Beneficial Microorganisms in Multicellular Life Forms

Eugene Rosenberg • Uri Gophna
Editors

Beneficial Microorganisms in Multicellular Life Forms

Springer

Editors
Eugene Rosenberg
University of Tel Aviv
Department of Molecular Microbiology
and Biotechnology
Ramat Aviv
Israel
eros@post.tau.ac.il

Uri Gophna
University of Tel Aviv
Dept. of Microbiology and
Biotechnology
Ramat Aviv
Israel
UriGo@tauex.tau.ac.il

ISBN 978-3-642-21679-4 e-ISBN 978-3-642-21680-0
DOI 10.1007/978-3-642-21680-0
Springer Heidelberg Dordrecht London New York

Library of Congress Control Number: 2011937068

Printed on acid-free paper

Springer is part of Springer Science+Business Media (www.springer.com)

Preface

Symbioses between microorganisms and animals and plants have been studied for more than 100 years. Until recently, these studies have concentrated on a primary symbiont and its host, e.g., *Rhizobia* species and their respective legumes, *Buchnera* and aphids, and *Vibrio fischeri* and squid. With the advent of molecular (culture-independent) techniques in microbiology during the last 15 years, it is now clear that all animals and plants contain hundreds or thousands of different microbial symbionts. In many cases the number of symbiotic microorganisms and their combined genetic information far exceed that of their hosts. For example, it has been estimated that the diverse bacterial symbionts in the human gut contain 200 fold more unique genes than the human host.

The diverse types of symbioses between microorganisms and eukaryotes have received growing attention in the last few years with regard to many different features of their complex interactions, such as the diversity and abundance of the symbionts, the type of advantage (or harm) the partners experience, how the interaction is initiated and maintained, and in recent years the role of the microorganisms in the evolution of the holobiont (host plus symbionts). These points taken together suggest that the genetic wealth of diverse microbial symbionts can play an important role both in adaptation and in evolution of higher organisms.

It is now clear that it is impossible to understand the health of plants, animals, and man without taking into consideration their microbiota. For example, it has recently been shown that the diverse symbionts in the human gut play a major role in obesity, priming the immune system, resistance to pathogenic bacteria, angiogenesis, fiber breakdown, and vitamin synthesis. There are also recent data on the importance of diverse symbionts in plant productivity, salt tolerance and mineral uptake, and the health of animals as diverse as sponges, corals, insects, and mammals. There is clearly a potential to combat diseases in animals, plants, and man by manipulating symbionts, i.e., by using probiotics (introduction of beneficial microorganisms) and prebiotics (alteration of the diet, or other environmental conditions, to encourage beneficial microorganisms). The success of these therapeutic approaches will depend upon a fundamental understanding of host–microbiota interactions – the major theme of this book.

The chapters in this book were derived from an international meeting on the Role of Microorganisms in the Adaptation and Evolution of Animals (including Man) and Plants which took place in Ein Gedi, Israel, in March 2011. The meeting was supported by the Bat-Sheva de Rothschild Foundation. We thank the Foundation for sponsoring the meeting. It was a pleasure to work with Ursula Gramm of Springer-Verlag, who made our job easier by providing all the necessary instructions. Finally, to the authors, thank you all.

Tel Aviv

Eugene Rosenberg
Uri Gophna

Contents

Part I Insect–Microbe Symbioses

1 **Microbial Symbioses in the Digestive Tract of Lower Termites** 3
Andreas Brune

2 **Insect "Symbiology" Is Coming of Age, Bridging Between Bench and Field** 27
Edouard Jurkevitch

3 **Chironomids and *Vibrio cholerae*** 43
Malka Halpern

4 **Role of Bacteria in Mating Preference in *Drosophila melanogaster*** 57
Gil Sharon, Daniel Segal, and Eugene Rosenberg

Part II Plant–Microbe Symbioses

5 **Legume–Microbe Symbioses** 73
Masayuki Sugawara and Michael J. Sadowsky

6 **Plant Growth Promotion by Rhizosphere Bacteria Through Direct Effects** 89
Yael Helman, Saul Burdman, and Yaacov Okon

7 **Rhizosphere Microorganisms** 105
Dror Minz and Maya Ofek

8 Microbial Protection Against Plant Disease 123
Eddie Cytryn and Max Kolton

Part III Coral–Microbe Symbioses

9 Bacterial Symbionts of Corals and Symbiodinium 139
Kim B. Ritchie

10 Coral-Associated Heterotrophic Protists 151
L. Arotsker, E. Kramarsky-Winter, and A. Kushmaro

11 Effect of Ocean Acidification on the Coral Microbial Community 163
Dalit Meron, Lena Hazanov, Maoz Fine, and Ehud Banin

Part IV Microbes in Mammalian Health and Disease

12 Toward the Educated Design of Bacterial Communities 177
Shiri Freilich and Eytan Ruppin

13 Oral Microbes in Health and Disease 189
Gilad Bachrach, Marina Faerman, Ofir Ginesin, Amir Eini, Asaf Sol, and Shunit Coppenhagen-Glazer

14 The Role of the Rumen Microbiota in Determining the Feed Efficiency of Dairy Cows 203
Itzhak Mizrahi

15 The Intestinal Microbiota and Intestinal Disease: Irritable Bowel Syndrome 211
Nirit Keren and Uri Gophna

16 Intestinal Microbiota and Intestinal Disease: Inflammatory Bowel Diseases 223
Amir Kovacs and Uri Gophna

17 A Role for Bacteria in the Development of Autoimmunity for Type 1 Diabetes 231
Adriana Giongo and Eric W. Triplett

18 Impact of Intestinal Microbial Communities upon Health 243
Harry J. Flint, Sylvia H. Duncan, and Petra Louis

19 Commensalism Versus Virulence 253
Dvora Biran, Anat Parket, and Eliora Z. Ron

20 Prebiotics: Modulators of the Human Gut Microflora 265
Uri Lesmes

21 Host Genetics and Gut Microbiota 281
Keren Buhnik-Rosenblau, Yael Danin-Poleg, and Yechezkel Kashi

Part V Evolution by Symbiosis

22 Microbial Symbiont Transmission: Basic Principles and Dark Sides 299
Silvia Bulgheresi

23 Hydra Go Bacterial 313
Thomas C.G. Bosch, Friederike Anton-Erxleben, René Augustin, Sören Franzenburg, and Sebastian Fraune

24 The Hologenome Concept 323
Eugene Rosenberg and Ilana Zilber-Rosenberg

Index 341

Part I
Insect–Microbe Symbioses

Chapter 1
Microbial Symbioses in the Digestive Tract of Lower Termites

Andreas Brune

Abstract The symbiotic gut microbiota of termites plays important roles in lignocellulose digestion and host nutrition. In contrast to the higher (evolutionarily advanced) termites, whose gut microbiota is largely prokaryotic, the capacity of lower (primitive) termites to digest wood depends on the cellulolytic gut flagellates housed in their enlarged hindgut paunch. The flagellates initiate a microbial feeding chain driven by the primary fermentations of carbohydrates to short-chain fatty acids, the major energy source of the host. Hydrogen, a central intermediate in the hindgut fermentations, is efficiently recycled by homoacetogenic spirochetes. They prevail over methanogenic archaea, which are restricted to particular microniches at the hindgut wall or within the flagellates. The spatial separation of microbial populations and metabolic activities gives rise to steep gradients of metabolites. The continuous influx of oxygen into the hindgut affects microbial metabolism in the microoxic periphery, and the anoxic status of the gut center is maintained only by the rapid reduction of oxygen by both aerobic and anaerobic microorganisms. Moreover, the gut microbiota also compensates for the low nitrogen content of wood by fixing atmospheric nitrogen, assimilating ammonia, providing essential amino acids and vitamins, and efficiently recycling nitrogenous wastes. The microorganisms responsible for these reactions are mostly unknown, but recent studies have implicated the bacterial symbionts of termite gut flagellates in these processes. These symbionts specifically colonize either the surface or the cytoplasm of the flagellates and represent novel bacterial lineages that occur exclusively in the hindgut of termites, often cospeciating with their respective hosts. Genome information indicates that the uncultivated symbionts of flagellates play a major role in the nitrogen metabolism of this tripartite symbiosis.

A. Brune (✉)
Department of Biogeochemistry, Max Planck Institute for Terrestrial Microbiology, Karl-von-Frisch-Strasse 10, Marburg 35043, Germany
e-mail: brune@mpi-marburg.mpg.de

E. Rosenberg and U. Gophna (eds.), *Beneficial Microorganisms in Multicellular Life Forms*, DOI 10.1007/978-3-642-21680-0_1,

1.1 Introduction

Like other animals thriving on a fiber-rich diet, termites (order Isoptera) are associated with microbial symbionts that aid in digestion (Brune 2009). While all termite species harbor prokaryotic symbionts in their intestinal tract, only the lower (primitive) termites live in a unique symbiosis that encompasses microorganisms from all three domains of life (Bacteria, Archaea, and Eukarya). In contrast to higher (evolutionarily advanced) termites, whose microbiota is mostly prokaryotic, lower termites harbor a dense assemblage of cellulolytic protists—the major agents of fiber digestion.

Symbiotic digestion takes place in the enlarged hindgut, which represents a microbial bioreactor that transforms plant fiber to short-chain fatty acids, the major energy source for the host, but also substantial amounts of methane, rendering termites a globally important source of this greenhouse gas (Fig. 1.1). Although the microbial processes resemble those in the rumen, termites differ from ruminants in many aspects, including their ability to digest highly lignified plant fiber. A particularly important property of termite guts is their small size, which creates steep gradients of oxygen and other metabolites and makes the gut wall a much more important habitat than in the rumen.

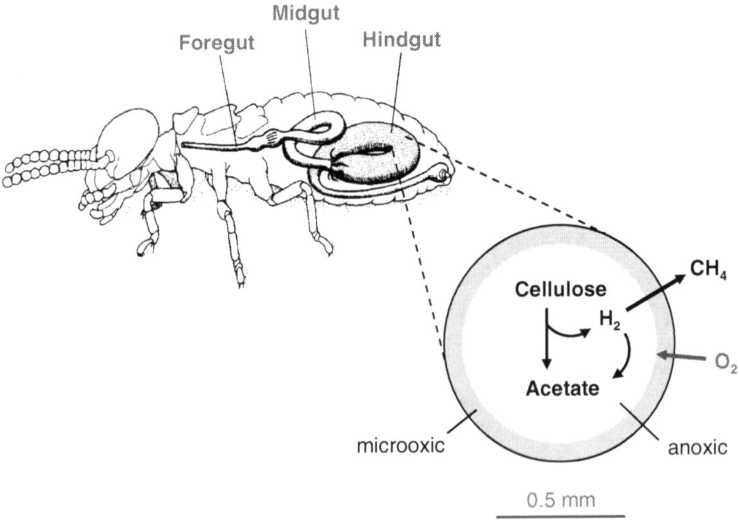

Fig. 1.1 The hindgut of termites is a microbial bioreactor that transforms lignocellulose to acetate and other short-chain fatty acids, which are the major energy source of the host. Hydrogen is an important intermediate that drives reductive acetogenesis and methanogenesis. The anoxic status of the hindgut lumen is maintained by the microorganisms colonizing the microoxic hindgut periphery, which consume the oxygen that constantly diffuses into the gut (From Brune and Ohkuma 2011)

The microbial symbionts of termites also make important contributions to host nutrition. In view of the essential role of the gut microbiota in symbiotic digestion, its is easily overlooked that microbial processes also compensate for both the low nitrogen content of wood and its poor nutritional value by fixing atmospheric nitrogen, assimilating ammonia, providing essential amino acids and vitamins, and efficiently recycling nitrogenous wastes.

This chapter will provide an overview of the tripartite symbiosis of flagellates, bacteria, and archaea in lower termites. It does not cover the diverse aspects of symbiotic digestion in higher termites, such as the breakdown of cellulose and hemicelluloses in the absence of gut flagellates (Tokuda and Watanabe 2007), the digestion of soil organic matter in soil- and humus-feeding groups with their extremely alkaline guts (Bignell 2006, 2011), or the fascinating symbiosis between fungus-cultivating termites and their basidiomycete partners (Nobre et al. 2011), and will only briefly touch on the production of endogenous cellulases by termites (Lo et al. 2011). Readers interested in these topics or in more details on the subject of symbiotic digestion in lower termites are referred to other comprehensive reviews of the literature (e.g., Brune 2010a, b; Hongoh and Ohkuma 2010; Brune and Ohkuma 2011; Hongoh 2011; Ohkuma and Brune 2011) and numerous excellent articles mentioned in the following sections.

1.2 Phylogenetic Diversity of the Gut Microbiota

The bulk of the microbiota in the hindgut paunch consists of unicellular eukaryotes—a highly diverse group of flagellates that are unique to lower termites. The sheer volume of their intracellular compartments and the enormous area of their cellular surfaces provide an abundance of habitats for the prokaryotic microorganisms. A single flagellate cell may harbor thousands of endosymbiotic and ectosymbiotic bacteria, and the majority of the bacteria in lower termites are associated with the gut flagellates. Also, the surface of the hindgut cuticle and the larger (filamentous) members of the bacterial community offer ample opportunity for bacterial attachment and are also the typical habitat of most of the archaea in the hindgut. Finally, life in the luminal fluid is open to any microorganisms that are able to maintain their position and prevent washout from the hindgut either by rapid growth or by their own locomotory power.

1.2.1 Gut Flagellates of Lower Termites

The perhaps most important event in the evolutionary history of termites was the establishment of the flagellates in the hindgut paunch, by which an ancestral, presumably subsocial, cockroach gained the capacity to efficiently digest cellulose. Wood-feeding cockroaches of the genus *Cryptocercus*, which are a sister group of termites and are considered the most primitive branch of the termite lineage, represent living fossils of such transition forms (Lo et al. 2000; Grimaldi and Engel 2005; Inward et al. 2007). The essential role of gut flagellates in cellulose digestion by lower termites had been documented already by the elegant work of Cleveland in the 1920s (see Breznak and Brune 1994). If the flagellates are removed from the gut, termites continue to feed on wood but die of starvation within a few weeks. However, if such Defaunation termites are then switched to a starch diet or are re-inoculated during contact with normally faunated nestmates, they maintain or regain their viability.

Neither the provenance of the symbionts nor the phylogenetic origin of the flagellates is clear. Termite gut flagellates comprise several monophyletic lineages among the "Excavata," a deep-branching supergroup of Eukarya, most of them in the phylum Parabasalia and one the phylum Preaxostyla (order Oxymonadida; see Hampl et al. 2009; Cepicka et al. 2010; Ohkuma and Brune 2011). Members of these lineages are unique to termites, and the characteristic assemblages of flagellates in different termite species can differ substantially in diversity. While *Hodotermopsis japonica* harbors at least 19 different flagellate species, only 3 species are present in *Coptotermes formosanus* (Inoue et al. 2000). Most intriguingly, the cellulolytic flagellates were completely lost again in the higher termites (family Termitidae), which harbor an essentially prokaryotic microbial community in their hindguts, and the source of cellulolytic activity in the hindgut of wood-feeding higher termites is far from clear (Brune 2007).

1.2.2 Bacterial Gut Microbiota

Ultrastructural studies have distinguished at least 20–30 different morphotypes of prokaryotic cells in each of several termite species investigated (see Brune 2006). The full extent of bacterial diversity, however, became apparent only with the application of molecular techniques. Numerous detailed studies of bacterial diversity in termite guts revealed the presence of hundreds of different phylotypes in a single termite species (typically using the threshold of 3% 16S rRNA sequence divergence in lieu of a species definition). Most communities are probably still undersampled because the few numerically predominant phylotypes are outnumbered by an enormous diversity of rare populations. Most phylotypes have no close relatives in public databases other than sequences derived from other termites, and when the molecular inventories are compared to the list of species isolated from such habitats, it is obvious that the majority of bacterial species in termite guts have not yet been cultivated (see Brune 2006; Ohkuma et al. 2006; Ohkuma and Brune 2011).

The gut microbiota of termites comprises species from a total of 15 bacterial phyla. By far the largest group in the hindgut of most wood-feeding termites—both in abundance and species richness—is the Spirochaetes (Breznak and Leadbetter 2006). The majority of spirochetes in termite guts belong to the genus *Treponema* and form a few immensely diverse but monophyletic groups, which must have been already acquired by the dictyopteran ancestor (Lilburn et al. 1999; Ohkuma et al. 1999; Iida et al. 2000; Hongoh et al. 2005). The enormous success of spirochetes in the colonization of the termite gut may be related to their high motility (a characteristic cork-screw mode of swimming), which allows them to maintain the most favorable position in the densely packed and highly dynamic gut environment without requiring attachment to the gut wall. Nevertheless, there are also many phylotypes that are regularly attached to the cell surface of specific gut flagellates (see Ohkuma and Brune 2011 for details).

The second most abundant bacterial phyla in lower termites are Bacteroidetes and Firmicutes (e.g., Hongoh et al. 2003, 2005; Yang et al. 2005). Both groups are highly diverse and form numerous termite-specific clusters, mostly within the orders Bacteroidales and Clostridiales. Also, members of Proteobacteria and Actinobacteria may be present in significant numbers. While some bacterial lineages seem to specifically colonize the gut wall (e.g., Nakajima et al. 2006), there are numerous lineages, especially among the Bacteroidetes, that represent specific symbionts of termite gut flagellates (see Ohkuma and Brune 2011). The endosymbionts of *Pseudotrichonympha grassii*, a large flagellate that dominates the bulk of the hindgut volume of *Coptotermes formosanus*, even outnumber the spirochetes in this lower termite (Noda et al. 2005; Shinzato et al. 2005). The functions of these endosymbionts and their cospeciation with *Pseudotrichonympha* species are discussed below.

Another bacterial lineage that is particularly abundant in many lower termites represents a deep-branching clade in the phylum Elusimicrobia. Originally discovered by Ohkuma and Kudo (Ohkuma and Kudo 1996) in 1996, the members of the "Termite group 1" are now called "Endomicrobia" because of their endosymbiotic association with many termite gut flagellates (Stingl et al. 2005). Endomicrobia are present in many gut flagellates (e.g., Ikeda-Ohtsubo et al. 2007; Ohkuma et al. 2007), and the endosymbionts of the large *Trichonympha* flagellates may amount to substantial populations in the hindgut (see Ohkuma and Brune 2011). Endomicrobia were most likely derived from putatively free-living relatives, which are present also in flagellate-free cockroaches and higher termites (Ohkuma et al. 2007; Ikeda-Ohtsubo et al. 2010). The cospeciation of Endomicrobia with *Trichonympha* species and their putative role in the symbiosis are discussed below.

1.2.3 Archaea Associated with Lower Termites

Methanogens are common among dictyopteran insects, and methanogenic archaea seem to be present in the hindgut of all termite species. However, archaeal rRNA amounts to only 1–2% of the total prokaryotic rRNA in the hindgut community (Brauman et al. 2001), and methane emission rates among wood-feeding termites are generally lower than among those with a humivorous lifestyle. In contrast to cockroaches and higher termites, which harbor several representatives of the orders

Methanobacteriales, Methanomicrobiales, and Methanosarcinales, and other deep-branching clades of uncultivated archaea, the archaeal community of lower termites is not very diverse. It consist almost exclusively of *Methanobrevibacter* species (order Methanobacteriales; see Brune 2010a, b).

Three members of the genus *Methanobrevibacter* have been isolated from *Reticulitermes flavipes* and are the only archaea from termite guts that have been obtained in pure culture (Leadbetter and Breznak 1996; Leadbetter et al. 1998). All of them possess a purely hydrogenotrophic metabolism and colonize the cuticle of the hindgut or the surface of filamentous bacteria colonizing the gut wall—the typical habitats of methanogens in lower termites. In several termites, specific lineages of uncultivated *Methanobrevibacter* spp. are associated with some of the smaller gut flagellates, but this phenomenon is much less common than the diverse bacterial associations (see Ohkuma 2008; Brune 2010a, b; Hongoh and Ohkuma 2010).

1.2.4 Associations of Gut Flagellates and Bacterial Symbionts

Since the majority of the bacterial cells in the hindgut of lower termites are associated with the gut flagellates and the relationships between bacteria and flagellates are typically highly specific (see Brune and Stingl 2005; Ohkuma 2008; Hongoh and Ohkuma 2010), it is not unexpected that many of the bacterial lineages identified in termite guts comprise flagellate-associated symbionts. Phylogenetic clustering suggests a common ancestry of the symbionts from closely related flagellates, and it has been speculated that once a bacterium has established a stable relationship with a gut flagellate, the partners in the symbiosis are likely to cospeciate. This is feasible in endosymbiotic systems, where the intracellular location of the bacterial symbionts combined with a considerable reduction of their genomes indicate that the symbiosis became obligate at least for the bacterial partner (see Ohkuma and Brune 2011).

A well-documented case is the association between endosymbiotic Bacteroidetes and flagellates of the genus *Pseudotrichonympha* in subterranean termites (Rhinotermitidae), which represents a unique case of cospeciation in a tripartite symbiosis. There is not only a strict congruence between the phylogenies of the "*Azobacteroides*" endosymbionts and their host flagellates, but also between the flagellates and their host termites (Noda et al. 2007). Another example of cospeciation is the symbiosis between endosymbiotic Endomicrobia and flagellates of the genus *Trichonympha*. Although the flagellates are spread over different termite families and multiple species may be present in the same termite gut, the symbiotic pairs have not been disrupted by host switches of the flagellate or the opportunities of the symbiont to switch between flagellates within the gut (Ikeda-Ohtsubo and Brune 2009).

While the basis of a purely vertical transmission is relatively easily envisaged for intracellular bacteria, the situation is different if the symbionts are attached to the external surfaces of their host. It has been underlined by several studies that certain species of ectosymbiotic Spirochetes and Bacteroidales can be associated with several protist species in the same host termite (e.g., Noda et al. 2003; Strassert et al. 2010). Nevertheless, strict cospeciation can be found also in an ectosymbiosis, as documented by the specific association of "*Armantifilum devescovinae*" with flagellates of the genus *Devescovina* in dry-wood termites (Kalotermitidae; Desai et al. 2010). Since there is no co-speciation between the *Devescovina* flagellates and their termite host, vertical transmission of the symbionts remains the most likely explanation for cospeciation also in this ectosymbiotic relationship.

A flagellate cell provides several niches for bacterial colonization, from both a structural and probably also a functional perspective, and multiple associations with different bacterial species seem to be more the rule than the exception. The symbionts are often from different bacterial phyla and each locate to different regions of the host cell, indicating the presence of specific interactions. There are many variations of this theme. The cells of *Trichonympha agilis* may simultaneously harbor at least two types of endosymbionts. One is "*Candidatus* Endomicrobium trichonymphae," the common endosymbiont of many *Trichonympha* species, the other a proteobacterium designated as "*Candidatus* Desulfovibrio trichonymphae" (Sato et al. 2009). While the former of the two symbiont populations occur mainly in the posterior part of the cell and presumably play a nutritional role (see below), the latter are located in the hydrogenosome-enriched anterior part and may be involved in the energy metabolism of the symbiotic pair. Also, a simultaneous association of flagellates with ectosymbiotic Bacteroidales or Spirochaetes and endosymbiotic Endomicrobia is not uncommon (e.g., Noda et al. 2003; Strassert et al. 2009; Desai et al. 2010), and the flagellate *Joenia annectens* may be simultaneously colonized by two only distantly related ectosymbiotic Bacteroidetes (Strassert et al. 2010).

An unusual symbiosis with multiple bacteria is present in the flagellate *Caduceia versatilis*, which is associated with both "*Armantifilum*"-like ectosymbiotic Bacteroidales and a second type of unusual ectosymbiont that belong to the phylum Synergistetes (Hongoh et al. 2007). The latter symbionts, which have been designated "*Candidatus* Tammella caduceiae," have been earlier demonstrated to be responsible for the motility of the flagellate host (Tamm 1982). An analogous case of motility symbiosis had been demonstrated already by Cleveland and Grimstone for the flagellate *Mixotricha paradoxa*, whose huge cell is not propelled by its few eukaryotic flagella but by the synchronous waving motion of ectosymbiotic spirochetes (Cleveland and Grimstone 1964). Also here, a second type of ectosymbiont is present, but it belongs to a different lineage of Bacteroidales from the ectosymbionts of *C. versatilis* (Wenzel et al. 2003). Despite several analogies in the morphology of the symbionts and the elaborate attachment structure in the cortex of the host cells, the two motility symbioses are clearly the result of evolutionary convergence.

It is not unusual that the ectosymbionts are intimately integrated into special attachment structures on the surface of the host. There are several reports where removal of ectosymbionts by antibiotics leads to a disintegration of the cortical attachment system and the loss of the characteristic shape of the host cell; this gave rise to the hypothesis that filamentous ectosymbionts may act as an exo-cytoskeleton (Leander and Keeling 2004; Radek and Nitsch 2007). In most cases, however, the biological basis for the specific associations between prokaryotes and gut flagellates is completely obscure (see Ohkuma 2008; Hongoh and Ohkuma 2010; Ohkuma and Brune 2011).

1.2.5 Coevolution of Microbiota and Termite Host

It is generally accepted that the present-day diversity of gut flagellates in lower termites is derived from a set of protists that had been acquired by their dictyopteran ancestor and subsequently coevolved with their host (Lo and Eggleton 2011). However, the evolutionary relationships between termites and gut flagellates are complicated by host switches, and the present-day assemblages of gut flagellates in different termite lineages bear clear indications of both the horizontal transfer of symbionts among members of different termite families and occasional losses of flagellates (Kitade 2004).

The cospeciaton of termites and gut flagellates is based on a reliable transfer of gut symbionts by proctodeal trophallaxis, a behavioral trait unique to termites and cockroaches of the genus *Cryptocercus*. It involves the transfer of a droplet of hindgut fluid from the rectum of one individual to the mouth of a nestmate, which not only distributes the symbionts among all colony members but also serves to transfer the community to the new colony via the alates. Proctodeal trophallaxis is considered a key element in the evolution of the wood-digestive capacity and eusociality in termites (Nalepa et al. 2001; Nalepa 2011).

There are numerous signs for cospeciation also in the many bacterial lineages that occur exclusively in the guts of termites (see Ohkuma and Brune 2011). Some of the lineages contain bacteria from all termite families, whereas others seem to be restricted to certain taxa. Since many of these lineages comprise symbionts of gut flagellates, it is possible that in many cases the apparent co-cladogenesis of termites and gut bacteria simply reflects the cospeciation of bacterial symbionts with their flagellate hosts. However, there are clear signs of co-cladogenesis also between lineages of other gut bacteria and flagellate-free hosts, and the case of "*Endomicrobium trichonymphae*" illustrates that certain lineages of flagellate symbionts must have been recruited from free-living gut bacteria long after the flagellate host had established its symbiosis with termites (Ikeda-Ohtsubo et al. 2010).

1.3 Roles of the Gut Microbiota in Digestion

The digestion of lignocellulose in termite guts involves both the host and its gut microbiota. Individual processes are separated by the transport of the digesta through the different gut compartments. Further compartmentation is accomplished by the segregation of wood particles into the digestive vacuoles of the gut

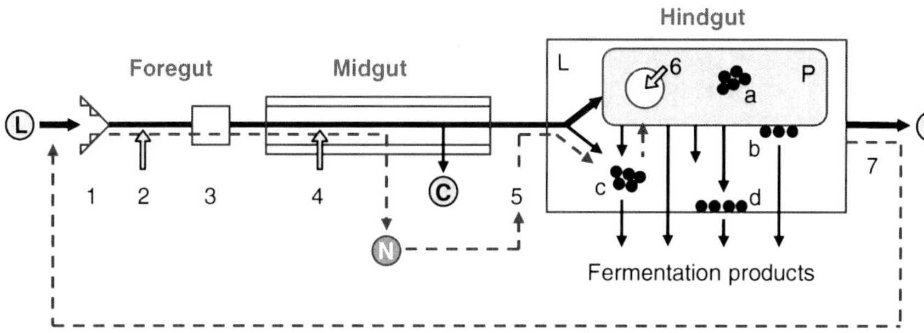

Fig. 1.2 Symbiotic digestion in a wood-feeding lower termite. The *bold lines* show the path of the insoluble material, the lignin-rich residues of which are released as feces; the thinner lines represent soluble degradation products that are eventually resorbed by the host. The *dashed lines* indicate the cycling of nitrogenous compounds. *Hollow arrows* mark the sites where cellulolytic enzymes are secreted. *Lower-case letters* refer to the different groups of prokaryotes in the hindgut, which are either endosymbionts (**a**) or ectosymbionts (**b**) of the gut flagellates, suspended in the gut lumen (**c**), or attached to the gut wall (**d**). The scheme has been simplified for the sake of clarity; not all possible interactions are shown. L, lignocellulose; *C* soluble carbohydrates, *N* nitrogenous compounds, *F* fecal matter, *L* gut lumen, *P* protozoa, *1*, mandibles; *2*, salivary glands; *3*, gizzard; *4*, midgut epithelium; *5*, Malpighian tubules; *6*, digestive vacuoles; *7*, proctodeal feeding (From Brune and Ohkuma 2011)

flagellates and the recycling of microbial biomass produced in the hindgut via proctodeal trophallaxis (Fig. 1.2).

The comminution by mandibles and gizzard tremendously increases the surface area of the ingested wood and is a prerequisite for symbiotic digestion in the hindgut compartment. However, the high efficiency with which cellulose and hemicellulose components are removed during gut passage seems to be based on more than just mechanical grinding. The microscopic-sized wood particles (ca. 20–100 μm) are mixed with enzymes contained in the saliva already in the anterior gut. Since passage of the digesta through the tubular midgut is very rapid and the endoglucanases of termites attack crystalline celluloses only inefficiently, the extent to which host enzymes contribute to cellulose digestion is not yet clear (see Watanabe and Tokuda 2009; Lo et al. 2011). Recent evidence also pointed out that the lignin component of wood may not be completely inert during gut passage but is subject to structural changes (Geib et al. 2008). It is possible that laccases and peroxidases produced by the insect and also abiotic processes such as iron-mediated hydroxyl radical formation improve digestibility of lignocellulose by chemical oxidation of lignin (see Tartar et al. 2009; Brune and Ohkuma 2011). However, concrete evidence for such speculations is so far lacking.

1.3.1 Cellulose Degradation by Gut Flagellates

Each of the different flagellate populations in the hindgut community of lower termites appears to have a specific role in lignocellulose digestion. Each population is specialized on different particle sizes of wood or on cellulose fractions of different crystallinity and degree of polymerization. Some flagellate species are essential for the digestion of hemicelluloses, which protect the cellulose fibers from enzymatic attack. Many of the smaller flagellates do not consume any wood particles and must be involved in the metabolism of soluble metabolites (see Inoue et al. 2000; Brugerolle and Radek 2006).

Parabasalid flagellates have been shown to produce a variety of exo-cellobiohydrolases and endoglucanases, the two components required for an efficient digestion of crystalline cellulose. A number of these enzymes have been expressed heterologously in other organisms and the recombinant proteins have been characterized biochemically (see Brune and Ohkuma 2011). Cellulases of glycosyl hydrolases family 7 seem to be innate to the flagellates, whereas those of other families seem to be of bacterial origin and most likely have been acquired by lateral gene transfer (Todaka et al. 2007, 2010).

The flagellates are also considered to be the source of hemicellulolytic activities in the hindgut of lower termites (Todaka et al. 2007; Zhou et al. 2007). The flagellate *Trichomitopsis termopsidis* from *Zootermopsis* spp., the only termite gut flagellate whose degradative capacities have been studied in pure culture, digests both microcrystalline cellulose and xylan, one of the major hemicelluloses of wood (Odelson and Breznak 1985a). Very little is known about the cellulolytic or hemicellulolytic activities of larger species of oxymonad flagellates, which also phagocytize wood. Many smaller gut flagellates seem to thrive exclusively on bacteria or soluble substrates. Bacterial degradation of plant fiber seems to be of little importance in lower termites. The immediate sequestration of wood particles into the digestive vacuoles of the gut flagellates probably renders them inaccessible to the prokaryotes in the gut fluid (see Brune and Ohkuma 2011).

1.3.2 Fermentation of Carbohydrates

Since the depolymerization products of plant polysaccharides are formed within the digestive vacuoles, the gut flagellates also have a central role in their subsequent fermentation. This was established about 70 years ago by the pioneering work of Robert E. Hungate, who identified hydrogen and acetate as the main fermentation products in flagellate suspensions of *Zootermopsis* spp. (see Breznak and Brune 1994). The bulk of the hydrogen seems to be formed by the large *Trichonympha* species inhabiting these termites, but exact stoichiometries of fermentation products are not available for these flagellates. The only quantitative study by Odelson and Breznak (Odelson and Breznak 1985b) documented that pure cultures of *Trichomitopsis termopsidis* ferment cellulose to 2 acetate, 2 CO_2 and $4H_2$ per glucose equivalent, which indicates that all redox equivalents formed in glycolysis are eventually released as H_2.

Hydrogen formation by eukaryotes usually takes place in hydrogenosomes. Although such organelles are a common feature of parabasalid flagellates

(Brugerolle and Radek 2006), it cannot be concluded that all species form the same fermentation products. The fermentation products of other parabasalids from termite guts have not been investigated, but distantly related *Trichomonas* spp. colonizing the urogenital tract of mammals have been shown to form substantial amounts of glycerol, lactate, and succinate in addition to H_2 (Müller 1988). Nothing is known about the fermentation products of oxymonads. Hydrogenosomes have not been reported for any oxymonad, but transcriptional analysis of the distantly related flagellate *Trimastix pyriformis* showed the presence of genes characteristic of such organelles (Hampl et al. 2008).

The contribution of bacteria to the fermentative processes in the hindgut of lower termites is far from clear. The lack of representative isolates makes it difficult to predict the metabolic activities of the different bacterial populations, and the microbial products accumulating in the hindgut fluid are not necessarily indicative of the nature of the primary fermentations. The restriction of polysaccharide degradation to the digestive vacuoles of the flagellates should limit the availability of soluble sugars in the gut fluid, and the turnover of free glucose seems to contribute only marginally to metabolic fluxes in *R. flavipes* hindguts (Tholen and Brune 2000). However, it is possible that soluble metabolites released by the fiber-digesting flagellates or stemming from midgut processes are important substrates for the bacteria and smaller protists (see Brune and Ohkuma 2011).

Some of the bacterial processes in the hindgut will be based on the reduced products of the primary fermentations, which are available in large quantities and give rise to secondary fermentations also in other anoxic habitats. Since such intermediates of microbial feeding chains usually do not accumulate to high concentrations, their importance is not easy to evidence. However, a rapid turnover of lactate to acetate was documented by microinjection of radioactive tracers into intact guts of *Reticulitermes* spp. (Tholen and Brune 2000; Pester and Brune 2007). The identity of the lactate-producing microorganisms is not clear, but it may be at least in part a fermentation product of the gut flagellates (see above). The microorganisms responsible for lactate consumption seem to be propionigenic bacteria located in the hindgut periphery because the intestinal fluxes through the lactate pool carbon shift from acetate to propionate production when oxygen is excluded (Tholen and Brune 2000). The bacteria responsible for these activities are not known, but uncultivated members of Bacteroidales and Propionibacteriaceae specifically colonizing the gut wall of *Reticulitermes* species are reasonable candidates (see Ohkuma and Brune 2011).

1.3.3 Methanogenesis and Reductive Acetogenesis

The hydrogen produced by the gut flagellates is subsequently consumed by hydrogenotrophic processes. Some lower termites show an enormous accumulation of hydrogen at the gut center, but hydrogen is a central metabolite in the microbial feeding chain also in species where it does not accumulate to high concentrations

(Ebert and Brune 1997; Pester and Brune 2007). In the absence of electron acceptors other than CO_2, the major processes responsible for the consumption of hydrogen in termite guts are methanogenesis and reductive acetogenesis.

Methane production in termite guts had been suspected already in the 1930s, but it took more than 40 years until the process was documented by Breznak and coworkers (see Brune 2010a for historical details). Most of the methanogenesis in lower termites is from H_2 and CO_2, which is corroborated by the purely hydrogenotrophic metabolism of all *Methanobrevibacter* species isolated from termite guts. More than 10 years later, however, Breznak and Switzer (1986) discovered that the guts of lower termites also showed high rates of reductive acetogenesis from H_2 and CO_2, an activity that was subsequently shown to be common among termite species from all major feeding guilds (Brauman et al. 1992).

The predominance of acetogenesis, reductive over methanogenesis in most wood-feeding termites coincides with an unusually high abundance of spirochetes in these taxa. The nature of this link became apparent when Leadbetter et al. (1999) isolated the first *Treponema* species from termite guts, which comprised the first spirochetes capable of reductive acetogenesis from H_2 and CO_2. Using the *fhs* gene (encoding a key enzyme of reductive acetogenesis) as a molecular marker, subsequent studies confirmed that spirochetes dominate reductive acetogenesis in lower termites, although not all *fhs* homologs present in the communities were expressed (Salmassi and Leadbetter 2003; Pester and Brune 2006). Moreover, microfluidic digital PCR of individual gut bacteria indicated that not all treponemes in a community carry an FTHFS homolog (Ottesen et al. 2006), and several of the spirochetes isolated from termite guts are not capable of reductive acetogenesis (Graber et al. 2004; Dröge et al. 2006, 2008). It is therefore likely that the diverse spirochetal communities in termite guts also have other functions. A metagenomic analysis of functional genes in the microbial community of a wood-feeding higher termite corroborated the predominance of spirochetes in reductive acetogenesis, but also implicated them in cellulose digestion (Warnecke et al. 2007).

The predominance of reductive acetogenesis over methanogenesis in termite guts had puzzled microbiologists for many years. Homoacetogens are usually outcompeted by methanogens in other anoxic environments. However, microsensor studies of *Reticulitermes flavipes* and *Zootermopsis nevadensis* revealed that the coexistence of both hydrogenotrophic processes in the termite gut is based on spatial separation (Ebert and Brune 1997; Tholen and Brune 2000; Pester and Brune 2007; Fig. 1.3). The activities of the hydrogen-producing gut flagellates causes a strong accumulation of hydrogen at the gut center and leads to steep hydrogen gradients toward the gut periphery. The methanogens, which are typically located only at the hindgut wall, are clearly hydrogen-limited, whereas the homoacetogens (presumably located exclusively within the gut lumen) are not. As long as methanogens are absent from the hindgut proper, reductive acetogenesis (presumably catalyzed by highly motile spirochetes) prevails, whereas methanogenesis is significantly increased in termites harboring gut flagellates with endosymbiotic methanogens (see Brune 2010a, b).

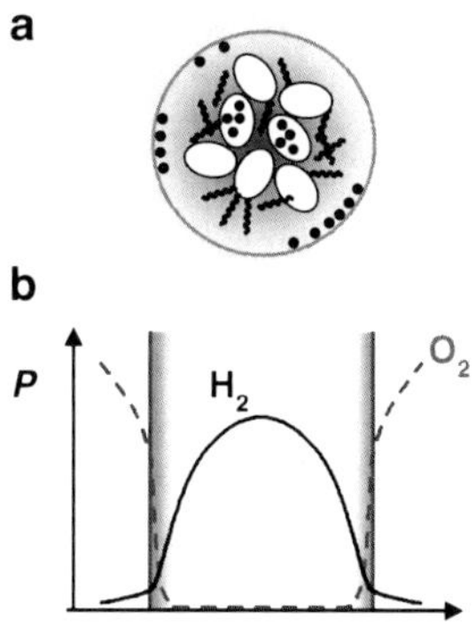

Fig. 1.3 Schematic cross section (**a**) of an agarose-embedded hindgut (*paunch region*) of a wood-feeding lower termite (*Reticulitermes* spp.), illustrating the location of methanogens (*filled circles*) attached to the hindgut wall and homoacetogenic spirochetes (spirals) within the gut proper. In some species, methanogens are also associated with the gut flagellates (*white ovals*). Radial profiles (**b**) of oxygen and hydrogen partial pressure (*P*) document that the respiratory activity of the gut microbiota maintains steep oxygen gradients (*dashed line*) within the gut periphery, rendering the center anoxic. Hydrogen (*solid line*) formed by the flagellates accumulates at the gut center but is consumed throughout the entire gut. The strong hydrogen sink below the gut wall is probably caused by methanogens, which prevent larger amounts of H_2 from escaping into the atmosphere (From Brune and Ohkuma 2011)

Steep metabolite gradients and rate-limiting metabolic fluxes must be expected also for any other metabolite whose sources and sinks are spatially separated within the gut. It is now well established that the microhabitats provided by the gut wall and the luminal fluid are colonized by specific lineages of prokaryotes, and there are also axial differences between the communities in different gut compartments. However, the perhaps most important factor causing a microscale heterogeneity of the community is the uneven distribution of different flagellate populations within the gut, each of them carrying unique populations of bacterial symbionts. The metabolic functions of the different microbial groups are slowly emerging.

1.3.4 Effects of Oxygen on Hindgut Metabolism

The introduction of microsensor techniques into termite gut research has fundamentally changed our concept of the metabolic processes in the termite–symbiont system (Brune et al. 1995). Originally considered to be purely anoxic fermenters, termite guts are now regarded as gradient systems characterized by the steady influx of oxygen across the gut wall (see Brune 1998; Brune et al. 2000). The steep oxygen gradients in the hindgut periphery indicate an efficient consumption of oxygen by the gut microbiota, which removes inflowing oxygen within fractions of a millimeter. Nevertheless, the minute dimensions of the hindgut dictate that a significant fraction of the gut microbiota will be exposed to oxygen, either continuously or for a limited time period (Brune et al. 2000; Brune and Ohkuma 2011).

Although the bacteria colonizing the hindgut wall remain to be characterized in a comprehensive manner, it appears that they comprise many aerotolerant, facultatively aerobic, or obligately aerobic representatives (see Brune 2006). The latter seem to comprise many microaerophiles specifically adapted to the hypoxic conditions of this habitat (Graber and Breznak 2005; Wertz and Breznak 2007). However, also the anaerobic bacteria isolated from termite guts are capable of reducing oxygen at astonishingly high rates (e.g., Boga et al. 2007) and shift their fermentations toward an increased acetate production when oxygen is present—a phenomenon relevant also under in situ conditions (Tholen and Brune 2000; Brune 2006). Even the notoriously oxygen-sensitive homoacetogens and methanogens are capable of hydrogen-dependent oxygen reduction (Boga and Brune 2003; Tholen et al. 2007). The ability of *Methanobrevibacter* species to remain metabolically active as long as the oxygen flux does not exceed their capacity for its removal may be the metabolic basis for their enigmatic colonization of the hindgut wall (see Brune 2010a for a detailed discussion).

It has been estimated that a substantial fraction of the respiratory activity of termites is attributable to the microbiota in their intestinal tract. However, the oxygen reduction potential of specific bacterial or archaeal populations colonizing the hindgut of lower termites does not suffice to explain the oxygen consumption of intact guts, and also the gut flagellates emerge as potential oxygen sinks (Tholen et al. 2007; Wertz and Breznak 2007). Since the latter have no mitochondria, the relative contributions of aerobic respiration and anaerobic (non-respiratory) processes to the oxygen-reducing activity of termite guts deserve to be further investigated.

In many environments, the oxic–anoxic interface is the habitat for aerobic methanotrophic bacteria. Although the counter-gradients of methane and oxygen in the hindgut periphery provide seemingly ideal conditions for aerobic methane oxidation (Brune et al. 2000), there is no evidence for the presence of methanotrophic bacteria or their activities in termite guts (Pester et al. 2007).

1.4 Roles of the Gut Microbiota in Host Nutrition

Based on the low nitrogen content of wood, xylophagous animals are strongly nitrogen-limited. Many wood-feeding termites show a preference for lignocellulosic substrates that are colonized by fungi, or complement their diet by taking up the hindgut content of nestmates. Termites living on sound wood acquire additional nitrogen through the nitrogen-fixing activity of their gut microbiota. The gut microbiota also ensures an efficient recycling of nitrogenous wastes within the system, which improves the nitrogen supply of the symbionts and promotes the overall performance of the digestive symbiosis (see Breznak 2000; Brune and Ohkuma 2011).

1.4.1 Upgrading and Recycling of Dietary Nitrogen

Insects have almost the same requirements for essential amino acids and vitamins as humans. Since wood is extremely poor in such nutrients, termites are under strong pressure to upgrade the quality of the nitrogenous components of their diet. The preference of lower termites for fungus-infested wood is well established (see Rouland-Lefèvre 2000), and the development of fungus gardening in higher termites is probably responsible for their enormous ecological success (Nobre et al. 2011). Most wood-feeding termites, however, supplement their diet with fresh hindgut content solicited from nestmates (proctodeal trophallaxis; see above), which not only serves to inoculate the gut of freshly molted individuals with a representative set of symbionts, but subjects the microbial biomass produced in the hindgut to the digestive activities of the host (see Brune and Ohkuma 2011; Fig. 1.2). By this strategy, termites solve the problem common to all hindgut fermenters—acquiring amino acids and other nutrients from their gut microbiota—in a manner analogous to that of rodents and lagomorphs, which consume cecum content to upgrade their diet.

The nitrogen cycle within the system is closed by the recycling of uric acid, the major waste product of nucleic acid and protein metabolism in insects. Uric acid secreted via the Malpighian tubules enters the gut hindgut, where it is immediately hydrolyzed by the hindgut microbiota, and ammonia is reassimilated into microbial biomass (see Breznak 2000). This strategy is reminiscent of nitrogen conservation in ruminants, which employ the ureolytic and assimilatory capacities of the rumen microbiota in the recycling of urea (Cotta and Russel 1997). Interestingly, all but the most primitive termites have lost the intracellular *Blattabacterium* symbionts common to the ancestral cockroaches, which are housed in the fat body and considered responsible for remobilizing uric acid during times of high nitrogen demand (Sabree et al. 2009).

1.4.2 Nitrogen Fixation by Gut Bacteria

No matter how efficiently nitrogen is recycled, it cannot increase the low nitrogen content of wood. Therefore, termites also exploit the nitrogen-fixing activity of their gut microbiota to compensate for the high C/N ratio of their diet. Since the independent discovery of nitrogen fixation in termite guts by Benemann (1973) and Breznak et al. (1973), a large number of studies have demonstrated the presence of dinitrogen-fixing bacteria, the incorporation of fixed N into termite biomass, and the widespread distribution of nitrogen fixation among termites of all families. Stable isotope analysis revealed that more than half of the nitrogen in certain termites may be derived from atmospheric N_2, and nitrogenase activities would be sufficient to double the nitrogen content of a colony within a few years (see Breznak 2000).

The capacity to fix dinitrogen is the exclusive domain of prokaryotes, and nitrogen-fixing bacteria have been isolated from various termites. However, there is little overlap between these isolates and the spectrum of potentially N_2-fixing microorganisms detected by cultivation-independent studies of the same environments (see Brune 2006). Analysis of the *nifH* gene sequences (a molecular marker for nitrogen fixation) obtained from various termite guts, mostly by Ohkuma and colleagues (see Brune and Ohkuma 2011), revealed that only few of the *nifH* genes were closely related to those of the nitrogen-fixing enterobacteria that are readily isolated from termite guts. Instead, most of the sequences belonged to hitherto uncultivated gut bacteria. Many of the sequences clustered with the *nifH* homologs of cultured termite gut spirochetes (Lilburn et al. 2001; Graber et al. 2004), underlining that related spirochetes may potentially contribute to nitrogen fixation in a variety of lower termites. Interestingly, many sequences from dry-wood termites clustered with so-called alternative nitrogenases that contain neither molybdenum nor vanadium cofactors; their preferential expression in *Neotermes koshunensis* may represent an adaption to the low molybdenum and vanadium content of the diet (Noda et al. 1999).

1.4.3 Nitrogen Metabolism of the Flagellate Symbionts

Although the processes surrounding the fixation, assimilation, and recycling of nitrogen in termite guts are reasonably well understood, the microorganisms responsible for these reactions are mostly unknown. In view of the fact that the majority of the termite gut microbiota remains uncultivated, further insights into the function of individual community members can be gained by genomic approaches. A pioneering metagenomic analysis of the luminal microbiota in the hindgut paunch of a higher termite provided important (and unexpected) insights into the metabolic potential of major phylogenetic groups (Warnecke et al. 2007; see also Brune 2007). However, it showed also the limitations of current metagenomic approaches when it comes to identifying specific links between genes located on different DNA fragments obtained from a highly diverse community. The enormous amount of flagellate DNA in hindgut preparations would add additional challenge and has so far discouraged any metagenomic studies of the gut microbiota in lower termites.

However, two recent studies have navigated around these problems in a most elegant manner. By amplifying just a tiny amount of genomic DNA extracted from a small number of bacteria collected from a single host cell, the authors were able to assemble the full genome sequences of two bacterial endosymbionts of termite gut flagellates (Hongoh et al. 2008a, b). Together, these studies documented the metabolic potential of the two largest populations of gut bacteria in the respective termites and provided the first evidence for a major role of the symbionts of flagellates in the nitrogen metabolism of the host termite.

One of the bacteria sequenced in this manner is "*Candidatus* Endomicrobium trichonymphae," the major endosymbiont of the flagellate *Trichonympha agilis* in *Reticulitermes speratus*. Genome analysis revealed that, despite a considerable reduction in genome size that is typical of obligate endosymbionts, "*E. trichonymphae*" seems to have retained a diversity of pathways for the biosynthesis of almost all amino acids and various cofactors (Hongoh et al. 2008a)—capacities that are lacking in a distantly related, free-living member of the *Elusimicrobia* phylum (Herlemann et al. 2009).

The second genome sequence is that of "*Candidatus* Azobacteroides pseudotrichonymphae," which colonizes the cytoplasm of flagellates of the genus *Pseudotrichonympha* and is the most abundant bacterium in the hindgut of *Coptotermes formosanus*. Genome analysis revealed that this uncultivated representative of the phylum Bacteroidetes not only possesses diverse pathways for the biosynthesis of amino acids and cofactors, but also has a complete set of genes necessary for nitrogen fixation, ammonium assimilation, and the dissimilation of urea—functions that were so far unprecedented among members of this phylum (Hongoh et al. 2008b).

Not much is known about the dietary requirements of the gut flagellates. In principle, they can complement their diet of wood particles by phagocytosis of hindgut bacteria. Although digestion of their own ectosymbionts has also been observed (Brugerolle and Radek 2006), it has not been reported that the endosymbionts are harvested in the same manner. Considering the enormous abundance of such endosymbionts in many of the large gut flagellates, they may play important roles in supplying their protist hosts—and thereby indirectly also the termite—with essential nutrients lacking in the lignocellulosic diet. Such a dependency may also be a part of the selective forces behind the cospeciation of the symbiotic pairs (see above).

1.5 Open Questions

Despite many advances in our understanding of the microbial processes in termite guts, there are still many aspects of symbiotic digestion that await further clarification. One of the most fundamental questions concerns the functional diversity of the gut flagellates. Since the pioneering studies of Hungate and Breznak, our understanding of the primary fermentations in the hindgut of lower termites has advanced only marginally. We still need to establish even the most fundamental metabolic properties of the different flagellate lineages, their respective fermentation products, and their interactions with bacterial partners. This concerns also their relationship to oxygen, whose continuous removal by the gut microbiota is a prerequisite for the fermentative breakdown of carbohydrates and thus the basis for the digestive symbiosis. Here, it is not at all clear to which extent prokaryotic and eukaryotic gut microorganisms are involved, and whether oxygen reduction is caused mainly by respiratory (aerobic) and non-respiratory (anaerobic) processes.

Another important field is the coevolution of termites and their gut microbiota and its functional implications. There are many signs of highly specific interactions leading to co-cladogenesis between specific phylogenetic groups, but the overall picture is far from clear. The situation is complicated by dynamic changes in the gut environment that are related to appearance or loss of major players and the need for new functions. In this context, is will be important to study the function of the flagellate symbionts in nitrogen metabolism in more detail. Also the role of gut bacteria in higher termites, which apparently replaced the fiber-digesting capacities of the gut flagellates, and the relative importance of host enzymes for cellulolytic process in foregut and midgut remain to be established. Lastly, also the question whether termites increase the digestibility of lignocellulose and humic substances by so far uncharacterized processes modifying the lignin fraction is a long-standing but unresolved issue with biotechnological implications.

References

Benemann JR (1973) Nitrogen fixation in termites. Science 181:164–165

Bignell DE (2006) Termites as soil engineers and soil processors. In: König H, Varma A (eds) Intestinal microorganisms of termites and other invertebrates. Springer, Berlin, pp 183–220

Bignell DE (2011) Morphology, physiology, biochemistry and functional design of the termite gut: an evolutionary wonderland. In: Bignell DE, Roisin Y, Lo N (eds) Biology of termites: a modern synthesis. Springer, Dordrecht, pp 375–412

Boga HI, Brune A (2003) Hydrogen-dependent oxygen reduction by homoacetogenic bacteria isolated from termite guts. Appl Environ Microbiol 69:779–786

Boga HI, Ji R, Ludwig W, Brune A (2007) *Sporotalea propionica* gen. nov. sp. nov., a hydrogen-oxidizing, oxygen-reducing, propionigenic firmicute from the intestinal tract of a soil-feeding termite. Arch Microbiol 187:15–27

Brauman A, Kane MD, Labat M, Breznak JA (1992) Genesis of acetate and methane by gut bacteria of nutritionally diverse termites. Science 257:1384–1387

Brauman A, Dore J, Eggleton P, Bignell D, Breznak JA, Kane MD (2001) Molecular phylogenetic profiling of prokaryotic communities in guts of termites with different feeding habits. FEMS Microbiol Ecol 35:27–36

Breznak JA (2000) Ecology of prokaryotic microbes in the guts of wood- and litter-feeding termites. In: Abe T, Bignell DE, Higashi M (eds) Termites: evolution, sociality, symbiosis, ecology. Kluwer, Dordrecht, pp 209–231

Breznak JA, Brune A (1994) Role of microorganisms in the digestion of lignocellulose by termites. Annu Rev Entomol 39:453–487

Breznak JA, Leadbetter JR (2006) Termite gut spirochetes. In: Dworkin M, Falkow S, Rosenberg E, Schleifer K-H, Stackebrandt E (eds) The prokaryotes, vol 7, 3rd edn, Proteobacteria: delta and epsilon subclasses. Deeply rooting bacteria. Springer, New York, pp 318–329

Breznak JA, Switzer JM (1986) Acetate synthesis from H_2 plus CO_2 by termite gut microbes. Appl Environ Microbiol 52:623–630

Breznak JA, Brill WJ, Mertins JW, Coppel HC (1973) Nitrogen fixation in termites. Nature 244:577–580

Brugerolle G, Radek R (2006) Symbiotic protozoa of termites. In: König H, Varma A (eds) Intestinal microorganisms of termites and other invertebrates. Springer, Berlin, pp 243–269

Brune A (1998) Termite guts: the world's smallest bioreactors. Trends Biotechnol 16:16–21

Brune A (2006) Symbiotic associations between termites and prokaryotes. In: Dworkin M, Falkow S, Rosenberg E (eds) The prokaryotes, vol 1, 3rd edn, Symbiotic associations, biotechnology, applied microbiology. Springer, New York, pp 439–474

Brune A (2007) Woodworker's digest. Nature 450:487–488

Brune A (2009) Symbionts aiding digestion. In: Resh VH, Cardé RT (eds) Encyclopedia of insects, 2nd edn. Academic, New York, pp 978–983

Brune A (2010a) Methanogenesis in the digestive tracts of insects. In: Timmis KN (ed) Handbook of hydrocarbon and lipid microbiology, vol 1. Springer, Heidelberg, pp 707–728

Brune A (2010b) Methanogens in the digestive tract of termites. In: Hackstein JHP (ed) (Endo) symbiotic methanogenic archaea. Springer, Heidelberg, pp 81–100

Brune A, Ohkuma M (2011) Role of the termite gut microbiota in symbiotic digestion. In: Bignell DE, Roisin Y, Lo N (eds) Biology of termites: a modern synthesis. Springer, Dordrecht, pp 439–475

Brune A, Stingl U (2005) Prokaryotic symbionts of termite gut flagellates: phylogenetic and metabolic implications of a tripartite symbiosis. In: Overmann J (ed) Molecular basis of symbiosis. Springer, Berlin, pp 39–60

Brune A, Emerson D, Breznak JA (1995) The termite gut microflora as an oxygen sink: microelectrode determination of oxygen and pH gradients in guts of lower and higher termites. Appl Environ Microbiol 61:2681–2687

Brune A, Frenzel P, Cypionka H (2000) Life at the oxic–anoxic interface: microbial activities and adaptations. FEMS Microbiol Rev 24:691–710

Cepicka I, Hampl V, Kulda J (2010) Critical taxonomic revision of parabasalids with description of one new genus and three new species. Protist 161:400–433

Cleveland LR, Grimstone AV (1964) The fine structure of the flagellate *Mixotricha paradoxa* and its associated micro-organisms. Proc R Soc Lond Ser B Biol Sci 159:668–686

Cotta MA, Russel JB (1997) Digestion of nitrogen in the rumen: a model for metabolism of nitrogen compounds in gastrointestinal environments. In: Mackie RI, White BA (eds) Gastrointestinal microbiology, vol 1. Chapman and Hall, New York, pp 380–423

Desai MS, Strassert JFH, Meuser K et al (2010) Strict cospeciation of devescovinid flagellates and *Bacteroidales* ectosymbionts in the gut of dry-wood termites (*Kalotermitidae*). Environ Microbiol 12:2120–2132

Dröge S, Fröhlich J, Radek R, König H (2006) *Spirochaeta coccoides* sp. nov., a novel coccoid spirochete from the hindgut of the termite *Neotermes castaneus*. Appl Environ Microbiol 72:392–397

Dröge S, Rachel R, Radek R, König H (2008) *Treponema isoptericolens* sp. nov., a novel spirochaete from the hindgut of the termite *Incisitermes tabogae*. Int J Syst Evol Microbiol 58:1079–1083

Ebert A, Brune A (1997) Hydrogen concentration profiles at the oxic-anoxic interface: a microsensor study of the hindgut of the wood-feeding lower termite *Reticulitermes flavipes* (Kollar). Appl Environ Microbiol 63:4039–4046

Geib SM, Filley TR, Hatcher PG et al (2008) Lignin degradation in wood-feeding insects. Proc Natl Acad Sci USA 105:12932–12937

Graber JR, Breznak JA (2005) Folate cross-feeding supports symbiotic homoacetogenic spirochetes. Appl Environ Microbiol 71:1883–1889

Graber JR, Leadbetter JR, Breznak JA (2004) Description of *Treponema azotonutricium* sp. nov. and *Treponema primitia* sp. nov., the first spirochetes isolated from termite guts. Appl Environ Microbiol 70:1315–1320

Grimaldi D, Engel MS (2005) Evolution of the insects. Cambridge University Press, New York, 772 pp

Hampl V, Silberman JD, Stechmann A et al (2008) Genetic evidence for a mitochondriate ancestry in the 'amitochondriate' flagellate *Trimastix pyriformis*. PLoS One 3:e1383

Hampl V, Hug L, Leigh JW, Dacks JB, Lang BF, Simpson AG, Roger AJ (2009) Phylogenomic analyses support the monophyly of Excavata and resolve relationships among eukaryotic "supergroups". Proc Natl Acad Sci USA 106:3859–3864

Herlemann DPR, Geissinger O, Ikeda-Ohtsubo W et al (2009) Genome analysis of "Elusimicrobium minutum," the first cultivated representative of the phylum "Elsimicrobia" (formerly Termite Group 1). Appl Environ Microbiol 75:2841–2849

Hongoh Y (2011) Toward the functional analysis of uncultivable, symbiotic microorganisms in the termite gut. Cell Mol Life Sci 68:1311–1325

Hongoh Y, Ohkuma M (2010) Termite gut flagellates and their methanogenic and eubacterial symbionts. In: Hackstein JHP (ed) (Endo)symbiotic methanogenic archaea. Springer, Heidelberg, pp 55–79

Hongoh Y, Ohkuma M, Kudo T (2003) Molecular analysis of bacterial microbiota in the gut of the termite *Reticulitermes speratus* (Isoptera; Rhinotermitidae). FEMS Microbiol Ecol 44:231–242

Hongoh Y, Deevong P, Inoue T et al (2005) Intra- and interspecific comparisons of bacterial diversity and community structure support coevolution of gut microbiota and termite host. Appl Environ Microbiol 71:6590–6599

Hongoh Y, Sato T, Dolan MF et al (2007) The motility symbiont of the termite gut flagellate *Caduceia versatilis* is a member of the "Synergistes" group. Appl Environ Microbiol 73:6270–6276

Hongoh Y, Sharma VK, Prakash T et al (2008a) Complete genome of the uncultured Termite Group 1 bacteria in a single host protist cell. Proc Natl Acad Sci USA 105:5555–5560

Hongoh Y, Sharma VK, Prakash T et al (2008b) Genome of an endosymbiont coupling N_2 fixation to cellulolysis within protist cells in termite gut. Science 322:1108–1109

Iida T, Ohkuma M, Ohtoko K, Kudo T (2000) Symbiotic spirochetes in the termite hindgut: phylogenetic identification of ectosymbiotic spirochetes of oxymonad protists. FEMS Microbiol Ecol 34:17–26

Ikeda-Ohtsubo W, Brune A (2009) Cospeciation of termite gut flagellates and their bacterial endosymbionts: *Trichonympha* species and '*Candidatus* Endomicrobium trichonymphae'. Mol Ecol 18:332–342

Ikeda-Ohtsubo W, Desai M, Stingl U, Brune A (2007) Phylogenetic diversity of '*Endomicrobia*' and their specific affiliation with termite gut flagellates. Microbiology 153:3458–3465

Ikeda-Ohtsubo W, Faivre N, Brune A (2010) Putatively free-living '*Endomicrobia*' — ancestors of the intracellular symbionts of termite gut flagellates? Environ Microbiol Rep 2:554–559

Inoue T, Kitade O, Yoshimura T, Yamaoka I (2000) Symbiotic associations with protists. In: Abe T, Bignell DE, Higashi M (eds) Termites: evolution, sociality, symbiosis, ecology. Kluwer, Dordrecht, pp 275–288

Inward D, Beccaloni G, Eggleton P (2007) Death of an order: a comprehensive molecular phylogenetic study confirms that termites are eusocial cockroaches. Biol Lett 3:331–335

Kitade O (2004) Comparison of symbiotic flagellate faunae between termites and a wood-feeding cockroach of the genus *Cryptocercus*. Microbiol Environ 19:215–220

Leadbetter JR, Breznak JA (1996) Physiological ecology of *Methanobrevibacter cuticularis* sp. nov. and *Methanobrevibacter curvatus* sp. nov., isolated from the hindgut of the termite *Reticulitermes flavipes*. Appl Environ Microbiol 62:3620–3631

Leadbetter JR, Crosby LD, Breznak JA (1998) *Methanobrevibacter filiformis* sp. nov., a filamentous methanogen from termite hindguts. Arch Microbiol 169:287–292

Leadbetter JR, Schmidt TM, Graber JR, Breznak JA (1999) Acetogenesis from H_2 plus CO_2 by spirochetes from termite guts. Science 283:686–689

Leander BS, Keeling PJ (2004) Symbiotic innovation in the oxymonad *Streblomastix strix*. J Eukaryot Microbiol 51:291–300

Lilburn TG, Schmidt TM, Breznak JA (1999) Phylogenetic diversity of termite gut spirochaetes. Environ Microbiol 1:331–345

Lilburn TG, Kim KS, Ostrom NE et al (2001) Nitrogen fixation by symbiotic and free-living spirochetes. Science 292:2495–2498

Lo N, Eggleton P (2011) Termite phylogenetics and co-cladogenesis with symbionts. In: Bignell DE, Roisin Y, Lo N (eds) Biology of termites: a modern synthesis. Springer, Dordrecht, pp 27–50

Lo N, Tokuda G, Watanabe H et al (2000) Evidence from multiple gene sequences indicates that termites evolved from wood-feeding cockroaches. Curr Biol 10:801–804

Lo N, Watanabe H, Tokuda G (2011) Evolution and function of endogenous termite cellulases. In: Bignell DE, Roisin Y, Lo N (eds) Biology of termites: a modern synthesis. Springer, Dordrecht, pp 51–67

Müller M (1988) Energy metabolism of protozoa without mitochondria. Annu Rev Microbiol 42:465–488

Nakajima H, Hongoh Y, Noda S et al (2006) Phylogenetic and morphological diversity of Bacteroidales members associated with the gut wall of termites. Biosci Biotechnol Biochem 70:211–218

Nalepa CA (2011) Altricial development in wood-feeding cockroaches: the key antecedent of termite eusociality. In: Bignell DE, Roisin Y, Lo N (eds) Biology of termites: a modern synthesis. Springer, Dordrecht, pp 69–95

Nalepa CA, Bignell DE, Bandi C (2001) Detritivory, coprophagy, and the evolution of digestive mutualisms in Dictyoptera. Insect Soc 48:194–201

Nobre T, Rouland-Lefèvre C, Aanen DK (2011) Comparative biology of fungus cultivation in termites and ants. In: Bignell DE, Roisin Y, Lo N (eds) Biology of termites: a modern synthesis. Springer, Dordrecht, pp 193–210

Noda S, Ohkuma M, Usami R et al (1999) Culture-independent characterization of a gene responsible for nitrogen fixation in the symbiotic microbial community in the gut of the termite *Neotermes koshunensis*. Appl Environ Microbiol 65:4935–4942

Noda S, Ohkuma M, Yamada A et al (2003) Phylogenetic position and in situ identification of ectosymbiotic spirochetes on protists in the termite gut. Appl Environ Microbiol 69: 625–633

Noda S, Iida T, Kitade S et al (2005) Endosymbiotic Bacteroidales bacteria of the flagellated protist *Pseudotrichonympha grassii* in the gut of the termite *Coptotermes formosanus*. Appl Environ Microbiol 71:8811–8817

Noda S, Kitade O, Inoue T, Kawai M, Kanuka M, Hiroshima K, Hongoh Y, Constantino R, Uys V, Zhong J, Kudo T, Ohkuma M (2007) Cospeciation in the triplex symbiosis of termite gut protists (*Pseudotrichonympha* spp.), their hosts, and their bacterial endosymbionts. Mol Ecol 16:1257–1266

Odelson DA, Breznak JA (1985a) Cellulase and other polymer-hydrolyzing activities of *Trichomitopsis termopsidis*, a symbiotic protozoan from termites. Appl Environ Microbiol 49:622–626

Odelson DA, Breznak JA (1985b) Nutrition and growth characteristics of *Trichomitopsis termopsidis*, a cellulolytic protozoan from termites. Appl Environ Microbiol 49:614–621

Ohkuma M (2008) Symbioses of flagellates and prokaryotes in the gut of lower termites. Trends Microbiol 16:345–352

Ohkuma M, Brune A (2011) Diversity, structure, and evolution of the termite gut microbial community. In: Bignell DE, Roisin Y, Lo N (eds) Biology of termites: a modern synthesis. Springer, Dordrecht, pp 413–438

Ohkuma M, Kudo T (1996) Phylogenetic diversity of the intestinal bacterial community in the termite *Reticulitermes speratus*. Appl Environ Microbiol 62:461–468

Ohkuma M, Iida T, Kudo T (1999) Phylogenetic relationships of symbiotic spirochetes in the gut of diverse termites. FEMS Microbiol Lett 181:123–129

Ohkuma M, Hongoh Y, Kudo T (2006) Diversity and molecular analyses of yet-uncultured microorganisms. In: König H, Varma A (eds) Intestinal microorganisms of termites and other invertebrates. Springer, Heidelberg, pp 303–317

Ohkuma M, Sato T, Noda S et al (2007) The candidate phylum 'Termite Group 1' of bacteria: phylogenetic diversity, distribution, and endosymbiont members of various gut flagellated protists. FEMS Microbiol Ecol 60:467–476

Ottesen EA, Hong JW, Quake SR, Leadbetter JR (2006) Microfluidic digital PCR enables multigene analysis of individual environmental bacteria. Science 314:1464–1467

Pester M, Brune A (2006) Expression profiles of *fhs* (FTHFS) genes support the hypothesis that spirochaetes dominate reductive acetogenesis in the hindgut of lower termites. Environ Microbiol 8:1261–1270

Pester M, Brune A (2007) Hydrogen is the central free intermediate during lignocellulose degradation by termite gut symbionts. ISME J 1:551–565

Pester M, Tholen A, Friedrich MW, Brune A (2007) Methane oxidation in termite hindguts: absence of evidence and evidence of absence. Appl Environ Microbiol 73:2024–2028

Radek R, Nitsch G (2007) Ectobiotic spirochetes of flagellates from the termite *Mastotermes darwiniensis*: attachment and cyst formation. Eur J Protistol 43:281–294

Rouland-Lefèvre C (2000) Symbiosis with fungi. In: Abe T, Bignell DE, Higashi M (eds) Termites: evolution, sociality, symbiosis, ecology. Kluwer, Dordrecht, pp 289–306

Sabree ZL, Kambhampati S, Moran NA (2009) Nitrogen recycling and nutritional provisioning by *Blattabacterium*, the cockroach endosymbiont. Proc Natl Acad Sci USA 106:19521–19526

Salmassi TM, Leadbetter JR (2003) Molecular aspects of CO_2-reductive acetogenesis in cultivated spirochetes and the gut community of the termite *Zootermopsis angusticollis*. Microbiology 149:2529–2537

Sato T, Hongoh Y, Noda S et al (2009) *Candidatus* Desulfovibrio trichonymphae, a novel intracellular symbiont of the flagellate *Trichonympha agilis* in termite gut. Environ Microbiol 11:1007–1015

Shinzato N, Muramatsu M, Matsui T, Watanabe Y (2005) Molecular phylogenetic diversity of the bacterial community in the gut of the termite *Coptotermes formosanus*. Biosci Biotechnol Biochem 69:1145–1155

Stingl U, Radek R, Yang H, Brune A (2005) "Endomicrobia": cytoplasmic symbionts of termite gut protozoa form a separate phylum of prokaryotes. Appl Environ Microbiol 71:1473–1479

Strassert JFH, Desai MS, Brune A, Radek R (2009) The true diversity of devescovinid flagellates in the termite *Incisitermes marginipennis*. Protist 160:522–535

Strassert JFH, Desai MS, Radek R, Brune A (2010) Identification and localization of the multiple bacterial symbionts of the termite gut flagellate *Joenia annectens*. Microbiology 156:2068–2079

Tamm SL (1982) Flagellated epibiotic bacteria propel a eucaryotic cell. J Cell Biol 94:697–709

Tartar A, Wheeler MM, Zhou X et al (2009) Parallel metatranscriptome analyses of host and symbiont gene expression in the gut of the termite *Reticulitermes flavipes*. Biotechnol Biofuels 2:25

Tholen A, Brune A (2000) Impact of oxygen on metabolic fluxes and *in situ* rates of reductive acetogenesis in the hindgut of the wood-feeding termite *Reticulitermes flavipes*. Environ Microbiol 2:436–449

Tholen A, Pester M, Brune A (2007) Simultaneous methanogenesis and oxygen reduction by *Methanobrevibacter cuticularis* at low oxygen fluxes. FEMS Microbiol Ecol 62:303–312

Todaka N, Moriya S, Saita K, Hondo T, Kiuchi I, Takasu H, Ohkuma M, Piero C, Hayashizaki Y, Kudo T (2007) Environmental cDNA analysis of the genes involved in lignocellulose digestion in the symbiotic protist community of *Reticulitermes speratus*. FEMS Microbiol Ecol 59:592–599

Todaka N, Inoue T, Saita K, Ohkuma M, Nalepa CA, Lenz M, Kudo T, Moriya S (2010) Phylogenetic analysis of cellulolytic enzyme genes from representative lineages of termites and a related cockroach. PLoS One 5:e8636

Tokuda G, Watanabe H (2007) Hidden cellulases in termites: revision of an old hypothesis. Biol Lett 3:336–339

Warnecke F, Luginbühl P, Ivanova N, Ghassemian M, Richardson TH et al (2007) Metagenomic and functional analysis of hindgut microbiota of a wood-feeding higher termite. Nature 450:560–565

Watanabe H, Tokuda G (2009) Cellulolytic systems in insects. Annu Rev Entomol 55:609–632

Wenzel M, Radek R, Brugerolle G, König H (2003) Identification of the ectosymbiotic bacteria of *Mixotricha paradoxa* involved in movement symbiosis. Eur J Protistol 39:11–23

Wertz JT, Breznak JA (2007) Physiological ecology of *Stenoxybacter acetivorans*, an obligate microaerophile in termite guts. Appl Environ Microbiol 73:6829–6841

Yang H, Schmitt-Wagner D, Stingl U, Brune A (2005) Niche heterogeneity determines bacterial community structure in the termite gut (*Reticulitermes santonensis*). Environ Microbiol 7:916–932

Zhou X, Smith JA, Oi FM, Koehler PG, Bennett GW, Scharf ME (2007) Correlation of cellulase gene expression and cellulolytic activity throughout the gut of the termite *Reticulitermes flavipes*. Gene 395:29–39

Chapter 2
Insect "Symbiology" Is Coming of Age, Bridging Between Bench and Field

Edouard Jurkevitch

Abstract Insects are major contributors to natural as well as to man-managed ecosystems and largely bear on human affairs, both positively and negatively. They have varied microbial communities that provide them with nutritional benefits, enhanced defenses, and improved stress resistance. This review centers on nutritional contributions of gut symbionts to their insect hosts. It explores the complex relationships between symbiont, host nutrition, and host life cycle, summarizing some of the important developments in symbiotic science. It particularly focuses on Tephritidae fruit flies that cause enormous agricultural damage. It exposes the present knowledge on the symbionts of the tephritids *Ceratitis capitata* (the Mediterranean fruit fly) and *Bactrocera oleae* (the Olive fly) and recapitulates recent advances in the improvement of the performance of mass-reared sterile males used in sterile insect technique to combat the medfly.

2.1 Introduction

The founding myth of the San people in the Kalahari Desert is that the first human grew from a seed planted in a mantis by a bee. The Minoans believed in the Bee goddess, and in the Homeric myth, the Thriaie speak the truth after eating honey. Ancient Egyptians made dung beetles sacred and used bee wax as seals and in ship building; wax has also been lighting homes for centuries, and the silk fiber tied East to West through interminable roads for more than 3,000 years; butterflies and caterpillars were shown as symbols of transience and impermanence of earthly things: All these tell of the close association between insects and man. Insect lovers cannot but marvel at the contrast between the general apprehension against these

E. Jurkevitch (✉)
Department of Plant Pathology and Microbiology, Faculty of Agriculture, Food and Environment, The Hebrew University of Jerusalem, Rehovot 76100, Israel
e-mail: jurkevi@agri.huji.ac.il

E. Rosenberg and U. Gophna (eds.), *Beneficial Microorganisms in Multicellular Life Forms*, 27
DOI 10.1007/978-3-642-21680-0_2, © Springer-Verlag Berlin Heidelberg 2011

arthropods we see today, and the reverence in which our ancestors seemed to hold them, leaving us mostly a positive, or at least revering view of these wonderful creatures.

Insects bear on human affairs. It was recently calculated that the economic value of pollination worldwide amounts to €153 billion (at a year 2005 value), or 9.5% of the total value of the world agricultural production used for human food (Gallai et al. 2009). Insects also provide enormous biocontrol, decomposition that increases soil fertility, and seed dispersion services estimated at tens of billions of dollars annually. Negative aspects of insects' impact on humans include diseases and agricultural pests. Insects and other pests destroy an estimated 40% of all crops worldwide, despite the yearly application of about three million tons of pesticides at a cost of more than $25 billions (Pimentel 2009). Insects are also vectors of human, other animals', and plant diseases, including, naming a few, *Plasmodium* spp. (malaria), transmitted by mosquitoes, *Trypanosoma* spp. (tsetse), transmitted by *Glossina* flies, the tobacco mosaic virus and the tomato yellow leaf curl virus, both transmitted by the white fly *Bemisia* spp. Thus, in human perspective, insect activities range from positive to negative. Most of the pollinating activities provided by insects are essential, and may be defined as irreplaceable, or obligate, both for agricultural production and for the sustenance of ecosystems. Others may be facultative, such as natural biocontrol in agricultural fields, that can be replaced by chemicals – of course with economic, environmental, and health costs.

Insects may interact with the microbes, the microbiota they carry, in similar ways: they may enter cooperative interactions or suffer from detrimental ones. These interactions may be obligate or facultative.

Cooperative, mutualistic interactions based on nutritional contributions will be the topic of this review. I will review various insect-symbiont systems and parameters that may affect the type of symbiosis and will emphasize the role of the microbiota in pest fruit flies, insects that negatively affect humans. The chapter will end with a view on how microbes may be manipulated to reduce the damage toll of such nefarious insects.

2.2 P and S Symbionts

Associations between an insect host and its symbionts may be obligate or facultative. The greater the benefits accrued, i.e., the more host fitness parameters (including development, longevity, and reproductive fitness) are dependent upon the symbiont, the more stringent the association. In such cases, symbionts are mostly called primary (P) symbionts. Secondary symbionts (S) are mostly dispensable or even deleterious (Hypsa and Novakova 2009). While P symbionts may share long evolutionary histories with their hosts and are usually transmitted from parent to progeny, S symbionts may be more recent and are transmitted in a more "relaxed" manner, i.e., largely horizontally (Moya et al. 2008). Yet, heritable associations tend to become mutualistic, and the host is severely harmed by the elimination of the symbiont (Baumann et al. 2006). Moreover, S symbionts can evolve to become obligate partners and establish a microbial consortium with the P symbiont or replace it (Wu et al. 2006; Toju et al. 2010).

Symbionts that are strictly transmitted vertically are obligate symbionts but not all obligate symbionts are solely transmitted vertically (Bright and Bulgheresi 2010). Strict symbionts are mostly intracellular endosymbionts, and are lodged within bacteriocytes (also called mycetomes).

2.3 Dietary Effects in Symbiotic Interactions

Symbionts can affect their hosts in many ways. Here, I will restrict to some aspects of dietary effects. Bacterial symbionts can improve diet quality when engaging in a mutualistic relationship and provision the host for nutrients lacking in its diet. Yet, the nature of the diet and the life cycle of the host seem to affect the type of symbiosis. Estes et al. (2009) proposed to interpret symbiotic relationships in insects along three variables defining a gradient of variability of the internal environment: (1) The host diet (monophagous versus polyphagous); (2) the host life cycle (hemimetabolous versus holometabolous); and (3) the location of the endosymbiont (intracellular versus extracellular).

Intracellular symbionts in holometabolous insects may face more challenges than intracellular symbionts of hemimetabolous insects: While both have to reach the progeny and pass through ecdysis, the former encounter the two, often ecologically largely different life stages of their hosts. To be conserved, the symbiont should be relevant to both stages: mutualistic in both, or mutualistic at one stage and parasitic at the other; or hitchhiking on particularities of the host's life cycle.

We will explore symbiosis along these lines, examining a few cases. This review does not intend to cover all aspects of dietary symbiotic interactions that certainly extend beyond straightforward nutrient provisioning but to provide a few lines of thought on some of their ecological and evolutionary aspects.

2.4 Nutrition-Based Symbiosis in Hemimetabolous Insects

Aphids and sharpshooters, sap-feeding insects, are at the one end of the spectrum: they feed on plant sap, a monotonous, nutrient deficient diet; they are hemimetabolous, with the nymphs and adults sharing the same habitat; and many aphids are monophagous, their endosymbionts are intracellular and as such are typically transmitted by transovarial mechanisms (Moran et al. 2008). Aphids, but also psyllids, whiteflies, mealybugs, and stinkbugs feed on phloem sap, a diet rich in sugar but poor in essential amino acids (Sandström and Pettersson 1994). Xylem sap provides an even poorer diet in sugar and amino acids and vitamins to sharpshooters. Data from metabolic, ecological, genetic, and genomic experiments performed on these insects support the assumption that P symbionts provide essential missing nutrients to their hosts: For example, the P symbiont *Buchnera* spp. provides essential amino acids to their aphid symbionts (Douglas 2006); a remarkable duo of complementary capabilities is found in sharpshooters where the γ-proteobacterium *Baumannia cicadellinicola* supplies vitamins and cofactors and the Bacteroidetes *Sulcia muelleri* synthesizes essential amino acids (Wu et al. 2006). Both symbionts can be classified as P symbionts.

In sap-feeding stinkbugs, the symbionts are obligate (Hosokawa et al. 2006) and may also provide their hosts with nutritional supplements (Fukatsu and Hosokawa 2009). However, in contrast to aphids, these insects are mostly polyphagous, feeding on more than one plant species, thus enjoying a wider variety of resources (Kuechler et al. 2011), hold their bacterial symbionts in the lumen of midgut crypts, and transfer them to progeny by postnatal transmission mechanisms. Other polyphagous stinkbugs include the broad-headed *Riptortus clavatus* (Heteroptera: Alydidae) that feed on seeds and seedpods. Their *Burkholderia* symbionts fill midgut crypts, are not vertically transmitted but are environmentally acquired by the nymph at each generation (Kikuchi et al. 2007). Similar findings were obtained from 39 phytophagous species in the Lygaeoidea and Coreoidea stinkbugs (Kikuchi et al. 2011). The symbionts may provide vitamins and essential amino acids and fix nitrogen.

In sharp contrast to the sap-sucking stinkbug, the stinkbug *Cimex lectularius* (the bedbug, Cimicidae) solely feeds on blood. Blood is a resource low in vitamins of the B complex, as well as in several cofactors, and a *Wolbachia* symbiont provisions for the missing nutrients. Its elimination leads to retarded growth and sterility. *Wolbachia* usually affect their host reproduction, and in some cases were found to conditionally improve host fitness (Weeks et al. 2007; Brownlie et al. 2009). Yet, in the bedbug, *Wolbachia* seems to have shifted roles: it is found in specific mycetomes; is vertically transmitted by a transovarial mechanism; and has become an obligate mutualist possibly involved in vitamin biosynthesis (Hosokawa et al. 2010). Since over 60% of insect species may carry *Wolbachia,* the bedbug case may not be an exception (Hosokawa et al. 2010). This suggests that specific symbionts may not be as functionally compartmentalized as previously thought and may evolve to fill rather different niches in different hosts.

Termites are the subjects of intensive research as they offer fascinating insights into the role, effects, and evolution of symbiosis. All termites feed on nitrogen deficient diets, living off the decomposition of lignocellulose. Lower termites feed only on wood; higher termites obtain their food also from lignocellulose, either directly or indirectly (see below). The hindgut of lower termites supports large populations of anaerobic flagellates most of which harbor numerous epibiotic and intracellular prokaryotic endosymbionts: In *Coptotermes formosanus,* about 70% of the total rRNA of the gut community is found within the gut protists (Ohkuma 2008) with free-living prokaryotes (including those attached to the gut wall) making up a few percent of the total population (Ohkuma 2003). For example, in the cellulolytic protist *Pseudotrichonympha grassii* of *C. formosanus,* an intracellular Bacteroidales ("*Candidatus* Azobacteroides pseudotrichonymphae") endosymbiont accounts for the majority of the bacterial cells (Hongoh et al. 2008a). In *Reticulitermes speratus*, Elusimicrobiota (formerly TG1) bacteria hosted within the cellulolytic flagellate *Trichonympha agilis* represent about 10% of the gut prokaryotic population (Hongoh et al. 2008b). The former bacterium can fix nitrogen, and both synthesize amino acids and co-factors while utilizing cellulose degradation products as a carbon source and producing acetate, the energy and carbon source of the termite (Hongoh et al. 2008a, b).

Higher termites lack protozoan symbionts but still retain an enlarged hindgut, which host diverse prokaryotes. In wood feeding higher termites such as *Nasutitermes*, these bacteria provide cellulose and hemicellulose degrading activities (Tokuda et al. 2005), acetogenesis (Breznak and Switzer 1986), and potentially nitrogen fixation (Warnecke et al. 2007). Yet, the higher termites in general display only low nitrogen fixation activity and vary their diet by extensive foraging (Ohkuma et al. 1999). Most higher termites are humivorous, feeding on soil, using plant tissue, fungal hyphae, and other microorganisms, but mainly humus, as food sources (Brune 2006). Humus constitutes the stable organic fraction in soil. It is composed of fulvic and humic acids, humins, lignin-like, and peptidic material and may include sugar amines, nucleic acids, phospholipids, vitamins, sulpholipids, and polysaccharides, providing carbon and energy to the termites (Ji and Brune 2001). Consequently, soil eating termites may not need to obtain additional nitrogen by nitrogen fixation as the C:N of the material they ingest is quite low (9–14) (the termite body's C:N = 4.3–6.9) (Tayasu et al. 1997). In fact, in place of the primary fermentation of carbohydrates that feeds the other termites, amino acids extracted from the humus peptidic fraction may constitute a major nutritive resource (Ohkuma and Brune 2006; Ji et al. 2000). In the context of this discussion, it is to be noted that soil is very heterogeneous, significantly varying in composition over time and space and can thus vary in nutritive value.

The feeding behavior of some wood feeding lower (like *Zootermopsis*, *Incisitermes* (*Kalotermes*), and *Reticulitermes*) and higher termites (Termitinae, Apotermitinae, and Nasutermitinae) is also noticeable: They use rotting (thus colonized by fungi) wood or highly decayed wood mixed with soil, (Bustamante and Martius 1998; Bignell and Egelton 2000), and thereby obtain a more diverse and richer diet. Lastly, Macrotermitine termites grow a fungal comb of *Termitomyces* (Agaricales, Tricholomataceae) fungi. The fungus acts as an exosymbiont, decomposing pre-digested plant material (e.g., wood, dry grass, and leaf litter) and the fungal nodules as well as the mature parts of the fungus comb are used as a food source for the termite (Brune 2006; Wood and Thomas 1989). Although both fungus and termite contribute glycosyl hydrolases, glucanases, and glucosidases, the gut bacteria also appear to participate in the digestion of the lignocellulose material (Liu et al. 2011).

We thus can summarize that in termites symbiotic interactions span from intracellular (lower, wood, and decaying wood eating termites that hold bacteria-filled protozoa) to extracellular (higher, decaying wood, and soil eating termites in which the bacteria are in the gut lumen) to exosymbiontic, with part of the microbiota acting as an "extended holobiont" (i.e. higher, fungal-garden tending termites). These associations are obligate (Sands 1969; Eutick et al. 1978): for instance, protozoan and prokaryotic symbionts are transmitted externally by coprophagy or trophollaxis. Finally, and most importantly, symbionts not only provide missing or limiting nutrients but also take part in food digestion.

2.5 Nutrition-Based Symbiosis in Holometabolous Insects

Tsetse flies belong to the Diptera, a phylum comprising about 20% of the insect diversity with more than 150,000 described species (Yeates and Wiegmann 2005). Diptera are holometabolous, i.e., they exhibit a larval and an imago stage, and experience complete metamorphosis during pupation between the two. This makes it possible for larvae and adults to differentially respond to selective pressures, to develop specific adaptations and lifestyles, to be ecologically different. Tsetse flies belong to the genus *Glossina*, feed on the blood of vertebrate animals, and are the vectors of trypanosomes. Although ontogeny includes three larval stages, pupation and an adult life, the tsetse life cycle is peculiar: a single egg is fertilized and retained within the uterus, where the larva develops. The larva is fed with a blood-derived diet very rich in proteins and lipids that is elaborated in the milk gland. It is the only food the larva eats, until the third instar when it crawls into the soil, stops feeding, and pupates to develop an adult body (www.fao.org/docrep/009/p5178e/P5178E04.htm). Therefore, while formally holometabolous, adult and larva do not experience great differences in nutrition, directly and indirectly feeding on blood, respectively. As in the bedbug, a bacterial symbiont, here named *Wigglesworthia,* may provide the deficient vitamins required for the fly's nutrition and fecundity (Akman et al. 2002). *Wigglesworthia* is an obligate P symbiont hosted in a bacteriome and therefore intracellular, yet it is transmitted extracellularly to the intrauterine larva via the mother's milk (Cheng and Aksoy 1999; Pais et al. 2008). Tsetse has a second symbiont (and *Wolbachia* as well, not discussed here), *Sodalis glossinidius,* found in midgut cells, the hemolymph, and other tissues, excluding the ovaries (Baumann et al. 2006). *S. glossinidus* appears to be evolving from a free-living to a mutualistic life style affects its host' s longevity and (Toh et al. 2006; Pais et al. 2008).

The family Tephritidae are dipterans. It is constituted of about 5,000 species in about 500 recognized genera (Evenhuis et al. 2008). A minority of these insects, circa 70 species, are known agricultural pests, inflicting significant damage to agricultural production. The family Tephritidae is well known for multiple invasions that cause important economic problems in fruit or vegetable crops in tropical and subtropical areas worldwide (Duyck et al. 2004). During invasion, newcomers can partially exclude and/or displace established fly populations. Many other dipterans may cause minor damage or are potentially harmful (White and Elson–Harris 1992).

The most deleterious genera *Anastrepha*, *Ceratitis*, *Bactrocera*, *Dacus*, and *Rhagoletis* infest fruits and cause billions of dollars of direct (crop loss) damage annually but much more when control measures are included. In addition, in the sub-family Tephritinae, the larvae feed on Asteraceae flower heads and often induce formation of galls (Headrick and Goeden 1994). The diversity of the Tephritidae is reflected in their anatomical adaptations: Many of these flies have digestive tracks containing specialized cavities or organs within which bacterial symbionts are hosted (Stammer 1929; Mazzon et al. 2008). The largely dominant populations in the Tephritidae belong to the Enterobacteriaceae, yet they differ between fly species (Jurkevitch 2011). Certain symbionts are readily culturable while others still remain uncultured.

Many of the fruit flies are polyphagous, feeding on different fruits and other food sources. Some of the fruit flies, such as *Ceratitis capitata*, the Mediterranean fruit fly (the medfly) can use a large collection of fruits as hosts and therefore affect many different crops. These fruits vary in protein ($0.86 \pm 0.59\%$ dry weight, dw) and carbohydrate, mostly sugar, content ($13.7 \pm 13.7\%$, dw) (means obtained from data on 37 fruits at: www.thefruitpages.com/contents.shtml; www.nal.usda.gov/fnic/foodcomp/search/). Consequently, the fruit habitat that medfly larvae experience varies but is invariably low in protein and high in sugar. Other fruit flies, such as *Bactrocera oleae* (the olive fruit fly), are specific and oviposit only in a particular host, in this case the olive fruit. Mature olives (the habitat of the developing larva) also contain little protein (circa ~3% dw) but a lot of oil (10–50% dw), carbohydrates (20–25, with sugar 4–10%, dw), and sterols (3–5% dw) (Zamora et al. 2001; Maestro Duran 1990). In contrast, at the adult stage, both flies appear to feed on sugar rich diets such as fruit juices, honeydew, nectar, and fruit and plant exudates, as well as on microorganisms (Drew and Yuval 2000), and occasionally on bird droppings and pollen (Christenson and Foote 1960; Drew and Yuval 2000). Thus, the medfly and the olive fly's diets are rather different at the larval stage and less so at the adult stage. As a result, it can be expected that if symbiotic associations occur during larval development and during adult life, they should be adapted to the different requirements of their hosts. According to the holobiont model (Zilber–Rosenberg and Rosenberg 2008), changes in adaptive traits may be associated with changes in the microbiota: A different dominant symbiont or a different community between the two stages – and between the medfly and olive larvae may thus be expected.

The microbiota associated with the medfly is essentially composed of Enterobacteriaceae and include *Citrobacter*, *Enterobacter*, *Klebsiella*, *Pantoea*, and *Pectobacterium* as dominant genera (Behar et al. 2005, 2008a; Ben Ami et al. 2010) and of *Erwinia dacicola*, *Enterobacter* sp. or *Acetobacter tropicalis* in the olive fly (Capuzzo et al. 2005; Estes et al. 2009; Kounatidis et al. 2009). Additionally, a cryptic but stable community of pseudomonads is found at very low concentrations in the medfly's gut (Behar et al. 2008b).

In the medfly, most of the populations are nitrogen fixers, pectinolytic or both, providing atmospheric nitrogen to adults and larvae and degrading the pectin in the infested fruits, potentially increasing sugar availability to the growing larva (Behar et al. 2005, 2008a). Thus, in the fruit, the symbionts can act externally to the insect's body. In addition, the gut microbiota may help recycle nitrogenous waste products into usable compounds such as uric acid and ammonia. Uricase, an enzyme that degrades uric acid into allantoin, may be produced by *Enterobacter* sp. present in the medfly's gut (Lauzon et al., 2000). Allantoin can be further processed to urea. In turn, Enterobacteriaceae like *K. oxytoca* or *Enterobacter gergoviae* can transform urea into ammonia (Zinder and Dworkin 2000).

The medfly's microbiota was shown to affect fitness parameters. Antibiotic treated males (thus bearing a much smaller gut microbiota) exhibited reduced mating competitiveness, as measured in a mating latency test; in females, the oviposition rate was significantly altered by antibiotic treatment (Ben Yosef et al. 2008a).

In the olive fly, the contribution of the gut bacteria in providing essential amino acids to the adult was demonstrated (Ben Yosef et al. 2010). It was postulated that bacteria associated with the larva actively detoxify the antimicrobial compounds produced at high concentrations by the green olive (Amiot et al. 1989; Ryan et al. 1999).

Although the same species are present at different stages of the medfly life, their abundance is significantly altered during ontogeny as reflected in the shift in community structure from parent to egg, larva, pupa, and adult (Behar et al. 2008a). The medfly symbionts appear to be extracellular at all stages and to be distributed in all sections of the gastrointestinal track. Gut colonization occurs stepwise, as bacterial foci expand in the gut during larval growth, and after pupation, during adulthood (unpublished data).

In the olive fly, the apparently same *E. dacicola* symbionts are intracellular at the larval stage, residing in the esophageal bulb, and extracellular, found in the intestinal lumen of the adult, transiting from the former to the latter as bacterial masses (Capuzzo et al. 2005; Estes et al. 2009). Thus, as in the medfly, the same community is found at both stages but the specific populations are found in different organs during the larval and adult stage and may differ in their relative abundances.

In both flies, egg smearing during passage in the ovipositor seems to be the mechanism by which the symbionts are transmitted to the next generation (Estes et al. 2009; Lauzon et al. 2009; Behar et al. 2008a; unpublished results). How can vertical transmission be guaranteed when nutritional needs and therefore the contribution of symbionts may change during the life cycle of the host?

Versatile, multivalent contributions by the same symbiont to both stages, as may occur in the olive fly, can offer a means to assure transmission to the next generation. "Ecological hitchhiking" may be another strategy: as seen above, the same species are found at all stages of the medfly's life cycle but differ in abundance. During oviposition, a mixed inoculum is introduced into the fruit, where it is differentially amplified. Larval, and then adult colonization provide subsequent bottlenecks that further and sequentially alter community structure (Behar et al. 2008a). The relative roles of competition, cooperation, local adaptation to the life stage, interactions with the immune system, or else, to shape the community are not known. Yet, the diversity at the strain level, the variability between individuals and the large population size of the gut microbiota (Behar et al. 2005, 2008a, b) help assure that a diverse community may graduate to the next generation. Stochastic effects may lead to eggs bearing different inocula. This, in turn, can provide an advantage when the environment in which the larva develops may change from generation to generation, as in a highly polyphagous insect like the medfly. Holding such a diverse community at the adult stage may come with a cost: Antibiotic-treated, nutritionally stressed flies lived significantly shorter lives than control flies or flies fed a full diet (with or without antibiotics) (Ben Yosef et al. 2008b).

The gut of Lepidopteran larvae do not contain specialized structures such as diverticula, and the contribution of microorganisms to nutrition and digestion was thought to be negligible (Appel 1994). In a recent study, a large (3.10^{11} cfu.ml^{-1}) community of anaerobic bacteria was found in the gut of *Bombyx mori* larvae, the silkworm that feeds on mulberry leaves (Anand et al. 2010), suggesting that they may play a role at the larval stage. A few isolates were characterized and classified as Enterobacteriaceae, mainly *Klebsiella, Citrobacter,* and *Erwinia*. Most were able to degrade cellulose, xanthan, and pectin (Anand et al. 2010). Yet, the abundance of the carbohydrate degraders in relation to the total culturable community was relatively small, so their contribution to digestion of the carbohydrates is uncertain. It is interesting that the microbiology of such a historically, culturally, and economically important insect has not received more attention.

Predatory holometabolous insects may be at the other end of the symbiotic spectrum: they exhibit clearly separated two-stage life cycles and feed on a complete diet. Intracellular host reproduction manipulators are present in predatory (including holometabolous and hemimetabolous) insects (Dunn and Stabb 2005; Hoy and Jeyaprakash 2005; Dedeine et al. 2001). However, the existence of diet upgrading in such insects has been questioned (Gulian and Cranston 2010). Nevertheless, bacteria may contribute an extended version of a digestion service to their hosts. In Myrmeleon, the ferocious appearing larvae are called antlions, and live off the body fluid of their prey for 1–3 years. The prey is paralyzed by an injected homologue of the GroEL chaperone that acts as a toxin and is produced by *Enterobacter aerogenes* in the predator's salivary glands (Yoshida et al. 2001). Interestingly, antlions have a discontinuous gut in which the midgut ends as a blind sac rather than being connected to the hindgut, and Enterobacteriaceae largely dominate this gastrointestinal track (Dunn and Stabb 2005). Enterobacteriaceae as well as *Bacillus* spp. were isolated from the crop and from the regurgating fluid and were shown to be insecticidal. Furthermore, bacterial cells may be injected into the prey by the predator (Egami et al. 2009; Nishiwaki et al. 2007) and proteases secreted by these bacteria may be involved in the liquification process of the prey content, i.e., they participate in extra-oral digestion (Egami et al. 2009). It remains to be seen what relation the relatively short-lived (20–45 days) adult antlions that feed on small flies, fruit, and honeydew (Burton and Burton 2002) entertain with their yet unknown bacteria.

2.6 Putting Symbionts to Work Against Insect Pests

In the medfly, the effects of the microbiota on longevity are intricate. As seen above, gut bacteria inferred a cost to nutritionally-stressed hosts (Ben Yosef et al. 2008b). However, the addition of a mixture of Enterobacteriaceae at high concentration to sugar-fed flies increased longevity. Substituting this inoculum with a low concentration of a *Pseudomonas aeruginosa* strain isolated from the fly gut lead to rapid death (Behar et al. 2008b) suggesting that the growth of these populations has to be tightly restricted. Symbionts have been shown to protect their insect hosts against pathogens (Brownlie and Johnson 2009) but they also have a role in keeping a proper balance between components of the gut microbiota. Commensal bacteria and the host immune system work in concert to modulate immune tolerance to commensal bacteria and respond to bacterial infection (Lhocine et al. 2008; Ryu et al. 2008). In the medfly, the dominant enterobacteria may help control the size of the pseudomonad community that however may not be essentially detrimental. In a sterile insect technique (SIT) facility mass-producing sterile males, a high level of pseudomonads were found in healthy flies (Ben Ami et al. 2010). SIT is an environmentally friendly approach that calls for the release of sterile males in the field to compete for mating with females against wild males, thereby reducing pest density (Hendrichs et al. 1995). SIT can lead to eradication under certain conditions (Vreysen et al. 2000). A main drawback of SIT has been the rather low mating performance of sterile males that compete poorly against wild males. Sterile males have to be frequently introduced in large numbers into the field, resulting in high operational costs.

In SIT, pupae are irradiated to render the male sterile, and this was shown to significantly alter the composition of the insect gut's microbiota (Ben Ami et al. 2010). In not irradiated males, *Klebsiella* spp. constituted about 20% of the total gut population in contrast to only 4% in irradiated males. Ben Ami et al. (2010) demonstrated that under laboratory conditions, the introduction of a *Klebsiell*a strain isolated from a wild medfly into the irradiated male improved the fly's mating performance in a mating latency test, illustrating that gut bacteria can affect male attractiveness. How females gauge males and which cues are affected by the male microbiota is unknown. Sharon et al. (2010) showed that in *Drosophila*, flies performed assortative mating based on the type of gut bacteria they had, and the female flies also displayed different cuticular hydrocarbon sex pheromones. Yet, in the medfly, males are the ones that produce sex hormones suggesting that different mechanisms affecting mating exist in different flies.

In a recent paper, Gavriel et al. (2011) described scaling up of the probiotic approach developed by Ben Ami et al. (2010). Probiotic-treated sterile flies successfully competed against control, untreated flies or against field-caught wild flies for mating with females, in settings ranging from 100 l tents to ~8 m^3 cages with an enclosed citrus tree mimicking field conditions. Mating with probiotic amended flies also significantly reduced the frequency of female remating with wild flies.

These studies offer a first proof that manipulation of a gut symbiont may be used to improve pest control.

The field of insect-symbiont interactions is yielding fascinating discoveries. At least some of these findings will certainly find their way to help improve the human lot.

References

Akman L, Yamashita A, Watanabe H, Oshima K, Shiba T, Hattori M, Aksoy S (2002) Genome sequence of the endocellular obligate symbiont of tsetse flies, *Wigglesworthia glossinidia*. Nat Genet 32:402–407

Amiot M-J, Fleuriet A, Macheix J-J (1989) Accumulation of oleuropein derivatives during olive maturation. Phytochemistry 28:67–69

Anand AA, Vennison JS, Sankar GS, Prabhu IGD, Vasan TP, Raghuraman GT, Geffrey JC, Vendan ES (2010) Isolation and characterization of bacteria from the gut of *Bombyx mori* that degrade cellulose, xylan, pectin and starch and their impact on digestion. J Insect Sci 10:1–20

Appel HM (1994) The chewing herbivore gut lumen: physicochemical conditions and their impact on plant nutrients, allelochemicals and insect pathogens. In: Bernays EA (ed) Insect–plant interactions, vol 1. CRC Press, Boca Raton, pp 209–221

Baumann P, Moran NA, Baumann L (2006) Bacteriocyte-associated endosymbionts of insects. In: Dworkin M, Rosenberg E, Schleifer K-H, Stackebrandt E (eds) The prokaryotes: symbiotic associations, biotechnology, applied microbiology, vol 1. Springer, New York, pp 403–438

Behar A, Yuval B, Jurkevitch E (2005) Enterobacteria-mediated nitrogen fixation in natural populations of the fruit fly *Ceratitis capitata*. Mol Ecol 14:2637–2643

Behar A, Jurkevitch E, Yuval B (2008a) Bringing back the fruit into fruit fly-bacteria interactions. Mol Ecol 17:1375–1386

Behar A, Yuval B, Jurkevitch E (2008b) Gut bacterial communities in the Mediterranean fruit fly (*Ceratitis capitata*) and their impact on host longevity. J Insect Physiol 54:1377–1383

Ben Ami E, Yuval B, Jurkevitch E (2010) Manipulation of the microbiota of mass-reared Mediterranean fruit flies *Ceratitis capitata* (Diptera: Tephritidae) improves sterile male sexual performance. ISME J 4:28–37

Ben Yosef M, Jurkevitch E, Yuval B (2008a) Effect of bacteria on nutritional status and reproductive success of the Mediterranean fruit fly *Ceratitis capitata*. Physiol Entomol 33:145–154

Ben Yosef M, Behar A, Jurkevitch E, Yuval B (2008b) Bacteria–diet interactions affect longevity in the medfly – *Ceratitis capitata*. J Appl Entomol 132:690–694

Ben Yosef M, Aharon Y, Jurkevitch E, Yuval B (2010) Give us the tools and we will do the job: symbiotic bacteria affect olive fly fitness in a diet dependent fashion. Proc R Soc B Biol Sci 277:1545–1552

Bignell DE, Egelton P (2000) Termites in ecosystems. In: Abe T, Bignel BE, Higashi M (eds) Termites: evolution, sociality, symbiosis, ecology. Kluwer Academic, Dordrecht, pp 363–388

Breznak J, Switzer J (1986) Acetate synthesis from H_2 plus CO_2 by termite gut microbes. Appl Environ Microbiol 52:623–630

Bright M, Bulgheresi S (2010) A complex journey: transmission of microbial symbionts. Nat Rev Microbiol 8:218–230

Brownlie JC, Johnson KN (2009) Symbiont-mediated protection in insect hosts. Trends Microbiol 17:348–354

Brownlie JC, Cass BN, Riegler M, Witsenburg JJ, Iturbe-Ormaetxe I, McGraw EA, O'Neill SL (2009) Evidence for metabolic provisioning by a common invertebrate endosymbiont, *Wolbachia pipientis*, during periods of nutritional stress. PLoS Pathog 5:e1000368

Brune A (2006) Symbiotic associations between termites and prokaryotes. In: Dworkin M, Falkow S, Rosenberg E, Schleifer K-H, Stackebrandt E (eds) The prokaryotes. Springer, New York, pp 439–474

Burton M, Burton R (2002) International wildlife encyclopedia. Marshall Cavendish Corporation, Tarrytown

Bustamante RNC, Martius C (1998) Nutritional preferences of wood-feeding termites inhabiting floodplain forests of the Amazon River. Brazil Acta Amazonica 28:301–307

Capuzzo C, Firrao G, Mazzon L, Squartini A, Girolami V (2005) 'Candidatus *Erwinia dacicola*', a coevolved symbiotic bacterium of the olive fly *Bactrocera oleae* (Gmelin). Int J Syst Evol Microbiol 55:1641–1647

Cheng Q, Aksoy S (1999) Tissue tropism, transmission and expression of foreign genes in vivo in midgut symbionts of tsetse flies. Insect Mol Biol 8:125–132

Christenson LD, Foote RH (1960) Biology of fruit flies. Annu Rev Entomol 5:171–92

Dedeine F, Vavre F, Fleury F, Loppin B, Hochberg ME, Boulétreau M (2001) Removing symbiotic *Wolbachia* bacteria specifically inhibits oogenesis in a parasitic wasp. Proc Natl Acad Sci USA 98:6247–6252

Douglas AE (2006) Phloem-sap feeding by animals: problems and solutions. J Exp Bot 57:747–754

Drew RAI, Yuval B (2000) The evolution of fruit fly feeding behavior. In: Aluja M, Norrbom A (eds) Fruit flies, phylogeny and evolution of behavior. CRC Press, Boca Raton, pp 731–749

Dunn AK, Stabb EV (2005) Culture-independent characterization of the microbiota of the ant lion *Myrmeleon mobilis* (Neuroptera: Myrmeleontidae). Appl Environ Microbiol 71:8784–8794

Duyck PF, David P, Quilici S (2004) A review of relationships between interspecific competition and invasions in fruit flies (Diptera: Tephritidae). Ecol Entomol 29:511–520

Egami I, Iiyama K, Zhang P, Chieda Y, Ino N, Hasegawa K, Lee JM, Kusakabe T, Yasunaga-Aoki C, Shimizu S (2009) Insecticidal bacterium isolated from an antlion larva from Munakata, Japan. J Appl Entomol 133:117–124

Estes AM, Hearn DJ, Bronstein JL, Pierson EA (2009) The olive fly endosymbiont, "*Candidatus Erwinia dacicola*," switches from an intracellular existence to an extracellular existence during host insect development. Appl Environ Microbiol 75:7097–7106

Eutick ML, Veivers P, O'Brien RW, Slaytor M (1978) Dependence of the higher termite, *Nasutitermes exitiosus* and the lower termite, *Coptotermes lacteus* on their gut flora. J Insect Physiol 24:363–368

Evenhuis NL, Pape T, Pont AC, Thompson FC (2008) Database of World Diptera. http://www.diptera.org/biosys.htm

Fukatsu T, Hosokawa T (2009) Capsule-transmitted obligate gut bacterium of plataspid stinkbugs: a novel model system for insect symbiosis studies. In: Bourtzis K, Miller TA (eds) Insect symbiosis, vol 3. CRC Press, Boca Raton

Gallai N, Salles J-M, Settele J, Vaissière BE (2009) Economic valuation of the vulnerability of world agriculture confronted with pollinator decline. Ecol Econ 68:810–821

Gavriel S, Jurkevitch E, Gazit Y, Yuval B (2011) Bacterially enriched diet improves sexual performance of sterile male Mediterranean fruit flies. J Appl Entomol. doi:10.1111/j.1439-0418.2010.01605.x

Gulian PJ, Cranston PS (2010) The insects: an outline of entomology. Wiley, London, pp. 565

Headrick DH, Goeden RD (1994) Reproductive behavior of California fruit flies and the classification and evolution of Tephritidae (Diptera) mating systems. Studia Dipterol 1:195–252

Hendrichs J, Franz G, Rendon P (1995) Increased effectiveness and applicability of the sterile insect technique through male-only releases for control of Mediterranean fruit flies during fruiting seasons. J Appl Entomol 119:371–377

Hongoh Y, Sharma VK, Prakash T, Noda S, Taylor TD, Kudo T, Sakaki Y, Toyoda A, Hattori M, Ohkuma M (2008a) Complete genome of the uncultured Termite Group 1 bacteria in a single host protist cell. Proc Natl Acad Sci 1–5:5555–5560

Hongoh Y, Sharma VK, Prakash T, Noda S, Taylor TD, Kudo T, Sakaki Y, Toyoda A, Hattori M, Ohkuma M (2008b) Genome of an endosymbiont coupling N_2 fixation to cellulolysis within protist cells in termite gut. Science 322:1108–1109

Hosokawa T, Kikuchi Y, Nikoh N, Shimada M, Fukatsu T (2006) Strict host-symbiont cospeciation and reductive genome evolution in insect gut bacteria. PLoS Biol 4:e337

Hosokawa T, Koga R, Kikuchi Y, Meng XY, Fukatsu T (2010) *Wolbachia* as a bacteriocyte-associated nutritional mutualist. Proc Natl Acad Sci 107:769–774

Hoy MA, Jeyaprakash A (2005) Microbial diversity in the predatory mite *Metaseiulus occidentalis* (Acari: Phytoseiidae) and its prey, *Tetranychus urticae* (Acari: Tetranychidae). Biol Control 32:427–441

Hypsa V, Novakova E (2009) Inset symbionts and molecular phylogenetics. In: Bourtzis K, Miller TA (eds) Insect symbiosis, vol 3. CRC Press, Boca Raton

Ji R, Brune A (2001) Transformation and mineralization of 14C-labeled cellulose, peptidoglycan, and protein by the soil-feeding termite. Biol Fertil Soils 33:166–174

Ji R, Kappler A, Brune A (2000) Transformation and mineralization of synthetic 14C-labeled humic model compounds by soil-feeding termites. Soil Biol Biochem 32:1281–1291

Jurkevitch E (2011) Riding the Trojan horse: combating pest insects with their own symbionts. Microb Biotech. doi:10.1111/j.1751-7915.2011.00249

Kikuchi Y, Hosokawa T, Fukatsu T (2007) Insect-microbe mutualism without vertical transmission: a stinkbug acquires a beneficial gut symbiont from the environment every generation. Appl Environ Microbiol 73:4308–4316

Kikuchi Y, Hosokawa T, Fukatsu T (2011) An ancient but promiscuous host-symbiont association between *Burkholderia* gut symbionts and their heteropteran hosts. ISME J 5:446–460

Kounatidis I, Crotti E, Sapountzis P, Sacchi L, Rizzi A, Chouaia B, Bandi C, Alma A, Daffonchio D, Mavragani-Tsipidou P, Bourtzis K (2009) *Acetobacter tropicalis* is a major symbiont of the olive fruit fly (*Bactrocera oleae*). Appl Environ Microbiol 75:3281–3288

Kuechler SM, Dettner K, Kehl S (2011) Characterization of an obligate intracellular bacterium in midgut epithelium of bulrush bug Chilacis typhae (Heteroptera, Lygaeidae, Artheneinae). Appl Environ Microbiol. doi:10.1128/AEM.02983-10

Lauzon CR, Sjogren RE, Prokopy RJ (2000) Enzymatic capabilities of bacteria associated with apple maggot flies: a postulated role in attraction. J Chem Ecol 26:953–967

Lauzon CR, McCombs SD, Potter SE, Peabody NC (2009) Establishment and vertical passage of *Enterobacter* (*Pantoea*) *agglomerans* and *Klebsiella pneumoniae* through all life stages of the Mediterranean fruit fly (Diptera: Tephritidae). Ann Entomol Soc Am 102:85–95

Lhocine N, Ribeiro PS, Buchon N, Wepf A, Wilson R, Tenev T, Lemaitre B, Gstaiger M, Meier P, Leulier F (2008) PIMS modulates immune tolerance by negatively regulating Drosophila innate immune signaling. Cell Host Microbe 4:147–158

Liu N, Yan X, Zhang M, Xie L, Wang Q, Huang Y, Zhou X, Wang S, Zhou Z (2011) Microbiome of fungus-growing termites: a new reservoir for lignocellulase genes. Appl Environ Microbiol 77:48–56

Maestro Duran R (1990) Relationship between the composition and ripening of the olive and the quality of the oil. Acta Hortic 290:441–451

Mazzon L, Piscedda A, Simonato M, Martinez-Sanudo I, Squartini A, Girolami V (2008) Presence of specific symbiotic bacteria in flies of the subfamily Tephritinae (Diptera Tephritidae) and their phylogenetic relationships: proposal of '*Candidatus Stammerula tephritidis*'. Int J Syst Evol Microbiol 58:1277–1287

Moran NA, McCutcheon JP, Nakabachi A (2008) Genomics and evolution of heritable bacterial symbionts. Annu Rev Genet 42:165–190

Moya A, Pereto J, Gil R, Latorre A (2008) Learning how to live together: genomic insights into prokaryote-animal symbioses. Nat Rev Genet 9:218–229

Nishiwaki H, Ito K, Shimomura M, Nakashima K, Matsuda K (2007) Insecticidal bacteria isolated from predatory larvae of the antlion species *Myrmeleon bore* (Neuroptera: Myrmeleontidae). J Invertebr Pathol 96:80–88

Ohkuma M, Noda S, Kudo T (1999) Phylogenetic diversity of nitrogen fixation genes in the symbiotic microbial community in the gut of diverse termites. Appl Environ Microbiol 65:4926–4934

Ohkuma M (2003) Termite symbiotic systems: efficient bio-recycling of lignocellulose. Appl Microbiol Biotechnol 61:1–9

Ohkuma M (2008) Symbioses of flagellates and prokaryotes in the gut of lower termites. Trends Microbiol 16:345–352

Ohkuma M, Brune A (2006) Role of the termite microbiota in symbiotic digestion. In: Bignell DE, Voisin Y, Lo N (eds) Biology of termites: a modern synthesis. Springer, Dordrecht, pp 439–476

Pais R, Lohs C, Wu Y, Wang J, Aksoy S (2008) The obligate mutualist *Wigglesworthia glossinidia* influences reproduction, digestion, and immunity processes of its host, the Tsetse fly. Appl Environ Microbiol 74:5965–5974

Pimentel D (2009) Pesticides and pest control. In: Peshin R, Dhawan AK (eds) Integrated pest management: innovation-development process, vol 1. Springer, Dordrecht

Ryan D, Robards K, Lavee S (1999) Accumulation of oleuropein derivatives during olive maturation. Int J Food Sci Technol 34:265–274

Ryu J-H, Kim S-H, Lee H-Y, Bai JY, Nam Y-D, Bae J-W, Lee DG, Shin SC, Ha E-M, Lee W-J (2008) Innate immune homeostasis by the homeobox gene *Caudal* and commensal-gut mutualism in *Drosophila*. Science 319:777–782

Sands WA (1969) The association of termites and fungi. In: Krishna K, Weesner FM (eds) Biology of termites. Academic, New York, pp 495–524

Sandström J, Pettersson J (1994) Amino acid composition of phloem sap and the relation to intraspecific variation in pea aphid (*Acyrthosiphon pisum*) performance. J Insect Physiol 40:947–955

Sharon G, Segal D, Ringo JM, Hefetz A, Zilber-Rosenberg I, Rosenberg E (2010) Commensal bacteria play a role in mating preference of *Drosophila melanogaster*. Proceedings of the National Academy of Sciences. /cgi/doi/10.1073/pnas.1009906107

Stammer HJ (1929) Die bakteriensymbiose der trypetiden (Diptera). Zoomorphology 15:481–523

Tayasu I, Abe T, Eggleton P, Bignell DE (1997) Nitrogen and carbon isotope ratios in termites: an indicator of trophic habit along the gradient from wood-feeding to soil-feeding. Ecol Entomol 22:343–351

Toh H, Weiss BL, Perkin SAH, Yamashita A, Oshima K, Hattori M, Aksoy S (2006) Massive genome erosion and functional adaptations provide insights into the symbiotic lifestyle of *Sodalis glossinidius* in the tsetse host. Genome Res 16:149–156

Toju H, Hosokawa T, Koga R, Nikoh N, Meng XY, Kimura N, Fukatsu T (2010) "Candidatus Curculioniphilus buchneri," a novel clade of bacterial endocellular symbionts from weevils of the genus Curculio. Appl Environ Microbiol 76:275–282

Tokuda G, Lo N, Watanabe H (2005) Marked variations in patterns of cellulase activity against crystalline- vs carboxymethyl-cellulose in the digestive systems of diverse, wood-feeding termites. Physiol Entomol 30:372–80

Vreysen MJ, Saleh KM, Ali MY, Abdulla AM, Zhu ZR, Juma KG, Dyck VA, Msangi AR, Mkonyi AR, Feldmann HU (2000) *Glossina austeni* (Diptera: Glossinidae) eradicated on the Island of Unguja, Zanzibar, using the sterile insect technique. J Econ Entomol 93:123–135

Warnecke F, Luginbuhl P, Ivanova N, Ghassemian M, Richardson T et al (2007) Metagenomic and functional analysis of hindgut microbiota of a wood-feeding higher termite. Nature 450:560–565

Weeks AR, Turelli M, Harcombe WR, Reynolds KT, Hoffmann AA (2007) From parasite to mutualist: rapid evolution of *Wolbachia* in natural populations of *Drosophila*. PLoS Biol 5:e114

White IM, Elson-Harris MM (1992) Fruit flies of economic significance: their identification and bionomics. CAB International, Wallingford

Wood TG, Thomas RJ (1989) The mutualistic association between Macrotermitinae and Termitomyces. In: Wilding N, Collins NM, Hammond PM, Webber JF (eds) Insect-fungus interaction. Academic, London, pp 69–92
Wu D, Daugherty SC, Van Aken SE, Pai GH, Watkins KL, Khouri H, Tallon LJ, Zaborsky JM, Dunbar HE, Tran PL, Moran NA, Eisen JA (2006) Metabolic complementarity and genomics of the dual bacterial symbiosis of sharpshooters. PLoS Biol 4:e188
Yeates DK, Wiegmann BM (2005) Phylogeny and evolution of Diptera: recent insight and new perspective. In: Yeates DK, Wiegmann BM (eds) The evolutionary biology of flies. Colombia University Press, New York, pp 14–44
Yoshida N, Oeda K, Watanabe E, Mikami T, Fukita Y, Nishimura K, Komai K, Matsuda K (2001) Protein function: chaperonin turned insect toxin. Nature 411:44–44
Zamora R, Alaiz M, Hidalgo FJ (2001) Influence of cultivar and fruit ripening on olive (*Olea europaea*) fruit protein content, composition, and antioxidant activity. J Agric Food Chem 49:4267–4270
Zilber-Rosenberg I, Rosenberg E (2008) Role of microorganisms in the evolution of animals and plants: the hologenome theory of evolution. FEMS Microbiol Rev 32:723–735
Zinder DE, Dworkin M (2000) Morphological and physiological diversity. In: Dworkin M, Falkow S, Rosenberg E, Schleifer K-H, Stackebrandt E (eds) The prokaryotes. Springer, New York, pp 185–220

Chapter 3
Chironomids and *Vibrio cholerae*

Malka Halpern

Abstract Chironomids (non-biting midges) are the most widely dispersed freshwater insects. Females deposit egg masses at the water's edge, each egg mass contains hundreds of eggs embedded in a gelatinous matrix. They undergo complete metamorphosis of four life stages: eggs, larvae, pupae, and adults. Non O1/O139 serogroups of *V. cholerae* inhabit all the four life stages of chironomids. haemagglutinin protease (HAP), an extracellular enzyme of *V. cholerae*, degrades the gelatinous matrix of chironomid egg masses and prevents the eggs from hatching. Chironomid populations and the *V. cholerae* in their egg masses were found to exhibit host–pathogen population dynamics. Quorum-sensing signals, which are known to commonly establish a successful infection in pathogenic bacteria, repress the cholera toxin expression in *V. cholerae* and upregulate HAP production. Thus, HAP regulation by quorum sensing should be understood with regard to the bacterial role in chironomids rather than humans. Further research is needed to understand the role of cholera toxin in the environmental existence of *V. cholerae*. Other endogenous bacterial species that inhabit the egg mass may also use the degraded gelatinous matrix as a nutritive source for their growth. By doing so, they probably control the population levels of *V. cholerae* in the egg mass. Vice versa, *V. cholerae* support the maintenance of other endogenic bacteria in the egg mass by secreting HAP. The interaction between *V. cholerae* and chironomids is probably a complicated mutualistic relationship rather than a simple host–pathogen interaction.

M. Halpern (✉)
Department of Biology and Environment, Faculty of Natural Sciences, University of Haifa, Oranim, Tivon 36006, Israel
e-mail: mhalpern@research.haifa.ac.il

E. Rosenberg and U. Gophna (eds.), *Beneficial Microorganisms in Multicellular Life Forms*,
DOI 10.1007/978-3-642-21680-0_3,

3.1 Introduction

3.1.1 Chironomids

Chironomids (*Diptera*; *Chironomidae*), known also as non-biting midges, are the most widely dispersed aquatic insects (Armitage et al. 1995). They undergo a complete metamorphosis of four life stages. Three stages (eggs, larvae, and pupae) take place in the water, the adults emerge into the air (Fig. 3.1). Female midges deposit eggs in jelly-like masses that float on the water surface. Each egg mass contains hundreds of eggs (Fig. 3.1a) (Nolte 1993). Upon hatching, the young larvae swim to the bottom to find a suitable site where they will construct a silken tube (Fig. 3.1b). The larvae pass four molts before they transform into a pupae. Adult midges do not feed and only live long enough to mate and lay egg masses. Broza and Halpern (2001) found that chironomid egg masses may serve as a natural reservoir of *Vibrio cholerae*.

3.1.2 **Vibrio cholerae** *and Cholera*

Cholera is an epidemic and life-threatening disease caused by *V. cholerae*, a Gram negative, motile curved rod, which is commonly associated with water. Serogroups O1 and O139 of *V. cholerae* are currently believed to be the only ones causing epidemics in humans (Kaper et al. 1995; Colwell 1996; Sack et al. 2004). However, non-O1 and non-O139 strains have also been associated with occasional outbreaks of cholera (Aldova et al. 1968; Kaper et al. 1995; Ko et al. 1998). Pathogenic strains of *V. cholerae* produce the cholera enterotoxin (CT) which causes acute dehydration in humans (Butler and Camilli 2005). Human infection with *V. cholerae* begins with the ingestion of contaminated food or water containing the bacterium. *V. cholerae* colonizes the small intestine and secrets CT which act as an A-B type

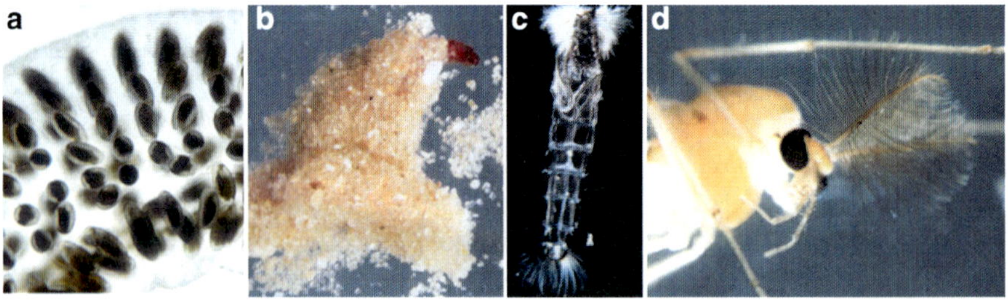

Fig. 3.1 Developmental stages of *Chironomus* sp. (**a**) Egg mass. The eggs are arranged in a row, folded into loops to form a spiral, and embedded in a thick gelatinous cylinder. (Original magnification × 40). (**b**) Larva dwelling in a sand tube. The head of the midge is seen outside the tube. (**c**) Pupal exuvium. (**d**) Male adult

toxin leading to ADP-ribosylation of a small G protein and constitutive activation of adenylate cyclase, thus giving rise to increased levels of cyclic AMP within the host cell. This results in rapid efflux of chloride ions and water from host intestinal cells and leads to severe dehydration (Kaper et al. 1995).

V. cholerae is a part of the normal flora and ecology of the surface water of our planet. Humans consuming unchlorinated surface water contaminated with pathogenic *V. cholerae* for drinking, may be infected at high rates (Kaper et al. 1995). In the aquatic environment, *V. cholerae* has been associated especially with zooplankton, particularly copepods (Huq et al. 1983; Colwell and Huq 2001; Thomas et al. 2006) and chironomids (Broza and Halpern 2001; Halpern et al. 2004, 2006, 2007a; Broza et al. 2005, 2008; Senderovich et al. 2008). Recent evidence points at fish as intermediate reservoirs of *V. cholerae* (Senderovich et al. 2010). As fish carrying the bacteria swim from one location to another, they serve as vectors on a small scale. Moreover, fish are consumed by waterbirds, which may disseminate the bacteria on a global scale (Halpern et al. 2008).

3.2 *V. cholerae* Inhabits All Chironomid Life Stages

3.2.1 Egg Masses

Chironomus sp. egg masses (Fig. 3.1a) harbor *V. cholerae* and act as its sole carbon source, thereby providing a possible natural reservoir for the cholera bacterium. Two hundred floating *Chironomus* egg masses (Fig. 3.1a) collected from a waste stabilization pond disintegrated overnight into thousands of individual eggs, most of which did not hatch (Broza and Halpern 2001). *Vibrio cholerae* serogroup O9 was isolated from the egg masses and was found to be the cause of the egg mass destruction and hatching prevention. *V. cholerae* is able to use chironomid egg masses as a sole nutritive source for growth (Broza and Halpern 2001). Hence, chironomid egg masses facilitate *V. cholerae* survival and multiplication in freshwater bodies.

To further support the observation made by Broza and Halpern (2001) during long-term monitoring of the *Vibrio*–chironomids association, three freshwater habitats were sampled in Israel: a spring, a polluted river, and an anaerobic waste stabilization pond (Halpern et al. 2004). Ca. 200 isolates of *V. cholerae* non-O1 and non-O139 were isolated from chironomid egg masses that were sampled from the different freshwater habitats. Ninety isolates belonged to 35 different serogroups (Halpern et al. 2004; Broza et al. 2005). The heterogeneity found in the bacterial isolates implies that there may be similarity in the ecological niche of all *V. cholerae* serogroups, non-O1, non-O139, as well as O1 and O139. More *V. cholerae* isolates were identified from chironomid egg masses in two more surveys in Israel (Halpern et al. 2006; Senderovich et al. 2008). However, these isolates were not identified by their specific serogroups. In sum, hundreds of *V. cholerae* isolates were identified from chironomid egg masses. All *V. cholerae* isolates that were tested for their ability to degrade chironomid egg masses have demonstrated this ability (Broza and Halpern 2001; Halpern et al. 2003, 2006, 2007a; Senderovich et al. 2008). Even though the pathogenic serogroups have not been identified from chironomids until today, it can be concluded that environmental *V. cholerae* in chironomids are highly variable.

3.2.1.1 VBNC *V. cholerae*

Chironomid egg masses harbor *V. cholerae* not only in a viable and culturable state but also in a viable but non-culturable (VBNC) form (Halpern et al. 2007a). Using the FISH technique, it was demonstrated that the average number of total *V. cholerae* cells per egg mass was $3.9 \pm 1.0 \times 10^4$. Less than 0.5% of the *V. cholerae* cells that were counted using a FISH specific probe for *V. cholerae* yielded colonies, demonstrating that most *V. cholerae* cells inhabiting the egg mass are in the VBNC state.

3.2.2 *Larvae, Pupae, and Adults*

V. cholerae was identified from chironomid larvae and pupae that were collected from a waste stabilization pond in Israel (Halpern, unpublished data). In an environmental study that was conducted during June 2006 in India, *V. cholerae* was identified from chironomid egg masses, larvae, exuviae (floating cuticle made of chitin that cover the pupae) and adults (Fig. 3.1) (Broza et al. 2008; Raz et al. 2010). These demonstrate that *V. cholerae* inhabit not only the egg masses but also other life stages of the insect.

Male and female adult flying chironomids were caught in the air during March–October 2001, in two sites neighboring waste stabilization ponds (WSP) in Israel. The adults carried viable non-O1 non-O139 serogroups of *V. cholerae* (Broza et al. 2005). Twelve different serogroups were identified from these *V. cholerae* isolates. In laboratory experiments, flying adult midges that emerged from *V. cholerae* O1 or *V. cholerae* O139 contaminated water, transferred the green fluorescent protein (GFP)-tagged pathogenic bacteria from one laboratory flasks to another (Broza et al. 2005, 2008). The spots on adults' bodies where the GFP-tagged bacteria were seen most frequently were the intersegmental membranes and the special pads of *Dipterans* legs, the pulvilum (Broza et al. 2005). The adherence of *V. cholerae* to chitin and the role of surface proteins in the process have been demonstrated (Tarsi and Pruzzo 1999) and may explain the mechanism by which these bacteria are transported by adult chironomids. The vector potential of adult flying chironomids was further demonstrated in a field experiment testing the efficacy of nets with varying mesh sizes in preventing *V. cholerae* colonization of artificial water bodies. The large-size mesh proved to be ineffective at preventing *V. cholerae* colonization of a water body, while the small-mesh-size net, which adequately prevented chironomid access, completely eliminated the detection of *V. cholerae* in the water above which it was spread. Simultaneously, *V. cholerae* was detected on flying adult chironomids and in egg masses laid in the accessible pools (Broza et al. 2005, 2008). Chironomid adults create huge swarms for mating, and during the mating or soon thereafter many of those adults are driven by winds off the ponds carrying environmental *V. cholerae* with them. Thus, aerial transfer by flying chironomids may play a role in the dissemination of *V. cholerae* (Broza et al. 2005, 2008; Halpern et al. 2008).

3.3 Virulence-Associated Genes in *V. cholerae* Isolated from Chironomids

It has to be emphasized that until now, cholera toxin (*ctxA*) has not been detected in any of the *V. cholerae* strains that were isolated from chironomids (Broza and Halpern 2001; Halpern et al. 2004, 2006, 2007a; Senderovich et al. 2008; Raz et al. 2010). However, three isolates from Zanzibar (serogroups O37, O37, O201) were characterized as producing the NAG-ST toxin (Halpern et al. 2004). The heat-stable enterotoxin NAG-ST is a virulence factor that has been associated with *V. cholerae* non-O1. It can cause diarrhea with symptoms that are indistinguishable from those of cholera. A NAG-ST positive isolate was also identified from chironomid exuviae during a local cholera outbreak in Vadodara, Gujarat, India (Raz et al. 2010).

The following virulence-associated genes were detected in two other field studies that examined *V. cholerae* isolates from chironomid egg masses in Israel: haemagglutinin protease (*hapA*, 100%); hemolysin (*hlyA*, 70%); and outer membrane protein (*ompU*, 28% and 16% respectively) (Halpern et al. 2006; Senderovich et al. 2008). Cholera toxin (*ctxA*) and the non-O1 heat-stable enterotoxin (*stn*/*sto*) were not detected in the above studies.

3.4 Haemagglutinin Protease: An Egg Mass Degrading Factor

The extracellular factor responsible for the degradation of chironomid egg masses was purified from the environmental strain *V. cholerae* O9 and the pathogenic strain O139 and was identified as the major secreted haemagglutinin protease of *V. cholerae* (Halpern et al. 2003). It was found that the purified HAP releases chironomid eggs from the gelatinous matrix and prevents the eggs from hatching. A comparison of the *V. cholerae* O1 strain and its HAP null mutant showed that the mutant could not degrade egg masses, demonstrating the role of HAP in egg mass degradation. Altogether, it was found that HAP has four different functional activities: The previously described haemagglutination and protease activities and two new activities – degradation of the glycoprotein matrix of the egg mass and prevention of the chironomid egg hatching. The classical biological activities of HAP (haemagglutination and proteolytic activities) were expressed as a distinct activity within the egg masses. The egg mass gelatinous matrix was found to consist of glycoproteins. In this context, the proteolytic activity of HAP is more properly referred to as a glycoproteolytic activity that might be connected to HAP's haemagglutination capabilities (Halpern et al. 2003). Benitez et al. (2001) showed that *V. cholerae* HAP is induced by nutrient limitation and is strongly repressed by glucose. This observation can explain the importance of HAP for *V. cholerae* in the environment with regard to utilization of egg masses. HAP may have a general role in enabling the bacteria to survive in the environment. Endogenic bacteria other than *V. cholerae* that inhabit the egg masses may also benefit from the utilization of the egg masses that is performed by the extracellular enzyme HAP of *V. cholerae*.

3.5 Seasonality of Chironomid Populations and *V. cholerae*

In a year-long survey of chironomid and *V. cholerae* population dynamics in Israel, chironomid populations were found to peak biannually (in spring and autumn) (Halpern et al. 2004, 2006). These peaks were followed by the presence of *V. cholerae* on the egg masses. Chironomid populations and the *V. cholerae* in their egg masses displayed host–pathogen population dynamics. The bacteria seemed to control the chironomid population density, most likely by secreting the HAP enzyme that degraded their egg masses, thus preventing their eggs from hatching (Halpern et al. 2003, 2004, 2006).

Cholera epidemics occur with seasonal regularity (Faruque et al. 2005). In India and Bangladesh, two peaks of major cholera outbreaks occur each year after the monsoons. A low peak of cholera cases occurs in March-May and a high peak in September-November (Samadi et al. 1983; Kaper et al. 1995; Sack et al. 2004). In Peru, epidemics are strictly confined to the warm season (Sack et al. 2004). Constantin de Magny et al. (2008) suggested that ocean and climate patterns are useful predictors of cholera epidemics. Although the seasonality of cholera in Bangladesh and elsewhere has been temporally associated with numerous physical and biological parameters, these factors do not directly cause epidemics, nor do they end them (Lipp et al. 2002).

V. cholerae can survive and multiply on chironomid egg masses (Broza and Halpern 2001; Halpern et al. 2003, 2004, 2006, 2007a; Senderovich et al. 2008; Raz et al. 2010). The bacteria have a seasonality pattern on chironomids (Halpern et al. 2004, 2006). It is likely that all *V. cholerae* serogroups, nonpathogenic as well as pathogenic, share the same ecological habitat (Lewin 1996). Pathogenic serogroups of *V. cholerae* are able to degrade chironomid egg masses and use them as a source of nutrients (Halpern et al. 2003) and can be transmitted from the larval to the adult stage (Broza et al. 2005). To determine the seasonal appearance of epidemic *V. cholerae* strains and outbreaks of cholera, a year-long survey, should be performed in an endemic area of the disease by monitoring the pathogenic O1 and O139 serogroups on chironomid egg masses. Such a study will determine whether dependent population dynamics exist between pathogenic *V. cholerae* and chironomids, as was found by Halpern et al. (2006) for the nonpathogenic serogroups. Chironomids, being the largest biomass in most aquatic systems, could provide an ideal monitoring tool of *V. cholerae*.

3.6 The Role of *V. cholerae* Quorum-Sensing Signals in the Relationships Between the Bacteria and the Insect

3.6.1 Quorum-Sensing Signals

In recent years, microbiologists realized that individual bacterial cells communicate with each other and coordinate group activities by releasing signal molecules into the environment (Fuqua and Greenberg 2002). The term quorum sensing is used to describe this phenomenon, whereby bacteria match up their behavior according to the signal molecules' composition and levels. At a particular threshold concentration, the signaling molecules bind to receptors on or in bacterial cells, leading to changes in gene expression in them. Quorum sensing allows bacteria to switch between two distinct gene expression programs: one that is favored at low cell densities for individually functioning, a-social behaviors, and another that is favored at high cell densities, for social group behaviors (Xavier and Bassler 2005; Ng and Bassler 2009). Quorum sensing can be intra- and inter-species specific. For intra-species quorum sensing, the emitter and responder are usually the same cell species. For inter-species quorum sensing the emitter and responder are usually not the same species.

Quorum-sensing signals mediate virulent factors in pathogenic bacteria. For example, to establish a successful infection, pathogenic bacteria become virulent only when they reach a certain local concentration in their host (Ng and Bassler 2009).

3.6.2 The Significance of Quorum-Sensing Signals for HAP Secretion Regulation

V. cholerae produces two quorum-sensing autoinducers and it responds to them by using parallel phosphorelay signaling systems (Miller et al. 2002). The two autoinducer molecules are termed CAI-1, (*S*)-3-hydroxytridecan-4-one, and AI-2, (2*S*,4*S*)-2-methyl-2,3,3,4-tetrahydroxytetrahydrofuran borate (Fig. 3.2) (Chen et al. 2002; Miller et al. 2002; Higgins et al. 2007).

The two autoinducer molecules CAI-1 (intra-species specific) and AI-2 (inter-species signal) act synergistically through the same response regulator, LuxO.

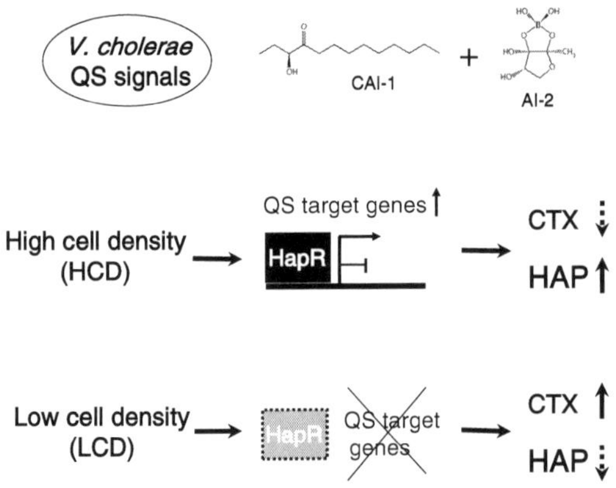

Fig. 3.2 Quorum-sensing signals and gene regulation in *V. cholerae. V. cholerae* produces two quorum-sensing (QS) autoinducer molecules; CAI-1 (intra-species specific) and AI-2 (inter-species signal). The two molecules act synergistically through the same response regulator, LuxO. When the cell density increases, quorum-sensing signals increase and the protease regulatory protein HapR is upregulated. As a result, haemagglutinin protease is upregulated and cholera toxin production is decreased. When cell density is low, quorum-sensing signals lower, HapR is downregulated and HAP is not produced. During low cell density the bacteria produces cholera toxin

In the low-cell-density state (i.e., when autoinducer levels are low), the autoinducer receptors function as kinases, and funnel phosphate to the response regulator, LuxO. Phosphorylated LuxO activates the expression of four genes encoding small regulatory RNAs (sRNAs) (Lenz et al. 2004). The sRNAs destabilize the mRNA encoding a major regulator of quorum sensing, HapR (Fig. 3.2) (Jobling and Holmes 1997; Lenz et al. 2004). When the cell density increases, the autoinducers accumulate, bind their cognate receptors, and switch the receptors to phosphatases. Phosphatase activity leads to dephosphorylation of LuxO and termination of the sRNAs genes expression. The mRNA encoding HapR is stabilized, and HapR protein is produced (Fig. 3.2) (Lenz et al. 2004; Miller et al. 2002). HapR is a DNA-binding transcription factor that initiates a program of gene expression that switches the cells from the individual, low-cell-density state to the high-cell-density state (Fig. 3.2). Under high-cell-density state, HapR upregulates the production of the extracellular enzyme, HAP, while under low-cell-density, Hap production is downregulated (Fig. 3.2). Quorum-sensing signals in *V. cholerae* repress the expression of the virulence factor cholera toxin. Cholera toxin is produced under low-cell-density when the quorum-sensing signals levels decrease (Fig. 3.2).

3.6.3 **V. cholerae** *Act as a Pathogen of Chironomids*

Higgins et al. (2007), Parsek (2007) and Ng and Bassler (2009), highlighted the dilemma that quorum sensing seems to repress *V. cholerae* virulence factor expression (CT production) in contrast to what has been observed for the virulence gene expression of other bacteria. Quorum-sensing signals in *V. cholerae* upregulate the production of an extracellular enzyme, HAP, which degrades chironomid egg masses and prevents the eggs from hatching, demonstrating that HAP is a virulence factor against chironomids (Halpern et al. 2003; Halpern 2010). Indeed, in a year-long survey *V. cholerae* and chironomids showed a pattern of predator–prey population dynamics (Halpern et al. 2004, 2006).

Quorum sensing in respect to *V. cholerae* virulence may have been misinterpreted. *V. cholerae* is a pathogen of chironomids and as such, its quorum-sensing system probably controls virulence gene expression with regard to its bacterial pathogenicity for this insect. Although two serogroups of this species are pathogenic to humans, all the serogroups, including the human pathogenic O1 and O139, are likely to be pathogenic to chironomids (Broza and Halpern 2001; Halpern et al. 2003, 2004, 2006), and their quorum-sensing system – especially with regard to HAP regulation – should be seen in this light. When chironomid populations reach high densities, *V. cholerae* on their egg masses attains high numbers, its quorum-sensing signals increase, and as a result HapR is upregulated. HapR activates HAP enzyme expression (Jobling and Holmes 1997; Zhu et al. 2002; Lin et al. 2005), which is then produced and secreted from *V. cholerae* cells. HAP degrades the egg masses, larvae do not emerge from the eggs and chironomid populations decrease (Fig. 3.3, A1–A6). At low chironomid population levels (Fig. 3.3, B1) the opposite process happens. *V. cholerae* cell density per egg mass is low, the autoinducers concentration is low, HapR is downregulated and HAP enzyme production is repressed. In the absence of the HAP enzyme, the egg masses are not affected, the eggs hatch into larvae, and chironomid populations thrive (Fig. 3.3, B1–B6) (Halpern 2010).

From a global point of view, chironomids are much more abundant in the environment than humans, so quorum-sensing signals of *V. cholerae* and the gene they regulate should be understood with regard to their pathogenic role in chironomids rather than humans.

3.7 Interactions Between *V. cholerae* and Chironomids' Microflora

3.7.1 Other Species of Endogenic Bacteria in Chironomids

Using culturable as well as FISH techniques (with a specific oligonucleotide probe for *V. cholerae*), it was shown that *V. cholerae* make up less than 0.5% of all the bacteria present in the egg masses (Halpern et al. 2007a). Species belonging to the following genera were cultured and identified from the egg masses: *Acinetobacter*, *Aeromonas*, *Brachymonas*, *Exiguobacterium*, *Klebsiella*, *Leucobacter*, *Oceanobacillus*, *Paracoccus*, *Pseudomonas*, *Rheinheimera*, and

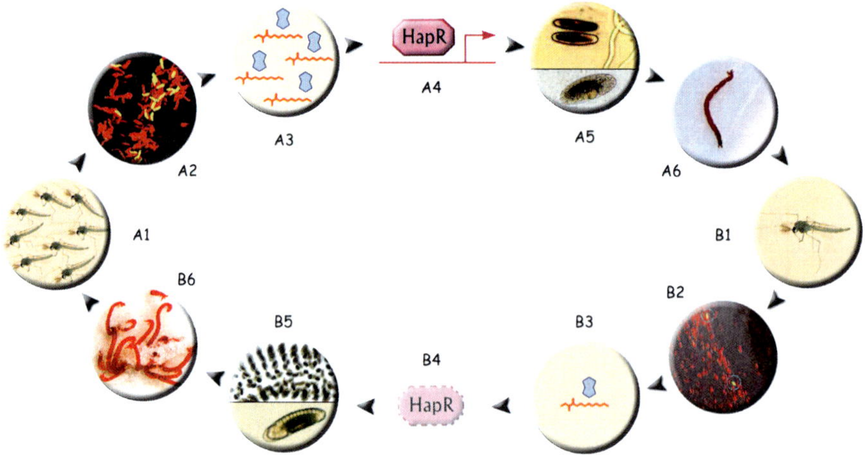

Fig. 3.3 Quorum-sensing system regulation of *V. cholerae* and its effect on chironomids and *V. cholerae* host–pathogen population dynamics*. *A1–A6* – High population densities of chironomids and *V. cholerae* cells lead to the up-regulation of the quorum-sensing signals. HAP is produced and as a result chironomid populations decrease. *B1–B6* – The opposite course. When chironomid and *V. cholerae* numbers are low, quorum-sensing signals are not produced, HAP is not formed and is not excreted, the egg masses are not affected and chironomid populations increase. *A1*. High chironomid density (Adults). *A2*. High *V. cholerae* cell density on the egg mass. Total bacteria stained with universal probe EUB 338 (*red*) *V. cholerae* stained with specific probe MAL2 (*yellow*). *A3*. Quorum-sensing signals (CAI-1 and AI-2) are upregulated. *A4*. Protease regulatory protein HapR is produced and upregulates the production of HAP enzyme. *A5*. HAP degrades the egg mass (*upper picture*) and prevents the eggs from hatching (a damaged egg, *lower part* of the picture). *A6*. Only few larvae hatch. The result of hatching prevention is a decrease in chironomid populations. *B1*. Low chironomid density. *B2*. Low *V. cholerae* cells density per egg mass. *B3*. Low quorum-sensing signals. *B4*. HapR and HAP are not produced. *B5*. The egg mass is not degraded (*upper picture*), eggs develop normally (*lower part* of the picture). *B6*. All larvae hatch from the eggs. The result is an increase in chironomid populations. *This figure is adopted from Halpern (2010)

Shewanella (Halpern et al. 2007a, b, 2009a, b; Raats and Halpern 2007; Senderovich et al. 2008). Hardly any isolate showed the ability to degrade the egg mass, and the ones that did, exhibited low efficiency compared to *V. cholerae*. Some of the identified species play a role in detoxifying metals or other toxic compounds (Ziemke et al. 1998; Srinath et al. 2001; Halpern et al. 2009a). What is the role that these bacteria play in the chironomids' life cycle? Chironomids display a wide range of tolerance toward pollution (Winner et al. 1980; Richardson and Kiffney 2000). However, the mechanism that protects the insect from pollutants is unknown. The presence of these species in chironomid egg masses may imply that these bacteria have a role in protecting chironomids from toxic compounds while enabling them to survive in polluted habitats. Yet, this role has still to be proven.

3.8 The Possible Dual Role of *V. cholerae* in Chironomids

V. cholerae is a stable resident in chironomids. However, its prevalence in the egg mass is very low (<0.5% of the bacterial community of the egg mass) (Halpern et al. 2007a). Although present in relatively low abundance, *V. cholerae* can degrade the egg masses by secreting the HAP enzyme, thus prevent eggs from hatching and may control the number of chironomids in the environment (Broza and Halpern 2001; Halpern et al. 2003, 2004, 2006; Halpern 2010). Other endogenous bacterial species that inhabit the egg masses may use the degraded gelatinous matrix for their own growth. By doing so, they probably control the population levels of *V. cholerae* in the egg mass. Vice versa, *V. cholerae* support the maintenance of endogenic bacteria in the egg mass by secreting HAP.

In conclusion: *V. cholerae* are pathogenic to chironomids on the one hand and support its endogenic bacteria on the other hand. Although in small numbers, *V. cholerae* is always present in chironomids. When its relative abundance is very low, it degrades only a very small portion of the egg mass and a minor harm is caused to the hatching larvae population. When the numbers of chironomid populations rise, *V. cholerae* abundance also goes up and then the pathogenic effect of the bacteria on the chironomids becomes dominant and controls chironomid populations.

The interaction between *V. cholerae* and the egg mass is probably a complicated mutualistic relationship rather than a simple host–pathogen interaction. More study is needed to understand this relationship.

3.9 Unresolved Questions and Future Research

- Are chironomids natural reservoirs of the pathogenic strains of *V. cholerae* as was found for the non O1/O139 strains?
  Up until today, the O1 or O139 serogroups of *V. cholerae* have not been isolated from chironomids. To determine the appearance of epidemic *V. cholerae* strains in different chironomid life stages, a year-long survey should be performed in an endemic area of the disease. If found, isolates should be compared to strains that cause cholera outbreaks. If a connection between isolates from chironomids and from cholera cases can be established, chironomid monitoring could be used as a predicting tool for cholera outbreaks.
- Does cholera toxin play a role in the *V. cholerae*-chironomid interactions?
  The HAP enzyme seems to play a role in the pathogenicity of *V. cholerae* toward chironomids. This enzyme is induced by quorum-sensing signals, while cholera toxin production occurs only when the bacterial population is low. The role of cholera toxin has to be examined in all the chironomid life stages. Understanding the function of cholera toxin in chironomids may improve our understanding of the appearance and disappearance of pathogenic strains in endemic areas.

- Are endogenic bacteria protecting the insect from pollutants?
  Chironomids are considered pollution tolerant. However, little is known about their protective mechanisms under these conditions. To answer the question whether the endogenic bacteria play a role in protecting the insect from pollutants that inhabit chironomid egg masses, Koch's postulates have to be applied to test if specific bacterial species inhabiting the egg masses or the larvae play a role in protecting chironomids from specific toxicants.

Acknowledgments I thank Prof. Simcha Lev-Yadun for his helpful comments on an earlier draft of the manuscript.

References

Aldova E, Laznickova K, Stepankova E, Lietava J (1968) Isolation of nonagglutinable *vibrios* from an enteritis outbreak in Czechoslovakia. J Infect Dis 118:25–31

Armitage P, Cranston PS, Pinder LCV (1995) The *Chironomidae*: the biology and ecology of non-biting midges. Chapman and Hall, London

Benitez JA, Silva AJ, Finkelstein RA (2001) Environmental signals controlling production of hemagglutinin/protease in *Vibrio cholerae*. Infect Immun 69:6549–6553

Broza M, Halpern M (2001) Chironomids egg masses and *Vibrio cholerae*. Nature 412:40

Broza M, Gancz H, Halpern M, Kashi Y (2005) Adult non-biting midges: possible windborne carriers of *Vibrio cholerae* non-O1 non-O139. Environ Microbiol 7:576–585

Broza M, Gancz H, Kashi Y (2008) The association between non-biting midges and *Vibrio cholerae*. Environ Microbiol 10:3193–3200

Butler SM, Camilli A (2005) Going against the grain: chemotaxis and infection in *Vibrio cholerae*. Nat Rev Microbiol 3:611–620

Chen X, SchauderS PN, Van Dorsselaer A, Pelczer I, Bassler BL, Hughson FM (2002) Structural identification of a bacterial quorum sensing signal containing boron. Nature 415:545–549

Colwell RR (1996) Global climate and infectious disease: the cholera paradigm. Science 274:2025–2031

Colwell RR, Huq A (2001) Marine ecosystems and cholera. Hydrobiologia 460:141–145

Constantin de Magny G, Murtugudde R, Sapiano MR, Nizam A, Brown CW, Busalacchi AJ, Yunus M, Nair GB, Gil AI, Lanata CF, Calkins J, Manna B, Rajendran K, Bhattacharya MK, Huq A, Sack RB, Colwell RR (2008) Environmental signatures associated with cholera epidemics. Proc Natl Acad Sci USA 105:17676–17681

Faruque SM, Naser IB, Islam MJ, Faruque AS, Ghosh AN, Nair GB, Sack DA, Mekalanos JJ (2005) Seasonal epidemics of cholera inversely correlate with the prevalence of environmental cholera phages. Proc Natl Acad Sci USA 102:1702–1707

Fuqua C, Greenberg EP (2002) Listening on Bacteria: acyl-homoserine lactone signaling. Nat Rev Mol Cell Biol 3:685–695

Halpern M (2010) Novel insights into Haemagglutinin Protease (HAP) Gene regulation in *Vibrio cholerae*. Mol Ecol 19:4108–4112

Halpern M, Izhaki I (2010) The environmental reservoirs and vector of *Vibrio cholerae*. In: Holmgren A, Borg G (eds) Handbook of disease outbreaks: prevention, detection and control. Nova, Hauppauge, pp 309–320

Halpern M, Gancz H, Broza M, Kashi Y (2003) *Vibrio cholerae* hemagglutinin/protease degrades chironomid egg masses. Appl Environ Microbiol 69:4200–4204

Halpern M, Broza YB, Mittler S, Arakawa E, Broza M (2004) Chironomid egg masses as a natural reservoir of *Vibrio cholerae* non-O1 and non-O139 in freshwater habitats. Microb Ecol 47:341–349

Halpern M, Raats D, Lavion R, Mittler S (2006) Dependent population dynamics between chironomids (non-biting midges) and *Vibrio cholerae*. FEMS Microbiol Ecol 55:98–104

Halpern M, Landsberg O, Raats D, Rosenberg E (2007a) Culturable and VBNC *Vibrio cholerae*; interactions with chironomid egg masses and their bacterial population. Microb Ecol 53:285–293

Halpern M, Senderovich Y, Snir S (2007b) *Rheinheimera chironomi* sp. nov., isolated from a chironomid (Diptera; Chironomidae) egg mass. Int J Syst Evol Microbiol 57:1872–1875

Halpern M, Senderovich Y, Izhaki I (2008) Waterfowl – the missing link in epidemic and pandemic cholera dissemination? PLoS Pathog 4:e1000173. doi:10.1371/journal.ppat.1000173

Halpern M, Shakéd T, Pukall R, Schumann P (2009a) *Leucobacter chironomi* sp. nov., a chromate resistant bacterium isolated from a chironomid egg mass. Int J Syst Evol Microbiol 59:665–670

Halpern M, Shakéd T, Schumann P (2009b) *Brachymonas chironomi* sp. nov., isolated from a chironomid egg mass, and emended description of the genus *Brachymonas*. Int J Syst Evol Microbiol 59:3025–3029

Higgins DA, Pomianek ME, Kraml CM, Taylor RK, Semmelhack MF, Bassler BL (2007) The major *Vibrio cholerae* autoinducer and its role in virulence factor production. Nature 450:883–886

Huq A, Small EB, West PA, Huq MI, Rahman R, Colwell RR (1983) Ecological relationships between *Vibrio cholerae* and planktonic crustacean copepods. Appl Environ Microbiol 45:275–283

Jobling MG, Holmes RK (1997) Characterization of *hapR*, a positive regulator of the *Vibrio cholerae* HA/protease gene *hap*, and its identification as a functional homologue of the *Vibrio harveyi luxR* gene. Mol Microbiol 26:1023–1034

Kaper JB, Morris JG, Levine MM (1995) Cholera. Clin Microbiol Rev 8:48–86

Ko WC, Chuang YC, Huang GC, Hsu SY (1998) Infections due to non-O1 *Vibrio cholerae* in southern Taiwan: predominance in cirrhotic patients. Clin Infect Dis 7:774–780

Lenz DH, Mok KC, Lilley BN, Kulkarni RV, Wingreen NS, Bassler BL (2004) The small RNA chaperone Hfq and multiple small RNAs control quorum sensing in *Vibrio harveyi* and *Vibrio cholerae*. Cell 118:69–82

Lewin SM (1996) Zoological microhabitats of *Vibrio cholerae*. In: Drasar BS, Forrest BD (eds) Cholera and the ecology of *Vibrio cholerae*. Chapman and Hall, London, pp 228–254

Lin W, Kovacikova G, Skorupski K (2005) Requirements for *Vibrio cholerae* HapR binding and transcriptional repression at the *hapR* promoter are distinct from those at the *aphA* promoter. J Bacteriol 187:3013–3019

Lipp EK, Huq A, Colwell RR (2002) Effects of global climate on infectious disease: the cholera model. Clin Microbiol Rev 15:757–770

Miller MB, Skorupski K, Lenz DH, Taylor RK, Bassler BL (2002) Parallel quorum sensing systems converge to regulate virulence in *Vibrio cholerae*. Cell 110:303–314

Ng W-L, Bassler BL (2009) Bacterial quorum sensing network architectures. Annu Rev Genet 43:197–222

Nolte U (1993) Egg masses of *Chironomidae* (*Diptera*). A review, including new observations and a preliminary key. Entomol Scand Suppl 43:5–75

Parsek MR (2007) Microbiology: bilingual bacteria. Nature 450:805–807

Raats D, Halpern M (2007) *Oceanobacillus chironomi* sp. nov., a halotolerant and facultative alkaliphilic species isolated from a chironomid egg mass. Int J Syst Evol Microbiol 57:255–259

Raz N, Danin-Poleg Y, Broza YY, Arakawa E, Ramakrishna BS, Broza M, Kashi Y (2010) Environmental monitoring of *Vibrio cholerae* using chironomids in India. Environ Microbiol Rep 2:96–103

Richardson JS, Kiffney PM (2000) Responses of a macroinvertebrate community from a pristine, southern British Columbia, Canada, stream to metals in experimental mesocosms. Environ Toxicol Chem 19:736–743

Sack DA, Sack RB, Nair GB, Siddique AK (2004) Cholera. Lancet 363:223–233

Samadi AR, Chowdhury MK, Huq MI, Khan MU (1983) Khan seasonality of classical and El tor cholera in Dhaka, Bangladesh: 17-year trends. Trans R Soc Trop Med Hyg 77:853–856

Senderovich Y, Gershtein Y, Halewa E, Halpern M (2008) *Vibrio cholerae* and *Aeromonas*: do they share a mutual host? ISME J 2:276–283

Senderovich Y, Izhaki I, Halpern M (2010) Fish as reservoirs and vectors of *Vibrio cholerae*. PLoS ONE 5:e8607. doi:10.1371/journal.pone.0008607

Srinath T, Khare S, Ramteke PW (2001) Isolation of hexavalent chromium-reducing Cr-tolerant facultative anaerobes from tannery effluent. J Gen Appl Microbiol 47:307–312

Tarsi R, Pruzzo C (1999) Role of surface proteins in *Vibrio cholerae* attachment to chitin. Appl Environ Microbiol 65:1348–1351

Thomas KU, Joseph N, Raveendran O, Nair S (2006) Salinity-induced survival strategy of *Vibrio cholerae* associated with copepods in Cochin backwaters. Mar Pollut Bull 52:1425–1430

Winner RW, Bossel MW, Farrel MP (1980) Insect community structure as an index of heavy-metal pollution in lotic ecosystems. Can J Fish Aquat Sci 37:647–655

Xavier KB, Bassler BL (2005) Interference with AI-2-mediated bacterial cell-cell communication. Nature 437:750–753

Zhu J, Miller MB, Vance RE, Dziejman M, Bassler BL, Mekalanos JJ (2002) Quorum-sensing regulators control virulence gene expression in *Vibrio cholerae*. Proc Natl Acad Sci USA 99:3129–3134

Ziemke F, Hofle MG, Lalucat J, Rossello-Mora R (1998) Reclassification of *Shewanella putrefaciens* Owen's genomic group II as *Shewanella baltica* sp. nov. Int J Syst Bacteriol 1:179–186

Chapter 4
Role of Bacteria in Mating Preference in *Drosophila melanogaster*

Gil Sharon, Daniel Segal, and Eugene Rosenberg

Abstract Assortative mating, considered to be an early event in speciation, has been studied for decades in the context of divergent adaptation. In *Drosophila* it is commonly attributed to genetic elements in the flies that exhibit assortative mating. However, some cases have been reported where the genetic basis for these differences was unclear. In light of the Hologenome Theory of Evolution (Zilber-Rosenberg and Rosenberg, 2008), we considered the microbiota of *Drosophila* as an additional element, acting together with its host to better adapt to a changing environment. The microbiota of any organism is closely linked to its host. Many of the impacts of the microbiota on its host are known. New evidence shows an interesting, previously unknown, role of the microbiota in influencing its host's behavior. In one case, as a result of adaptation to a new substrate, the microbiota changed with behavioral implications on its host flies. By changing its host's mating preference, the microbiota has the potential of driving the evolution of its host. In this chapter, the mating process in *Drosophila* will be reviewed within the framework of the hologenome theory of evolution. Some conclusions and speculations on how microbes and their *Drosophila* host interact will be presented.

4.1 Introduction

The concept of species has been debated for many years, since it was proposed by Darwin (1859) 150 years ago. The biological species concept, proposed by Dobzhansky (1937) and Mayr (1942), is widely accepted. It defines a species as "... a reproductive community of populations (reproductively isolated from others) that occupies a specific niche in nature" (Mayr 1982). According to this concept, the

---

G. Sharon (✉) • D. Segal • E. Rosenberg
Department of Molecular Microbiology and Biotechnology, Tel Aviv University, Ramat Aviv, Tel Aviv 69978, Israel
e-mail: gilsharo@post.tau.ac.il

E. Rosenberg and U. Gophna (eds.), *Beneficial Microorganisms in Multicellular Life Forms*, DOI 10.1007/978-3-642-21680-0_4,

first step towards the formation of a new species is the creation of a reproduction barrier or homospecific assortative mating (in which one community prefers to mate with its own members rather than the members of a different community). The adaptive divergence hypothesis posits that reproductive isolation is acquired as a by-product of genetic divergence, gradually and continuously, as spatially and temporally isolated populations adapt to local ecological conditions (Darwin 1859; Muller 1942). Under these conditions, speciation should take 10^4 (Darwin 1859) to 10^5 generations (Futuyma 1979). Strong selection increases the speed in which reproductive isolation is obtained (Levins 1968). According to these concepts, speciation is driven by either pre- or post-zygotic isolation mechanisms, namely mating isolation and hybrid sterility or inviability, respectively.

As mating isolation and speciation take many generations, it is a great challenge to study it experimentally (at least in higher organisms) and assortative mating has been used as a proxy for studying mechanisms by which mating isolation may arise. To date, many different organisms and model systems have been used to study assortative mating experimentally. Most if not all of these studies concluded that the basis for assortative mating is genetic.

The fruit fly, *Drosophila*, a common model system for studying genetics and evolution, has been used to investigate assortative mating in both natural and laboratory populations, showing assortative mating as a result of adaptive divergence, over long periods of time (Rice and Hostert 1993). Some reports showed that assortative mating may arise in relatively short time frames that are seemingly not sufficient for genetic differences to arise. Thus, one can argue that assortative mating must arise due to a mechanism that is not *sensu stricto* genetic.

In this chapter, the mating of *Drosophila melanogaster* will be reviewed, and mechanisms by which assortative mating develops will be discussed in light of the hologenome theory of evolution (Zilber-Rosenberg and Rosenberg 2008) as well as the recent data showing a bacterial role in diet-induced *Drosophila* assortative mating.

4.2 Mating in *Drosophila melanogaster*

Sexual reproduction is the key process at the base of the success of many higher organisms. It ensures the transfer of genes from the parents to their offspring and thus the continuum of the species. As such, mate choice is crucial; choosing the best mate possible will ensure the survival of an individual's offspring so that its genes are transferred to future generations. The process of mate choice is a sophisticated blend of various cues that elicits the "eligibility" of an individual to mate [e.g., size (Byrne and Rice 2006), odors (Howard and Blomquist 2005), etc.].

In *D. melanogaster*, two main cues are known (amongst others) and considered to play the key roles in courtship and mate choice: (1) Courtship song and (2) Cuticular hydrocarbons (CHs) (i.e., sex pheromones). This concert of cues is hard wired to the *Drosophila* genome (albeit it is also influenced by the environment) and both cues and receptors play crucial roles in the process of decision making (Dickson 2008).

4.2.1 Courtship

Courtship in *D. melanogaster* is a process that involves a sequential series of behaviors by both the male and female (Hall 1994). In general, males decide whom to court and females decide whether they should mate with their suitor or not (Dickson 2008). During courtship, an exchange of signals, mainly auditory and chemical, directs success or failure of the process. At the first step, the male notices the female and directs itself towards its future mate. He follows her and taps on her abdomen. The male then "serenades" the female by extending and vibrating its wings (one at a time). The male then extends its proboscis and licks the female's genitalia. After this "conversation" between the two, the male attempts to copulate with the female. If rejected, the male will rest for a few moments, and start the courtship sequence again, without tapping or licking. Although seemingly passive, females have a say at whether or not copulation will occur. First, a female has to be receptive to the male's courtship. Receptivity is conveyed by a general slowdown as well as taking a specific position so that mating is successful. A non-receptive female (e.g., one that has already copulated) will actively reject a courting male.

4.2.2 Signals in Mating

Communication is a key attribute of social organisms and is used to orchestrate many different social behaviors. Two main types of signals play a role in mating behavior in *D. melanogaster*: (1) an auditory signal, the "love song" produced by the vibration of the male's wing while courting a female and (2) an olfactory/gustatory signal that conveys many different attributes of a fly. These signals confer specificity (Ritchie et al. 1999) and allow for mating efforts to be made only where they are most likely to be fruitful.

4.2.2.1 Auditory Signals

The "love song" is an obligatory signal for mating. Female *Drosophila* will not mate with wingless males that cannot produce it (Liimatainen et al. 1992). This auditory signal is complex, and in *D. melanogaster* is composed of a "sine song" and a "pulse song" (reviewed by Tauber and Eberl 2003). Each part of this signal is characterized by many different variables such as the frequency of the song, the intervals between pulses, the number of pulses, and the length of the sinusoidal oscillation of the intervals. These variables confer specificity to each species' song during courtship and thus the courtship song is thought to contribute to mating isolation (Ritchie et al. 1999).

4.2.2.2 Pheromones

The role of CHs in the ecology and behavior of *Drosophila* spp., as well as in other insects, has been studied in the past few decades and their major role in mating and assortative mating has been studied extensively (reviewed by Howard and Blomquist 2005). Other than their major role in communication, CHs are crucial for resistance to desiccation (Rouault et al. 2004).

CHs are derived from fatty-acids and vary in length, methylation and saturation. The mature CHs are produced in specific cells, oenocytes (Billeter et al. 2009), and are secreted to the surface of the fly. One exception is the male-specific cis-vaccinyl-acetate, produced in the male's ejaculatory bulb (Butterworth 1969). Since the volatility of CHs is low, they can serve as signals for short distance (Ferveur 2005). The CH profile of *Drosophila* is generally stable, but depends on age and strain and is dependent on both genetic and environmental factors (reviewed by Ferveur 2005).

4.3 Assortative Mating in *Drosophila*

Drosophila choose their mate according to two different levels of variation in the fly. The first is general and common in each sex and usually relates to physical traits of the fly (male or female). For example, male and female *Drosophila* prefer a larger mate (Byrne and Rice 2006; Friberg and Arnqvist 2003). The second level is more specific and usually used for distinguishing which is the most "attractive" mate. These are more specific stimuli (e.g., courtship song and CHs, see above) that help flies make this important decision of whom is the best mate for them (a decision that may determine the potential fitness of future generations). The male's decision-making process is based on an intrinsic ability to discriminate males from females as well as an acquired (at least in part) skill of discriminating receptive and non-receptive females (e.g., females of the same species as opposed to females of a different species) (Dickson 2008). Once a male decides to court a specific female, it is mostly up to her whether or not they will copulate. Females decide mostly based on the male's courtship song (Rybak et al. 2002).

Assortative mating in *Drosophila* has been studied for decades and is commonly attributed to genetic elements in the flies that exhibit assortative mating. In many cases, these flies present different courtship songs (in males) and CH (in both males and females) profiles that stand at the base of this behavior. However, some cases have been reported where the genetic basis for these differences were unclear. This uncertainty originates mostly when these differences are observed in too short time scales to allow for genetic divergence to accumulate [as assortative mating is thought to arise as a by-product of adaptation (Rice and Hostert 1993)] and in sympatric populations where gene flow is possible.

4.3.1 The Genetic Basis for Assortative Mating in Drosophila

In *Drosophila* populations that diverged a long time ago, phenotypic variability, leading to assortative mating, can often be attributed to polymorphism in different genes. One interesting example in *D. melanogaster* is the case of Zimbabwe vs. cosmopolitan strains that exhibit strong homogamic assortative mating, that stem from the different CH profiles these flies present (Hollocher et al. 1997). Genetic analysis of this behavior rendered a remarkable result – all autosomal chromosomes were found to be important for this behavior (Wu et al. 1995). Capy et al. (2000) reported significant phenotypic and genetic differences in two sympatric populations of *D. melanogaster* in Congo that also show homogamic assortative mating. Another fascinating example is the "Evolution Canyon," Mount Carmel, Israel (Korol et al. 2000; Iliadi et al. 2001) where *Drosophila* populations collected from opposite slopes that differ in solar radiation, flora, temperature, and aridity (but separated by only 100–400 m), show homogamic assortative mating. The "Evolution Canyon" is an intriguing site that shows how ecological differentiation of natural sympatric populations influences behavior and evolution. Genetic differences between the *D. melanogaster* populations of the two slopes were found (Michalak et al. 2001).

As divergent selection has been studied in controlled laboratory experiments, many data have accumulated showing that by dividing a population and growing the parts under different conditions assortative mating arises. For example, Kilias et al. (1980) collected two geographically distinct populations of *D. melanogaster*. Each one was split into two populations – one reared under cold-dry-dark conditions while the other under warm-moist-light conditions. After 5 years (ca. 100–120 generations), divergently selected populations derived from either the same or different reared populations showed prezygotic isolation (~50% excess). Parallel selected populations that were reared separately, but under the same conditions, showed no isolation. In another study, De Oliveira and Cordeiro (1980) grew *Drosophila willistoni* at different pHs. After 26 generations, mating choice tests showed significant assortative mating between the populations. Dodd (1989) used year-old (ca. 35 generations) *Drosophila pseudoobscura* populations (collected at Bryce Canyon, USA) adapted to either starch or maltose (four populations of each). Out of 16 crosses between starch-adapted and maltose-adapted populations, ten showed significant behavioral isolation (i.e., mating preference). Before mating tests took place, each population was reared on the same cornmeal-molasses-yeast medium for one generation. When crossing starch-adapted flies with other starch-adapted flies or maltose-adapted with maltose-adapted, no preference was observed.

Divergent selection may lead to assortative mating. It is thought that assortative mating in these cases emerges via "incidental pleiotropy or genetic hitchhiking" (Rice and Hostert 1993). However, in a review by Andersson and Simmons (2006) the authors state: "As experimental evidence accumulated, mate choice became widely recognized, but the genetic mechanisms underlying its evolution remain the subject of debate". In light of the hologenome theory of evolution (Zilber-Rosenberg and Rosenberg 2008) it was hypothesized that indeed there is a genetic basis to this phenomenon, but it is the *Drosophila* hologenome rather than only the host genome that changes, through changes in the fly's microbiota. This hypothesis means that the first genetic component to change is the microbiota of the fly (that adapt quickly), inducing assortative mating that will later be fixed in the host genome by the known slower mechanisms of mutation and selection.

Polymorphism in various genes, in populations that have diverged a long time ago (as in the case of Zimbabwe vs. cosmopolitan *D. melanogaster*) has been found to be responsible to phenotypic polymorphism in courtship song and CH profile phenotypic polymorphism. One example is polymorphism in *desat2* (a female-specific desaturase gene involved in diene hydrocarbon biosynthesis) that alters the enzyme's specificity in Tai strain females (Dallerac et al. 2000). Likewise, the specificity of the courtship song has been shown to be controlled by the *period* gene (Alt et al. 1998). These examples (two of many) show that genetic polymorphisms control the two major stimuli in courtship.

4.4 The Role of Bacteria in Medium-Induced Assortative Mating

Many bacteria are known to live in close association with insects, and *Drosophila* is no different. Bacterial symbionts were shown to be essential for the well being of their hosts by aiding their development (e.g., Dedeine et al. 2001), aiding in harvesting necessary nutrients (e.g., Breznak et al. 1973; Blatch et al. 2010), and defending from potential pathogens (e.g., Jaenike et al. 2010). Other bacteria, however, are known to have deleterious effects on the host (e.g., Aronson et al. 1986). The nature of the association may vary: Primary symbionts are intracellular endosymbionts that are transmitted transovarianly whereas secondary symbionts are transmitted through the environment and reside on the surface of the fly and in its gut.

4.4.1 The Role of the Endosymbionts in Drosophila *Health and Mating*

An extensive study by Mateos et al. (2006) found only two confirmed endosymbionts in *Drosophila* – *Spiroplasma* and *Wolbachia*. In that study, 19 species were infected with *Wolbachia* and three with *Spiroplasma* out of 181 strains of *Drosophila* (25 species) examined.

4.4.1.1 Wolbachia

Wolbachia are alphaproteobacteria of the order Riketsialles. They are primary endosymbionts, common in arthropods and nematodes. These bacteria show different symbioses in different hosts – from complete mutualism in nematodes to parasitism in arthropods (reviewed by Werren et al. 2008). *Wolbachia* can induce various phenotypic effects in its hosts, some of which have profound impacts on mating behavior as well as the fitness of offspring: (1) Cytoplasmic incompatibility: A deleterious modification of sperm, induced by *Wolbachia*, is rescued only in *Wolbachia*-infected embryos while in non-infected embryos, or embryos that are infected with a different strain of *Wolbachia* will not develop properly (Werren et al. 2008). (2) Male-killing: During embryogenesis, *Wolbachia* kills males (that are of less importance in maintaining the infection in the population) to allow more food for females (Dyer and Jaenike 2004). (3) Assortative mating: Koukou et al. (2006) directly showed a bacterial role in inducing homospecific assortative mating. Long-term selection populations were subjected to divergent selection for tolerance of toxins in the food (heavy metals versus ethanol). After 30 years, these populations differed in sexual isolation and *Wolbachia* infection status. Removal of *Wolbachia* decreased levels of mate discrimination between the populations by about 50%. They wrote "The presence of *Wolbachia* (or another undetected bacterial associate) acts as an additive factor contributing to the level of pre-mating isolation..."

4.4.1.2 Spiroplasma

Besides *Wolbachia*, another genus of male-killing bacteria (in some *Drosophila* species) is *Spiroplasma*. These bacteria are helical and lack a cell wall (Williamson et al. 1998). *Spiroplasma* can infect different *Drosophila* species where they either induce male-killing (as in *D. melanogaster*) or have no effect on sex ratio (Montenegro et al. 2005; Haselkorn et al. 2009).

4.4.2 Commensal Bacteria in Drosophila

D. melanogaster is colonized by a diverse community of bacteria. These symbiotic bacteria span three major phyla – Proteobacteria, Bacteroidetes, and Firmicutes (Corby-harris et al. 2007; Cox and Gilmore 2007). Associated bacteria are undoubtedly important for their host as they play an important role in development and health. One seminal research by Brummel and colleagues (2004) showed that commensal bacteria in *D. melanogaster* increase the lifespan of their host. Beyond the immediate effect bacteria have on the fitness of their host, there is evidence of a bacterial role in increasing the host's mating success: a common method for biological pest control is the sterile insect technique (by irradiation). However, sterile Mediterranean fruit fly (Medfly) males show decreased abundances of Enterobacteria, and especially *Klebsiella* species. These flies show significantly lower mating success with wild females than non-irradiated male Medflies. Infecting these flies with *Klebsiella oxytoca* improved mating success of irradiated males (Gavriel et al. 2011; Ben Ami et al. 2010).

Recently, we have shown that mating preference can arise in *D. melanogaster* as a result of an adaptation to a new medium. This phenomenon was found to be dependent on the bacteria harbored on/in the fly (Sharon et al. 2010).

4.4.3 *The Role of* Lactobacillus plantarum *in Medium-Induced Assortative Mating*

This chapter has exemplified how research performed on assortative mating in *Drosophila* in the last few decades has shown that mating could be achieved by adapting *Drosophila* spp. to different environmental conditions (e.g., medium, pH, and heavy metals). However, it was not clear which mechanisms were responsible for these phenomena. The common assumption was that assortative mating is brought forth as a by product of selection for one trait or the other, where *Drosophila* genes that may confer mating preference "hitchhike" on other genes that are selected for by the environmental conditions. Recently, it has been shown that a bacterial component of the fly microbiota was responsible for inducing homogamic mating preference in wild-type Oregon-R *D. melanogaster* (Sharon et al. 2010). In light of the hologenome theory, it was hypothesized that the driving force for environmentally-induced mating preference is the different bacterial communities that arise in response to the changing environment. To test this hypothesis a population of *D. melanogaster* was divided and reared one part on a molasses-based medium and the other on a starch-based medium. When the isolated populations were mixed, "molasses flies" preferred to mate with "molasses flies" and "starch flies" preferred to mate with "starch flies." The mating preference appeared after only one generation and was maintained for at least 37 generations. Antibiotic treatment abolished mating preference, suggesting that the fly microbiota was responsible for the phenomenon. This was confirmed by infection experiments with microbiota obtained from the fly media (prior to antibiotic treatment). By using 16S rRNA gene clone libraries it was found that *L. plantarum* was an abundant bacterium in starch-bred flies (ten times more abundant than in molasses-bred files). Infection experiments of antibiotic-treated flies with pure cultures of *L. plantarum* isolated from "starch flies" showed that it can induce mating preference, and thus it was concluded that *L. plantarum* was responsible, at least in part, for the observed medium-induced mating preference.

Analyses of the CH bouquet of flies showed a significant difference between molasses- and starch-bred flies. These differences were reduced by antibiotic treatment, suggesting that symbiotic bacteria influence mating preference by changing the levels of specific CHs. The data suggest that host-associated bacteria play a role in adaptation and drive speciation of their hosts. These results raise many questions regarding various aspects of this phenomenon – from the mechanism by which *Lactobacilli* induce mating preference to its impact on how we view evolution, and from the microbial ecology of flies to their chemical ecology.

4.4.4 *The Possible Effect of Bacteria on* Drosophila *Adaptation, Speciation and Evolution, Within the Framework of the Hologenome Theory of Evolution*

Drosophila is an insect that moves around, looking for food, a good mate and a place to lay their eggs in order to establish the next generation. When considering how laboratory-controlled experiments (presented above) may reflect on how bacteria actually influence mating preference in nature, one has to take into account the lifestyle of these remarkable organisms. Once a female *Drosophila* copulates, it stores sperm and gradually uses them to fertilize its eggs before laying them. It then looks for a good substrate for feeding and oviposition. Upon laying eggs, the female fly deposits a "starter" culture of bacteria in order to ferment the medium to make it readily available for offspring larvae to feed on (Behar et al. 2008). We hypothesize that the deposited bacterial community now reshapes as a result of the new environment it is subjected to – some bacterial species multiply faster than others (amplification) while some can't cope with the new environment (decay). Bacteria already present in the substrates are introduced to the community (and are probably more adapted to it than others). The bacteria probably interact with each other – some produce antibiotics (competition) while some degrade a tough substrate and allow others to feed on the by products (cooperation) and others exchange genetic information (lateral gene transfer). Larvae feed on this developing bacterial community along with the decaying organic matter. The indigested bacteria now have to deal with a new environment, the gut. These bacteria are most probably the same ones to colonize the adult fly's gut and surface and thus are capable of influencing the mating behavior of the emerging flies (Sharon et al. 2010). In theory, these flies may now have a preference to mate with their own kind, the kind of flies that harbor the microbial community best adapted to the specific substrates. We further presume that as mating isolation becomes established, the genome of the *Drosophila* host slowly changes, to better interact with its new habitat as well as with its microbial community.

At present, the above scenario is only a speculation. However, consideration of various reports supports this hypothesis. One recently published example is the effect of preference for oviposition sites has on reproductive isolation of the beetle *Callosobruchus maculates*, where oviposition in a new substrate induced mating preference starting the second generation, which was lost within two generations if not selected for (Rova and Björklund 2011). The Hawaiian islands are the most diverse place on earth, in terms of the number of *Drosophila* species – approximately 1,000 different species have been described and are believed to originate from a single introduction of a *Drosophila* sp. to the islands (Boake 2005). The case of Hawaiian *Drosophila* and theories of how this amazing diversity emerged have been reviewed by Ritchie (2007) and by Boake (2005). A major theory trying to explain this diversity was suggested by Carson and Templeton (1984). It proposes founder-flush cycles upon migration from older islands to the newer ones. And it is believed that sexual isolation is at the basis of speciation in the Hawaiian Islands (Boake 2005). One might consider, however, that sexual isolation arose by changing microbial communities, rather than genetic differences between fly populations due to founder-flush cycles. This reasoning could also explain the above mentioned mating preferences in cosmopolitan vs. African *D. melanogaster* (Hollocher et al. 1997). However, since these species/strains diverged a long time ago, it is almost impossible to account for the present differences by bacterial communities alone.

4.5 Unresolved Questions and Future Research

Two major questions arise from the current knowledge of the influence bacteria have on *D. melanogaster*'s mating preference: (1) What is the mechanism by which mating preference is induced by commensal bacteria? (2) Do bacteria influence mating behavior of *Drosophila* not only in controlled laboratory experiments but also in nature?

In order to elucidate the mechanism, it is crucial to first localize *L. plantarum* in the fly. Although it is assumed to be present and act in the gut, this remains to be demonstrated. Fluorescent in-situ hybridization (FISH) is a widely used method in microbial ecology by which one can probe for a specific bacterium by targeting its 16S rRNA. As a first stage, it would be of interest to realize whether Lactobacilli reside in the gut, on the fly outer surface or in any other fly tissue. If the bacterium resides on the surface, it is likely that it either produces a volatile compound or it degrades fly-produced CHs to change the pheromonal bouquet presented by the fly. If it resides in the gut or in other tissues, the bacterium might alter gene expression that would be manifested in the pheromonal repertoire of the fly. Bacteria in the gut (or other tissues of the fly) may alter gene expression through the innate immune response, which is designed to interact with the mutualistic as well as pathogenic and commensal microbes the fly encounters. Downstream signaling may result in differential expression of specific CH biosynthesis genes. In a recent work, Richard et al. (2008) found that the innate immune response to bacterial components plays a role in CH secretion in honey bees.

In order to test how well the laboratory model system describes what actually happens in nature, a few different approaches can be utilized: (1) Using ecologically diverged populations that exhibit mating preference, and treating them with antibiotics to "cure" mating preference. If indeed mating preference can be eliminated by antibiotics, it would indicate a bacterial role in inducing mating preference in nature, and thus divergence. (2) By generalizing this phenomenon to construct mathematical models that can be tested, one might get an idea of how new species can evolve through induction of mating preference via changing microbial communities. (3) Introducing a natural population of flies to a new substrate, while the old substrate is still present and testing for mating preference as a proxy for divergence. The research of how bacteria induce mating preference and its effect on the evolution of new species is in its infancy and future research might help us understand how organisms of different kingdoms cooperate in order to make themselves better adapted to their environment and to transmit this adaptation to the next generation.

References

Alt S, Ringo J, Talyn B, Bray W, Dowse H (1998) The period gene controls courtship song cycles in *Drosophila melanogaster*. Anim Behav 56:87–97

Andersson M, Simmons LW (2006) Sexual selection and mate choice. Trends Ecol Evol 21(6):296–302

Aronson AI, Beckman W, Dunn P (1986) *Bacillus-thuringiensis* and related insect pathogens. Microbiol Rev 50(1):1–24

Behar A, Jurkevitch E, Yuval B (2008) Bringing back the fruit into fruit fly-bacteria interactions. Mol Ecol 17(5):1375–1386

Ben Ami E, Yuval B, Jurkevitch E (2010) Manipulation of the microbiota of mass-reared Mediterranean fruit flies *Ceratitis capitata* (Diptera: Tephritidae) improves sterile male sexual performance. Isme J 4(1):28–37

Billeter JC, Atallah J, Krupp JJ, Millar JG, Levine JD (2009) Specialized cells tag sexual and species identity in *Drosophila melanogaster*. Nature 461(7266):987–U250

Blatch SA, Meyer KW, Harrison JF (2010) Effects of dietary folic acid level and symbiotic folate production on fitness and development in the fruit fly Drosophila melanogaster. Fly 4(4):312–319

Boake CRB (2005) Sexual selection and speciation in Hawaiian *Drosophila*. Behav Genet 35(3):297–303

Breznak JA, Brill WJ, Mertins JW, Coppel HC (1973) Nitrogen fixation in termites. Nature 244(5418):577–580

Brummel T, Ching A, Seroude L, Simon AF, Benzer S (2004) *Drosophila* lifespan enhancement by exogenous bacteria. Proc Natl Acad Sci USA 101(35):12974–12979

Butterworth FM (1969) Lipids of *Drosophila* – a newly detected lipid in male. Science 163(3873):1356–1357

Byrne PG, Rice WR (2006) Evidence for adaptive male mate choice in the fruit fly *Drosophila melanogaster*. Proc R Soc B Biol Sci 273(1589):917–922

Capy P, Veuille M, Paillette M, Jallon JM, Vouidibio J, David JR (2000) Sexual isolation of genetically differentiated sympatric populations of *Drosophila melanogaster* in Brazzaville, Congo: the first step towards speciation? Heredity 84(4):468–475

Carson HL, Templeton AR (1984) Genetic revolutions in relation to speciation phenomena – the founding of new populations. Annu Rev Ecol Syst 15:97–131
Corby-Harris V, Pontaroli AC, Shimkets LJ, Bennetzen JL, Habel KE, Promislow DEL (2007) Geographical distribution and diversity of bacteria associated with natural populations of *Drosophila melanogaster*. Appl Environ Microbiol 73(11):3470–3479
Cox CR, Gilmore MS (2007) Native microbial colonization of *Drosophila melanogaster* and its use as a model of *Enterococcus faecalis* pathogenesis. Infect Immun 75(4):1565–1576
Dallerac R, Labeur C, Jallon JM, Knippie DC, Roelofs WL, Wicker-Thomas C (2000) A Delta 9 desaturase gene with a different substrate specificity is responsible for the cuticular diene hydrocarbon polymorphism in *Drosophila melanogaster*. Proc Natl Acad Sci USA 97(17):9449–9454
Darwin C (1859) On the origin of species by means of natural selection. J. Murray, London
De Oliveira AK, Cordeiro AR (1980) Adaptation of *Drosophila-willistoni* experimental populations to extreme Ph medium .2. Development of incipient reproductive isolation. Heredity 44(Feb):123–130
Dedeine F, Vavre F, Fleury F, Loppin B, Hochberg ME, Bouletreau M (2001) Removing symbiotic *Wolbachia* bacteria specifically inhibits oogenesis in a parasitic wasp. Proc Natl Acad Sci USA 98(11):6247–6252
Dickson BJ (2008) Wired for sex: the neurobiology of *drosophila* mating decisions. Science 322(5903):904–909
Dobzhansky TG (1937) Genetics and the origin of species, vol 11, Columbia biological series. Columbia University Press, New York
Dodd DMB (1989) Reproductive isolation as a consequence of adaptive divergence in *Drosophila-pseudoobscura*. Evolution 43(6):1308–1311
Dyer KA, Jaenike J (2004) Evolutionarily stable infection by a male-killing endosymbiont in *Drosophila innubila*: molecular evidence from the host and parasite genomes. Genetics 168 (3):1443–1455
Ferveur JF (2005) Cuticular hydrocarbons: their evolution and roles in *Drosophila* pheromonal communication. Behav Genet 35(3):279–295
Friberg U, Arnqvist G (2003) Fitness effects of female mate choice: preferred males are detrimental for *Drosophila melanogaster* females. J Evol Biol 16(5):797–811
Futuyma DJ (1979) Evolutionary biology, 1st edn. Sinauer Associates, Sunderland
Gavriel S, Jurkevitch E, Gazit Y, Yuval, B (2011) Bacterially enriched diet improves sexual performance of sterile male Mediterranean fruit flies. Journal of Applied Entomology, 135: no. doi: 10.1111/j.1439-0418.2010.01605.x
Hall JC (1994) The mating of a Fly. Science 264(5166):1702–1714
Haselkorn TS, Markow TA, Moran NA (2009) Multiple introductions of the Spiroplasma bacterial endosymbiont into *Drosophila*. Mol Ecol 18(6):1294–1305
Hollocher H, Ting CT, Pollack F, Wu CI (1997) Incipient speciation by sexual isolation in *Drosophila melanogaster*: variation in mating preference and correlation between sexes. Evolution 51(4):1175–1181
Howard RW, Blomquist GJ (2005) Ecological, behavioral, and biochemical aspects of insect hydrocarbons. Annu Rev Entomol 50(1):371–393. doi:10.1146/annurev.ento.50.071803.130359
Iliadi K, Iliadi N, Rashkovetsky E, Minkov I, Nevo E, Korol A (2001) Sexual and reproductive behaviour of Drosophila melanogaster from a microclimatically interslope differentiated population of 'Evolution Canyon' (Mount Carmel, Israel). Proc R Soc Lond Ser B Biol Sci 268(1483):2365–2374
Jaenike J, Unckless R, Cockburn SN, Boelio LM, Perlman SJ (2010) Adaptation via symbiosis: recent spread of a *Drosophila* defensive symbiont. Science 329(5988):212–215
Kilias G, Alahiotis SN, Pelecanos M (1980) A multifactorial genetic investigation of speciation theory using *Drosophila-melanogaster*. Evolution 34(4):730–737

Korol A, Rashkovetsky E, Iliadi K, Michalak P, Ronin Y, Nevo E (2000) Nonrandom mating in *Drosophila melanogaster* laboratory populations derived from closely adjacent ecologically contrasting slopes at "Evolution Canyon". Proc Natl Acad Sci USA 97(23):12637–12642

Koukou K, Pavlikaki H, Kilias G, Werren JH, Bourtzis K, Alahiotisi SN (2006) Influence of antibiotic treatment and *Wolbachia* curing on sexual isolation among *Drosophila melanogaster* cage populations. Evolution 60(1):87–96

Levins R (1968) Evolution in changing environments; some theoretical explorations, vol 2, Monographs in population biology. Princeton University Press, Princeton

Liimatainen J, Hoikkala A, Aspi J, Welbergen P (1992) Courtship in *Drosophila-montana* – the effects of male auditory signals on the behavior of flies. Anim Behav 43(1):35–48

Mateos M, Castrezana SJ, Nankivell BJ, Estes AM, Markow TA, Moran NA (2006) Heritable endosymbionts of Drosophila. Genetics 174(1):363–376

Mayr E (1942) Systematics and the origin of species from the viewpoint of a zoologist, Columbia biological series. No. XIII. Columbia University Press, New York

Mayr E (1982) The growth of biological thought: diversity, evolution, and inheritance. Belknap Press, Cambridge

Michalak P, Minkov I, Helin A, Lerman DN, Bettencourt BR, Feder ME, Korol AB, Nevo E (2001) Genetic evidence for adaptation-driven incipient speciation of *Drosophila melanogaster* along a microclimatic contrast in "Evolution Canyon," Israel. Proc Natl Acad Sci USA 98(23):13195–13200

Montenegro H, Solferini VN, Klaczko LB, Hurst GDD (2005) Male-killing *Spiroplasma* naturally infecting *Drosophila melanogaster*. Insect Mol Biol 14(3):281–287

Muller JJ (1942) Isolating mechanisms, evolution and temperature. Biol Symp 6:71–125

Rice WR, Hostert EE (1993) Laboratory experiments on speciation – what have we learned in 40 years. Evolution 47(6):1637–1653

Richard FJ, Aubert A, Grozinger CM (2008) Modulation of social interactions by immune stimulation in honey bee, *Apis mellifera*, workers. BMC Biol 6:50

Ritchie MG (2007) Sexual selection and speciation. Annu Rev Ecol Evol Syst 38(1):79–102

Ritchie MG, Halsey EJ, Gleason JM (1999) Drosophila song as a species-specific mating signal and the behavioural importance of Kyriacou & Hall cycles in *D-melanogaster* song. Anim Behav 58:649–657

Rouault JD, Marican C, Wicker-Thomas C, Jallon JM (2004) Relations between cuticular hydrocarbon (HC) polymorphism, resistance against desiccation and breeding temperature; a model for HC evolution in *D-melanogaster* and *D-simulans*. Genetica 120(1–3):195–212

Rova E, Björklund M (2011) Can preference for oviposition sites initiate reproductive isolation in *Callosobruchus maculates*. Plos One 6(1):e14628

Rybak F, Sureau G, Aubin T (2002) Functional coupling of acoustic and chemical signals in the courtship behaviour of the male *Drosophila melanogaster*. Proc R Soc Lond Ser B Biol Sci 269 (1492):695–701

Sharon G, Segal D, Ringo JM, Hefetz A, Zilber-Rosenberg I, Rosenberg E (2010) Commensal bacteria play a role in mating preference of *Drosophila melanogaster*. Proc Natl Acad Sci USA 107(46):20051–20056

Tauber E, Eberl DF (2003) Acoustic communication in *Drosophila*. Behav Process 64(2):197–210

Werren JH, Baldo L, Clark ME (2008) *Wolbachia*: master manipulators of invertebrate biology. Nat Rev Microbiol 6(10):741–751

Williamson DL, Whitcomb RF, Tully JG, Gasparich GE, Rose DL, Carle P, Bove JM, Hackett KJ, Adams JR, Henegar RB, Konai M, Chastel C, French FE (1998) Revised group classification of the genus *Spiroplasma*. Int J Syst Bacteriol 48:1–12

Wu CI et al (1995) Sexual isolation in *Drosophila melanogaster*: a possible case of incipient speciation. Proc Natl Acad Sci USA 92(7):2519–2523

Zilber-Rosenberg I, Rosenberg E (2008) Role of microorganisms in the evolution of animals and plants: the hologenome theory of evolution. FEMS Microbiol Rev 32(5):723–735

Part II
Plant–Microbe Symbioses

Chapter 5
Legume–Microbe Symbioses

Masayuki Sugawara and Michael J. Sadowsky

Abstract The world will most likely face severe food shortages in the not-too-distant future, in part due to rapid population growth. Based on all models, significant increases in crop production will be needed to maintain human nutrition levels and this will require increased yields in current crop areas and use of land areas now considered to be marginal. If sufficient water is available, nitrogen (N) acquisition and assimilation ranks second in importance for plant growth next to photosynthesis. The production of food is dependent on the availability of N for plant growth and much of this N is now supplied in a non-sustainable manner from fossil fuels. However, many diverse bacteria contribute to N_2 fixation (BNF) in soil and aquatic systems. In most agricultural systems, the primary source of biologically-fixed N occurs via the symbiotic interactions of legumes and soil bacteria collectively termed the rhizobia (including the genera *Allorhizobium*, *Azorhizobium*, *Rhizobium*, *Mesorhizobium*, *Ensifer* (formerly, *Sinorhizobium*) or *Bradyrhizobium*) and several new bacterial groups in the genera *Devosia*, *Methylobacterium*, *Ochrobactrum*, *Shinella*, *Cupriavidus*, and *Burkholderia*. These microbes have the ability to form N-fixing root nodules on roots/stems of legumes. Legumes may provide up of 35% of the world's protein intake, and symbiotically-fixed N may account for 90 Tg N year^{-1}. Over the last decade, it has been shown that the host plant and microbial genome both control metabolic processes that are essential to the proper functioning of the symbiotic interaction. Since N_2-fixing symbioses require the interaction of two evolutionarily unique organisms (plants and microbes) that interact in an appropriate environment, research aimed at improving this process is interdisciplinary by definition. Consequently, understanding and utilizing these important symbiotic systems requires studies in microbial genetics, genomics, microbial and plant physiology, ecology, and soil science. Recently, there have been several exciting developments in the basic and applied sciences

M. Sugawara • M.J. Sadowsky (✉)
University of Minnesota, St. Paul, MN 55108, USA
e-mail: sadowsky@umn.edu

E. Rosenberg and U. Gophna (eds.), *Beneficial Microorganisms in Multicellular Life Forms*,
DOI 10.1007/978-3-642-21680-0_5,

directed as this effort, with significant progress being made in bacterial and plant genomics, and in the genetic manipulation of the host plants to better understand the symbiotic process. While the symbiosis between rhizobia and its legume host has obvious importance for agricultural productivity, it also influences the global N cycle, is ecologically beneficial, and reduces use of our limited fossil fuel resources. The effective use of biological N fixation, via the application of rhizobia, results in better cropping systems that have a decreased impact on the environment through the decreased use of N fertilizers. For this reason, the ecological and genetic features of rhizobia have been studied extensively. Here, I present a discussion on the overall mechanisms by which legumes and microbes interact to fix atmospheric N and the genetics and genomics of these tightly linked symbiotic partners.

5.1 Introduction

Symbiotic interactions between microorganisms and plants are widespread among a variety of eukaryotic and prokaryotic taxa. Bacteria that associate with plants as symbiotic partners include symbiotic N-fixing rhizobia, actinorhizal bacteria (e.g., *Frankia* species), many endophytic bacteria, the cyanobacteria *Anabaena* and *Nostoc*, and several plant-growth promoting rhizobacteria (PGPR). While many of these symbioses have been of wide interest to both plant and microbial biologists, the interactions between legumes and rhizobia have been amongst the most extensively studied symbiotic systems over the last three decades.

Rhizobia are N-fixing bacteria that form root and/or stem nodules on leguminous plants (Fig. 5.1). Within the nodules, rhizobia convert atmospheric dinitrogen

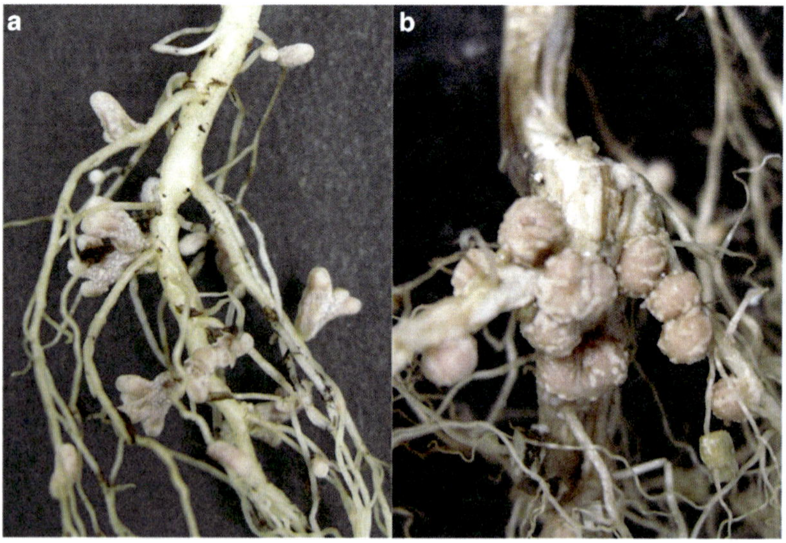

Fig. 5.1 Nodule types formed on leguminous plants. (**a**); Indeterminant nodules formed on *Medicago truncatula* roots inoculated with *Ensifer medicae*. (**b**); Determinant nodules formed on *Glycine max* (soybean) roots inoculated with *Bradyrhizobium japonicum*

(N_2) gas into ammonia (NH_3). This fixed N is subsequently assimilated by the host plant resulting in improved plant growth and productively, even under N-limiting environmental conditions. While this has obvious importance for agricultural productivity, it also influences the global N cycle, is ecologically beneficial, and reduces use of our limited fossil fuel resources.

N is one of the major limiting nutrient factors affecting plant growth in the biosphere. The application of N fertilizers often negatively impacts the environment through nitrate-mediated water pollution, and the evolution of a greenhouse gases, such as nitrous oxide (N_2O), into the atmosphere (Kinzig and Socorow 1994; Mosier et al. 1996). Thus, the effective use of biological N fixation, via the application of rhizobia, results in more ideal cropping systems that have a decreased impact on the environment (Bohlool et al. 1992). For this reason, the ecological and genetic features of rhizobia have been studied extensively.

5.2 Taxonomy of Rhizobia

The taxonomy of root and stem nodulating bacteria, collectively referred to as the rhizobia, is in a state of flux. Although this ever-changing taxonomy affects what the organisms are called and how they are distinguished, it has little impact on their true phylogenetic relationships, their use for enhancing crop productivity, their positive impact in the environment, and use as model systems to study symbioses. Sequence analysis of small subunit rRNA (SSU rRNA) supports the contention that most of these rhizobial species belong to the alpha-proteobacteria and are members of the genera *Allorhizobium*, *Azorhizobium*, *Rhizobium*, *Mesorhizobium*, *Ensifer* (formerly, *Sinorhizobium*) or *Bradyrhizobium* (Martinez-Romero and Caballero-Mellado 1996; Young and Haukka 1996). In addition to these divisions, recent research has shown that some root nodule bacteria, including *Devosia neptuniae* (Rivas et al. 2002), *Methylobacterium nodulans* (Sy et al. 2001), *Ochrobactrum lupini* (Trujillo et al. 2005) and *Ochrobactrum cytisi* (Zurdo-Piñeiro et al. 2007), *Shinella kummerowiae* (Lin et al. 2008) are members of the alpha-proteobacteria, whereas several beta-proteobacterial species, including *Cupriavidus* (formerly *Ralstonia*; Chen et al. 2001) and *Burkholderia* (Moulin et al. 2001; Vandamme et al. 2002), also have the ability to form N-fixing root nodules on some legumes.

5.3 Genome Architecture of Rhizobia

Several approaches have been used to define and study the involvement of whole bacterial genomes in the symbiotic process. As of this date, about 15 rhizobial genome sequences have been completed and are currently available online (Table 5.1), and approximately 150 more are currently being sequenced. In most of the genome-sequenced rhizobia, the genes involved in nodulation and N fixation

Table 5.1 Architecture of sequenced rhizobial genomes

Rhizobial strain	Number of replicons	Total size (bp)	Genbank accession number	Reference
Alpha-proteobacteria				
Azorhizobium caulinodans ORS571	1	5,369,772	AP009384	Lee et al. (2008)
Bradyrhizobium japonicum USDA110	1	9,105,828	BA000040	Kaneko et al. (2002)
Bradyrhizobium sp. BTAi1	2	8,493,513	CP000494, CP000495	Giraud et al. (2007)
Bradyrhizobium sp. ORS278	1	7,456,587	CU234118	Giraud et al. (2007)
Ensifer (*Sinorhizobium*) *medicae* WSM419	4	6,817,576	CP000738-CP000741	Reeve et al. (2010)
Ensifer (*Sinorhizobium*) *meliloti* 1021	3	6,691,694	AL591688, AL591985, AE006469	Galibert et al. (2001)
Mesorhizobium loti MAFF303099	3	7,596,297	AP002994-AP003017	Kaneko et al. (2000)
Methylobacterium nodulans ORS2060	8	8,839,022	CP001349-CP001356	Unpublished
Rhizobium etli CFN42	7	6,530,228	CP000133-CP000138	González et al. (2006)
Rhizobium etli CIAT652	4	6,448,048	CP001074-CP001077	Unpublished
Rhizobium leguminosarum bv. *trifolii* WSM1325	6	7,418,122	CP001622-CP001627	Unpublished
Rhizobium leguminosarum bv. *trifolii* WSM2304	5	6,872,702	CP001191-CP001195	Unpublished
Rhizobium leguminosarum bv. *viciae* 3841	7	7,751,309	AM236080-AM236086	Young et al. (2006)
Rhizobium sp. NGR234	3	6,891,900	CP000874, CP001389, U00090	Schmeisser et al. (2009)
Beta-proteobacteria				
Burkholderia phymatum STM815	4	8,676,562	CP001043-CP001046	Unpublished
Cupriavidus (*Ralstonia*) *taiwanensis* LMG19424	3	6,476,522	CU633749-CU633751	Amadou et al. (2008)

genes (see below) are clustered on large, and often self-transmissible megaplasmids (pSyms), or are located within large genomic islands, referred to as symbiotic islands (SIs). These features in many ways emphasize the accessory nature of the symbiosis-related genes and their ability to be acquired by microorganisms via horizontal gene transfer (MacLean et al. 2007).

A SI present in *Mesorhizobium loti* strain ICMP3153 was found capable of transforming non-symbiotic strains of *M. loti* into symbiotic counterparts (Sullivan and Ronson 1998). Large SIs of 611 and 681 kb have been found in *M. loti* MAFF303099 and *B. japonicum* strain USDA110, respectively (Kaneko et al. 2000, 2002). Although the wide-spread transmissibility of these islands has yet to be confirmed, the integration of SIs into the *M. loti* and *B. japonicum* genomes occurs within phe-tRNA and val-tRNA genes, respectively (Kaneko et al. 2000, 2002). The association of SIs, in both rhizobial species, with a phage-related integrase implies the SIs may have originated from the ancient integration of a bacteriophage. Beside the SI, 14 smaller genomic islands, with lower GC contents, were also found in the genome of *B. japonicum* USDA110 (Kaneko et al. 2002).

Since some of these genomic islands were found to be missing in several strains of *B. japonicum*, the 14 genomic islands were thought to be likely inserted into the ancestor genome of USDA110 via horizontal gene transfer events (Itakura et al. 2009).

In contrast to what has been found in the slower-growing bradyrhizobia, most of the genes involved in symbioses in genome-sequenced *Ensifer* and *(Sino)Rhizobium* strains are located on symbiotic plasmids, pSyms. In many cases, the pSyms have been shown to be transferred among bacteria via conjugation (Rao et al. 1994; Freiberg et al. 1997; Brom et al. 2004; Pérez-Mendoza et al. 2004).

5.4 The Nodulation Process of Rhizobia with Legume Plants

The nodulation process requires molecular communication between both symbiotic partners and involves the induction and repression of a large number of bacterial and plant genes. Free-living rhizobia infect and form N_2-fixing symbioses with legumes in a series of discrete stages or steps. In most instances, stages in the process include proliferation of rhizobia in the rhizosphere, recognition of host by rhizobia, attachment of rhizobia to susceptible root hair cells, root-hair curling and infection-thread formation, initiation of nodule primordium, and transformation of free-living rhizobia into N_2-fixing bacteroids in developed mature nodule.

More recent studies have shown that rhizobia infect their respective host plants and induce root or stem nodules using several different mechanisms. While infection through root hairs is commonly seen with most legumes (Hadri et al. 1998), rhizobia can also invade the host plant via wounds, cracks, or lesions caused by the emergence of secondary roots, as occurs in peanut and *Stylosanthes* (Boogerd and van Rossum 1997). In these cases, rhizobia spread intercellularly. There are also instances where the same rhizobial strain infects one legume through root hairs and another via cracks or wounds (Sen and Weaver 1988). Lastly, rhizobia may initiate infection of the host via cavities or openings produced through the emergence of adventitious root primordia on the stems of legumes such as *Sesbania*, *Aeschynomene*, *Neptunia*, and *Discolobium* (Boivin et al. 1997; Giraud et al. 2007).

In root hair infection, rhizobia attach to susceptible root hairs within minutes of inoculation or contact with the host plant. Rhizobial cells often attach perpendicular, in an end-on_manner, to the root hair cells. It has been suggested that adhesion is initially mediated by a calcium (Ca)-binding protein rhicadhesin, or by plant lectins, and subsequent bonding via the production of cellulose fibrils (Kijne 1992). It has also been hypothesized that rhizobia produce localized hydrolysis of the root hair cell wall (Kijne 1992). Subsequent penetration of rhizobia through the cell wall leads to root-hair curling, which may be visible 6–18 h after inoculation. Within the root hair, rhizobia are enclosed within a plant-derived infection thread, and move distally down the root hair towards the root cortex. Cell division in the root cortex, in advance of the approaching infection thread, leads to the production of nodule primordial (Kijne 1992). Spread of the infection thread among cells of the nodule primordium follows, with the release of rhizobia into host cortex by an endocytotic process. Rhizobia are not found free in the cytoplasm, but rather are surrounded by a host-derived peribacteroid membrane. This serves to compartmentalize the rhizobia into a symbiosome. One to several rhizobia can be localized to a single symbiosome, and this varies by host plant species.

Legume root nodules are usually visible 6–18 days after inoculation, but this varies considerably with the selection of bacterial strain, host cultivar, inoculant density and placement, and incubation temperature. The number of nodules produced on each legume host is tightly controlled by the host and rhizobial genotype, the efficiency of the symbiotic interaction, by environmental factors, such as the soil N level, and by the presence of existing nodules (Singleton and Stockinger 1983; Sagan and Gresshoff 1996; Caetano-Annoles 1997).

Nodule shape in legumes is determined by the host plant and is regulated by the pattern of cortical cell divisions within the root cortex. Two basic types of nodules are formed on legumes: determinant and indeterminant (Franssen et al. 1992). Indeterminant nodules are most commonly formed in symbioses between the fast-growing root-nodule bacteria, such as members of the genera *Ensifer* or *Rhizobium*, and temperate legumes (pea, clover, and alfalfa; Fig. 5.1a). In contrast, determinate nodules are typically induced by the bradyrhizobia and some rhizobial species strains, and are more common on tropical legumes, such as soybean and bean (Fig. 5.1b). Morphologically, indeterminate nodules have defined, persistent apical meristems and are elongated and sometimes lobed, whereas determinant nodules do not have persistent meristems and are usually round (Hadri et al. 1998; Fig. 5.1).

5.5 Nodulation Genes in Rhizobia

A large number of bacterial genes have been identified which are involved in the formation of nodules on leguminous plants. Collectively, more than 65 nodulation genes (designated *nod*, *noe*, and *nol*) have been identified in the rhizobia, although each strain may only contain a subset of these. A more complete description of the types and function of a majority of these genes can be found in several reviews (Bladergroen and Spaink 1998; Niner and Hirsch 1998; Schlaman et al. 1998; Long 2001; Kobayashi and Broughton 2008).

Several studies have shown that relatively few genes are required for nodulation of legumes (Long et al. 1985; Long 1989; Göttfert 1993; van Rhijn and Vanderleyden 1995). In *E. meliloti*, the symbiont of alfalfa, nodulation genes (located on an 8.5 kb fragment of the pSym) contain sequences necessary for the nodulation of a wide variety of legume hosts (Kondorosi et al. 1989; Truchet et al. 1991). These genes, referred to as "common nodulation" genes and designated *nodA*, *nodB* and *nodC*, have homologues in other fast- and slow-growing rhizobial species (Egelhoff and Long 1985; Long et al. 1985). The common nodulation genes are involved in biosynthesis of the chitin backbone of Nod-factor (see below), and organized in a similar cluster in most rhizobia (Long et al. 1985; Long 1989; van Rhijn and Vanderleyden 1995). A fourth gene, *nodD*, is regulatory and together with plant flavonoid signals (see below) activates transcription of other inducible nodulation genes (Long 1989; Martinez et al. 1990; van Brussel et al. 1990). *Rhizobium leguminosarm* bvs. *viceae* and *trifolii* have single copies of *nodD*, and the symbionts *E. meliloti*, *B. japonicum* and *M. loti* have multiple copies of *nodD* (Göttfert et al. 1986, 1990; Honma and Ausubel 1987; Kaneko et al. 2000). Other nodulation genes are involved in various modifications and secretion of Nod-factor, and transcriptional regulators of nodulation genes.

Intriguingly, *Bradyrhizobium* strains BTAi1 and ORS278, which induce nodules on both the root and stem of the aquatic legume *Aeschynomene,* do not possess functional *nodABC* genes (Giraud et al. 2007). This indicates that canonical *nodABC* genes and Nod-factor are not required for symbiosis for these strains, and implies that an alternative nodulation strategy exists in the *Bradyrhizobium–Aeschynomene* symbiotic interaction (Giraud et al. 2007; Bonaldi et al. 2010).

5.6 Signal Exchange and Induction of Nod Genes

Although the regulation of nodulation genes in the rhizobia is not fully understood, a lot is known about communication between rhizobia and a susceptible legume hosts. Several studies have firmly established that flavonoid signal molecules present in root and seed exudates are necessary for *nod* gene expression (Schlaman et al. 1998; Long 2001). Flavones, isoflavones, flavanols, flavanones, and closely related compounds, have been identified as *nod* gene inducers, and each is specific for a particular legume-*Rhizobium* interaction (Schlaman et al. 1998). This imparts host-specificity on the legume–microbe symbiosis, although flavonoid compounds are only one of several determinants of host specificity. For example, Spaink et al. (1991) reported the differential induction of *nodD* genes in various fast-growing rhizobia by a range of flavonoid compounds and exudates. The induction of nodulation genes requires the regulatory *nodD* gene product (Mulligan and Long 1985; Shearman et al. 1986; Long 1989), and current studies suggest that the inducer binds NodD, causing a change in its conformation (Kondorosi et al. 1988; Fisher and Long 1989, 1993; Kobayashi and Broughton 2008). The activated NodD in turn binds to a regulatory, promoter-like sequence, found upstream of rhizobial *nod* genes, the Nod-box (Horvath et al. 1986; Rostas et al. 1986; Shearman et al. 1986 ; Hong et al. 1987; Kondorosi et al. 1988).

Repressor proteins have also been shown to play a role in *nod* gene regulation (Kondorosi et al. 1988). A *nod* gene repressor encoded by the *nolR* gene has been identified in *E. meliloti* strain 41 (Kondorosi et al. 1989, 1991), and by the *nolA* and $nodD_2$ has been identified in *B. japonicum* strain USDA 110 (Sadowsky et al. 1991; Göttfert et al. 1992; Loh and Stacey 2003).

5.7 Extracellular Nodulation Factors

One of the principal functions of *nod* genes is the production of extracellular lipochitinoligosaccharide (LCO) molecules, also known as Nod factors (Carlson et al. 1993, 1994). These molecules, acting at hormonal-like 10^{-8}–10^{-9} M concentrations, can act to: (1) stimulate the plant to produce more *nod* gene inducers (van Brussel et al. 1990); (2) deform root hairs on homologous hosts (Banfalvi and Kondorosi 1989; Faucher et al. 1989); and (3) initiate cell division in the root cortex (Lerouge et al. 1990; Spaink et al. 1991; Sanjuan et al. 1992; Schultze et al. 1992). In *E. meliloti* these signal molecules are acetylated and sulfated glucosamine oligosaccharides (Lerouge et al. 1990), and similar molecules have been identified in other legume symbiotic systems (Pueppke 1996; Downie 1998 for a review). Numerous observations support the well-accepted theory that some nodulation genes control host specificity by decorating Nod factors with various substituents. Thus, the nodulation genes involved in modification of the Nod-factor also appear to impart host-specificity to the legume–microbe symbiosis. For example, the host-specific nodulation genes, such as *nodL* or *nodH*, specify the various Nod-factor substitutions. Mutations in these genes can result in changes in the rhizobial host range (Spaink et al. 1987; Kobayashi and Broughton 2008). Therefore, nod genes and Nod-factors play a central role in nodulation and host range determination. In *E. meliloti*, the *nodP*, *nodQ*, and *nodH* genes are involved in the sulfation of the Nod-factor reducing sugar (Faucher et al. 1989; Roche et al. 1991). Disruption of any of these genes affects host specificity. Each rhizobial strain usually produces only one or a few Nod factors, which are subsequently recognized by the host legume. However, *Rhizobium* spp. strain NGR234, which nodulates over 125 different legume species (Pueppke and Broughton 1999), produces diverse (more than 18) Nod factors, which vary in the substituents attached to a similar backbone structure (Price et al. 1992). Nod factors have been shown to bind to their cognate host receptors and are thought to initiate a programmed series of events on the plant host leading to nodule formation (Endre et al. 2002; Limpens et al. 2003; Madsen et al. 2003; Bersoult et al. 2005). Interestingly, purified Nod factors, which are structurally similar to those produced by the appropriate rhizobial symbiont, can induce nodules on their specific host plant in the absence of a bacterium (Downie 1998; Truchet et al. 1991; Schultze et al. 1992; Mergaert et al. 1993). The nodules produced, however, are devoid of bacteria and, as such, cannot fix N.

5.8 N Fixation in Rhizobia

N fixation is the natural process, either biological or abiotic, by which nitrogen gas (N_2) in the atmosphere is converted into ammonium (NH_4). Only bacteria containing the enzyme nitrogenase can reduce N_2 to ammonium. This is the only known enzyme that can carry out this energetically unfavorable reaction.

As described above, legume plants enter into a symbiotic interaction with N-fixing rhizobia resulting in the formation of root or stem nodules. Nitrogenase is very sensitive to inactivation by oxygen and within the nodules leghemoglobin maintains a low-oxygen concentration to protect enzyme function (nanomolar range) (Downie 2005).

Two major types of rhizobial N_2 fixation genes have been described, *nif* and *fix* genes. The *nif* genes are structurally and functionally related to those first described in the free-living diazotrophic bacterium *Klebsiella pneumoniae* (Kennedy 1989). As with the nodulation genes, a majority of the *nif* genes are plasmid borne and contiguous in the rhizobia. In contrast, nif genes are chromosomally-located in the bradyrhizobia and mesorhizobia. The *fix* genes are also involved in the N_2 fixation process, but have no similar structural or functional homologues in *K. pneumoniae*.

The enzyme complex nitrogenase is encoded by the *nifDK* and *nifH* genes. Nitrogenase consists of two protein subunits, a molybdenum-iron (MoFe) protein and an Fe-containing protein. These structural components of the nitrogenase enzyme complex are often referred to as subunits I and II, respectively. The *nifK* and *nifD* genes encode the MoFe protein subunits, wheras *nifH* encodes the Fe subunit protein. A FeMo cofactor (FeMo-Co) is required for activation of the MoFe protein and is assembled from the *nifB*, *nifV*, *nifN*, and *nifE* genes.

There are at least 20 *nif*-specific genes that are localized in about eight operons in *K. pneumoniae* (Dean and Jacobson 1992). Although the organization of *nif* genes in other organisms varies tremendously (Downie 1998), *nifHD* and *nifK* are conserved in disparate N_2-fixing organisms and rhizobia (Ruvkun and Ausubel 1980).

The gene products NifA and NifL control the regulation of all other *nif* genes. While NifA is positive activator of transcription of the *nif* operons, NifL is involved in negative control. In *K. pneumoniae,* and several other free-living diazotrophic microbes, *nif* gene expression is regulated by oxygen and N levels (Merrick 1992). Ammonia (NH_3) causes NifL to act as a negative regulator and prevents the activator function of NifA. This has been referred to as the N control system, and has been shown to regulate several enzymes that are capable of producing NH_3. Merrick (1992) and Dean and Jacobson (1992), and Kaminski et al. (1998) give excellent in-depth reviews of the structure and regulation of N_2 fixation in free-living and symbiotic bacteria.

5.9 Other Genes Involved in Nodualtion and Symbiotic N Fixation

Other plasmid- and chromosomally borne genes in the rhizobia have also been found to function indirectly in nodulation and symbiotic N_2-fixation. This includes genes involved in the biosynthesis of exopolysaccharides (*exo*), lipopolysaccharides (*lps*), and beta 1,2-glucans (*ndv*), those used for hydrogen uptake (*hup*), glutamine synthase (*glu*), dicarboxylate transport (*dct*), bacteroid development (*bacA*), type III and type IV secretion systems (*tts* and *virB*), purine biosynthesis (*pur*), inhibition of plant ethylene biosynthesis (*acdS, rtx*), and nodulation efficiency (*nfe*). These genes either directly or indirectly influence nodulation and symbiotic N fixation. Some review articles on the structure and function of these and other symbiosis-related genes are provided by Spaink (1995), Pueppke (1996), Long (1989), Sugawara et al. (2006) and Kobayashi and Broughton (2008).

5.10 Unresolved Questions and Future Research

While extensive research studies on legume-rhizobia interactions completed over the last three decades have provided valuable information on this agriculturally-important symbiotic system, many unresolved questions remain. Chief among these concerns are the competition for nodulation problem (Triplett and Sadowsky 1992). Competition for nodule occupancy occurs whenever two or more rhizobial strains have the opportunity for infection and nodulation of a susceptible legume plant. Nodulation of a target legume by unwanted strains has been an intractable issue to rhizobiologists for many years. Competition by indigenous, soil-borne, strains frequently limits enhanced plant growth and productivity following inoculation by a superior N_2-fixing inoculant strain. Several phenotypes, and in some cases genes, have been identified as playing an important role in nodulation competitiveness, including antibiosis, motility, nodulation speed, cell-surface characteristics, and nodulation efficiency. While it has been postulated that the degree of competitiveness is also due to tolerance of rhizobia to environmental stresses, such as desiccation, salinity, and temperature (Sadowsky 2005), it is apparent that multiple genetic loci and traits collectively contribute to competitive interactions in soil systems.

Recently, advances in several "omic" sciences have provided useful information about microbial and plant genes involved in the interaction of rhizobia with their cognate legume host plants (Stacey et al. 2006; MacLean et al. 2007). For example, oligonucleotide microarrays have been used to examine the global transcriptional responses of *B. japonicum* USDA 110 in the bacteroid state, and to growth under osmotic and desiccation stress conditions, when cultured in minimal and rich media, and under chemoautotrophic growth conditions (Chang et al. 2007; Cytryn et al. 2007; Franck et al. 2008). Based on these analyses, several new and well-characterized *Bradyrhizobium* genes have been shown to be specifically involved in tolerances to physiological stresses or to be responsive to growth conditions (Chang et al. 2007; Cytryn et al. 2007; Franck et al. 2008). Similar "omics" information obtained from legume hosts has also rapidly advanced the field (Gepts et al. 2005; Udvardi et al. 2005; Young and Udvardi 2009). This type of global information at the genomic and proteomic levels needs to be further integrated in order to have a better understanding of the legume–microbe symbioses (Delmotte et al. 2010). Despite these advances, and the currently more rapid advances in genome and RNA sequencing, the genomics-derived data needs to be coupled to traditional genetics and biochemical efforts. This, along with new bioinformatics tools, will allow researchers to rapidly define and utilize microbial and host genetic loci that are involved in nodulation, N fixation, and the tolerance to a large number of environmental stresses, and to better understand and ultimately improve the symbioses between rhizobia and legumes.

References

Amadou C, Pascal G, Mangenot S, Glew M, Bontemps C, Capela D, Carrère S, Cruveiller S, Dossat C, Lajus A, Marchetti M, Poinsot V, Rouy Z, Servin B, Saad M, Schenowitz C, Barbe V, Batut J, Médigue C, Masson-Boivin C (2008) Genome sequence of the beta-rhizobium *Cupriavidus taiwanensis* and comparative genomics of rhizobia. Genome Res 18:1472–1483

Banfalvi Z, Kondorosi A (1989) Production of root hair deformation factors by *Rhizobium meliloti* nodulation genes in *Escherichia coli*: *hsnD* (*nodH*) is involved in plant host-specific modification of the *nodABC* factor. Plant Mol Biol 13:1–12

Bersoult A, Camut S, Perhald A, Kereszt A, Kiss GB, Cullimore JV (2005) Expression of the *Medicago truncatula* DM12 gene suggests roles of the symbiotic nodulation receptor kinase in nodules and during early nodule development. Mol Plant Microbe Interact 18:869–876

Bladergroen MR, Spaink HP (1998) Genes and signal molecules involved in the Rhizobia-leguminoseae symbiosis. Curr Opin Plant Biol 1:353–359

Bohlool BB, Ladha JK, Garrity DP, George T (1992) Biological nitrogen fixation for sustainable agriculture: a perspective. Plant and Soil 141:1–11

Boivin C, Ndoye I, Molouba F, Delajudie P, Dupuy N, Dreyfus B (1997) Stem nodulation in legumes – diversity, mechanisms, and unusual characteristics. Critical Rev Plant Sci 16:1–30

Bonaldi K, Gourion B, Fardoux J, Hannibal L, Cartieaux F, Boursot M, Vallenet D, Chaintreuil C, Prin Y, Nouwen N, Giraud E (2010) Large-scale transposon mutagenesis of photosynthetic *Bradyrhizobium* sp. strain ORS278 reveals new genetic loci putatively important for Nod-independent symbiosis with *Aeschynomene indica*. Mol Plant Microbe Interact 23:760–770

Boogerd FC, van Rossum D (1997) Nodulation of groundnut by *Bradyrhizobium*: a simple infection process by crack entry. FEMS Microbiol Rev 21:5–27

Brom S, Girard L, Tun-Garrido C, los Garcia-de SA, Bustos P, González V, Romero D (2004) Transfer of the symbiotic plasmid of *Rhizobium etli* CFN42 requires cointegration with p42a, which may be mediated by site-specific recombination. J Bacteriol 186:7538–7548

Caetano-Annoles G (1997) Molecular dissection and improvement of the nodule symbiosis in legumes. Field Crops Res 53:47–68

Carlson RW, Sanjuan J, Bhat UR, Glushka J, Spaink HP, Wijfjes AHM, van Brussel AAN, Stokkermans TJW, Peters NK, Stacey G (1993) The structures and biological activities of the lipooligosaccharide nodulation signals produced by type I and II strains of *Bradyrhizobium japonicum*. J Biol Chem 268:18372–18381

Carlson RW, Price NPJ, Stacey G (1994) The biosynthesis of rhizobial lipo-oligosaccharide nodulation signal molecules. Mol Plant Microbe Interact 7:684–695

Chang WS, Franck WL, Cytryn E, Jeong S, Joshi T, Emerich DW, Sadowsky MJ, Xu D, Stacey G (2007) An oligonucleotide microarray resource for transcriptional profiling of *Bradyrhizobium japonicum*. Mol Plant Microbe Interact 20:1298–1307

Chen WM, Laevens S, Lee TM, Coenye T, De Vos P, Mergeay M, Vandamme P (2001) *Ralstonia taiwanensis* sp. nov., isolated from root nodules of *Mimosa* species and sputum of a cystic fibrosis patient. Int J Syst Evol Microbiol 51:1729–1735

Cytryn EJ, Sangurdekar DP, Streeter JG, Franck WL, Chang WS, Stacey G, Emerich DW, Joshi T, Xu D, Sadowsky MJ (2007) Transcriptional and physiological responses of *Bradyrhizobium japonicum* to desiccation-induced stress. J Bacteriol 189:6751–6762

Dean DR, Jacobson MR (1992) Bichemical genetics of nitrogenase. In: Stacey G, Burris B, Evans HJ (eds) Biological nitrogen fixation. Chapman and Hall, New York, pp 763–834

Delmotte N, Ahrens CH, Knief C, Qeli E, Koch M, Fischer HM, Vorholt JA, Hennecke H, Pessi G (2010) An integrated proteomics and transcriptomics reference data set provides new insights into the *Bradyrhizobium japonicum* bacteroid metabolism in soybean root nodules. Proteomics 7:1391–1400

Downie JA (1998) Functions of Rhizobial nodulation genes. In: Spaink HP, Kondorosi A, Hooykaas PJJ (eds) The Rhizobiaceae. Kluwer Academic Publishers, Dordrecht, pp 387–402

Downie JA (2005) Legume haemoglobins: symbiotic nitrogen fixation needs bloody nodules. Curr Biol 15:196–198

Egelhoff TT, Long SR (1985) *Rhizobium meliloti* nodulation genes: identification of *nodDABC* gene products, purification of *nodA* protein, and expression of *nodA* in *Rhizobium meliloti*. J Bacteriol 164:591–599

Endre G, Kereszt A, Kevei Z, Mihacea S, Kaló P, Kiss GB (2002) A receptor kinase gene regulating symbiotic nodule development. Nature 417:962–966

Faucher C, Camut S, Denarie J, Truchet G (1989) The *nodH* and *nodQ* host range genes of *Rhizobium meliloti* behave as avirulence genes in *R. legumiosarum* bv *viceae* and determine changes in the production of plant-specific extracellular signals. Mol Plant Microbe Interact 2:291–300

Fisher RF, Long SR (1989) DNA footprint analysis of the trascriptional activator proteins NodD1 and NodD3 on inducible *nod* gene promoters. J Bacteriol 171:5492–5502

Fisher RF, Long SR (1993) Interactions of NodD at the nod Box: NodD binds to two distinct sites on the same face of the helix and induces a bend in the DNA. J Mol Biol 233:336–348

Franck WL, Chang WS, Qiu J, Sugawara M, Sadowsky MJ, Smith SA, Stacey G (2008) Whole-genome transcriptional profiling of *Bradyrhizobium japonicum* during chemoautotrophic growth. J Bacteriol 190:6697–6705

Franssen HJ, Nap JP, Bisseling T (1992) Nodulins in root nodule development. In: Stacey G, Burris B, Evans HJ (eds) Biological nitrogen fixation. Chapman and Hall, New York, pp 598–624

Freiberg C, Fellay R, Bairoch A, Broughton WJ, Rosenthal A, Perret X (1997) Molecular basis of symbiosis between *Rhizobium* and legumes. Nature 387:394–401

Galibert F, Finan TM, Long SR, Puhler A, Abola P, Ampe F, Barloy-Hubler F, Barnett MJ, Becker A, Boistard P, Bothe G, Boutry M, Bowser L, Buhrmester J, Cadieu E, Capela D, Chain P, Cowie A, Davis RW, Dreano S, Federspiel NA, Fisher RF, Gloux S, Godrie T, Goffeau A, Golding B, Gouzy J, Gurjal M, Hernandez-Lucas I, Hong A, Huizar L, Hyman RW, Jones T, Kahn D, Kahn ML, Kalman S, Keating DH, Kiss E, Komp C, Lelaure V, Masuy D, Palm C, Peck MC, Pohl TM, Portetelle D, Purnelle B, Ramsperger U, Surzycki R, Thebault P, Vandenbol M, Vorholter FJ, Weidner S, Wells DH, Wong K, Yeh KC, Batut J (2001) The composite genome of the legume symbiont *Sinorhizobium meliloti*. Science 293:668–672

Gepts P, Beavis WD, Brummer EC, Shoemaker RC, Stalker HT, Weeden NF, Young ND (2005) Legumes as a model plant family. Genomics for food and feed report of the cross-legume advances through genomics conference. Plant Physiol 137:1228–1235

Giraud E, Moulin L, Vallenet D, Barbe V, Cytryn E, Avarre JC, Jaubert M, Simon D, Cartieaux F, Prin Y, Bena G, Hannibal L, Fardoux J, Kojadinovic M, Vuillet L, Lajus A, Cruveiller S, Rouy Z, Mangenot S, Segurens B, Dossat C, Franck WL, Chang WS, Saunders E, Bruce D, Richardson P, Normand P, Dreyfus B, Pignol D, Stacey G, Emerich D, Verméglio A, Médigue C, Sadowsky M (2007) Legumes symbioses: absence of Nod genes in photosynthetic bradyrhizobia. Science 316:1307–1312

González V, Santamaría RI, Bustos P, Hernández-González I, Medrano-Soto A, Moreno-Hagelsieb G, Janga SC, Ramírez MA, Jiménez-Jacinto V, Collado-Vides J, Dávila G (2006) The partitioned *Rhizobium etli* genome: genetic and metabolic redundancy in seven interacting replicons. Proc Natl Acad Sci USA 103:3834–3839

Göttfert M (1993) Regulation and function of rhizobial nodulation genes. FEMS Microbiol Lett 104:39–64

Göttfert M, Horvath B, Kondorosi E, Putnoky P, Rodriguez-Quniones F, Kondorosi A (1986) At least two *nodD* genes are necessary for efficient nodulation of alfalfa by *Rhizobium meliloti*. J Mol Biol 191:411–420

Göttfert M, Grob P, Hennecke H (1990) Proposed regulatory pathway encoded by the *nodV* and the *nodW* genes, determinants of host specificity in *Bradyrhizobium japonicum*. Proc Natl Acad Sci USA 87:2680–2684

Göttfert M, Holzhäuser D, Bäni D, Hennecke H (1992) Structural and functional analysis of two different *nodD* genes in *Bradyrhizobium japonicum* USDA110. Mol Plant Microbe Interact 5:257–265

Hadri AE, Spaink HP, Bisseling T, Brewin NJ (1998) Diversity of root nodulation and rhizobial infection processes. In: Spaink HP, Kondorosi A, Hooykaas PJJ (eds) The Rhizobiaceae. Kluwer Academic Publishers, Dordrecht, pp 348–360

Hong GF, Burn JE, Johnston AWB (1987) Evidence that DNA involved in the expression of nodulation (nod) genes in *Rhizobium* binds to the regulatory gene *nodD*. Nucl Acids Res 15:9677–9690

Honma MA, Ausubel FM (1987) *Rhizobium meliloti* has three functional copies of the *nodD* symbiotic regulatory element. Proc Natl Acad Sci USA 84:8558–8562

Horvath B, Kondorosi E, John M, Schmidt J, Török I, Györgypal Z, Barabas I, Wieneke U, Schell J, Kondorosi A (1986) Organization, structure, and symbiotic function of *Rhizobium meliloti* nodulation genes determining host specificity for alfalfa. Cell 46:335–343

Itakura M, Saeki K, Omori H, Yokoyama T, Kaneko T, Tabata S, Ohwada T, Tajima S, Uchiumi T, Honnma K, Fujita K, Iwata H, Saeki Y, Hara Y, Ikeda S, Eda S, Mitsui H, Minamisawa K (2009) Genomic comparison of *Bradyrhizobium japonicum* strains with different symbiotic nitrogen-fixing capabilities and other Bradyrhizobiaceae members. ISME J 3:326–339

Kaminski PA, Batutu J, Boistard P (1998) A survey of symbiotic nitrogen fixation by Rhizobia. In: Spaink HP, Kondorosi A, Hooykaas PJJ (eds) The Rhizobiaceae. Kluwer Academic Publishers, Dordrecht, pp 432–460

Kaneko T, Nakamura Y, Sato S, Asamizu E, Kato T, Sasamoto S, Watanabe A, Idesawa K, Ishikawa A, Kawashima K, Kimura T, Kishida Y, Kiyokawa C, Kohara M, Matsumoto M, Matsuno A, Mochizuki Y, Nakayama S, Nakazaki N, Shimpo S, Sugimoto M, Takeuchi C, Yamada M, Tabata S (2000) Complete genome structure of the nitrogen-fixing symbiotic bacterium *Mesorhizobium loti*. DNA Res 7:331–338

Kaneko T, Nakamura Y, Sato S, Minamisawa K, Uchiumi T, Sasamoto S, Watanabe A, Idesawa K, Iriguchi M, Kawashima K, Kohara M, Matsumoto M, Shimpo S, Tsuruoka H, Wada T, Yamada M, Tabata S (2002) Complete genomic sequence of nitrogen-fixing symbiotic bacterium *Bradyrhizobium japonicum* USDA110. DNA Res 9:189–197

Kennedy C (1989) The genetics of nitrogen fixation. In: Hopwood DA, Chater KE (eds) Genetics of bacterial diversity. Academic, New York, pp 107–127

Kijne JW (1992) The *Rhizobium* infection process. In: Stacey G, Burris B, Evans HJ (eds) Biological nitrogen fixation. Chapman and Hall, New York, pp 349–398

Kinzig AP, Socorow RH (1994) Is nitrogen fertilizeruse nearing a balance? Reply. Phys Today 47:24–35

Kobayashi H, Broughton WJ (2008) Fine-tuning of symbiotic genes in rhizobia: flavonoid signal transduction cascade. In: Dilworth MJ, James EK, Sprent JI, Newton WE (eds) Nitrogen-fixing leguminous symbioses. Springer, Dordrecht

Kondorosi E, Gyuris J, Schmidt J, John M, Duda E, Schell J, Kondorosi A (1988) Positive and negative control of nodulation genes in *Rhizobium meliloti* strain 41. In: Verma DPS, Palacios R (eds) Mol microbe-plant interact. APS Press, St. Paul, p 73

Kondorosi E, Gyuris J, Schmidt J, John M, Duda E, Hoffmann B, Schell J, Kondorosi A (1989) Positive and negative control of *nod* gene expression in *Rhizobium meliloti* is required for optimal nodulation. EMBO J 8:1331–1340

Kondorosi E, Pierre M, Cren M, Haumann U, Buiré M, Hoffmann B, Schell J, Kondorosi A (1991) Identification of NolR, a negatively transacting factor controlling the *nod* regulon in *Rhizobium meliloti*. J Mol Biol 222:885–896

Lee KB, De Backer P, Aono T, Liu CT, Suzuki S, Suzuki T, Kaneko T, Yamada M, Tabata S, Kupfer DM, Najar FZ, Wiley GB, Roe B, Binnewies TT, Ussery DW, D'Haeze W, Herder JD, Gevers D, Vereecke D, Holsters M, Oyaizu H (2008) The genome of the versatile nitrogen fixer *Azorhizobium caulinodans* ORS571. BMC Genomics 9:271

Lerouge P, Roche P, Faucher C, Maillet F, Truchet G, Promé JC, Dénarié J (1990) Symbiotic host specificity of *Rhizobium meliloti* is determined by a sulphated and acylated glucoisamine oligosaccharide signal. Nature 344:781–784

Limpens E, Franken C, Smit P, Willemse J, Bisseling T, Geurts R (2003) LysM domain receptor kinases regulating rhizobial Nod factor-induced infection. Science 302:630–633

Lin DX, Wang ET, Tang H, Han TX, He YR, Guan SH, Chen WX (2008) *Shinella kummerowiae* sp. nov., a symbiotic bacterium isolated from root nodules of the herbal legume *Kummerowia stipulacea*. Int J Syst Evol Microbiol 58:1409–1413

Loh J, Stacey G (2003) Nodulation gene regulation in *Bradyrhizobium japonicum*: a unique integration of global regulatory circuits. Appl Environ Microbiol 69:10–17

Long SR (1989) *Rhizobium*-legume nodulation: life together in the underground. Cell 56:203–214

Long SR (2001) Genes and signals in the *Rhizobium*-legume symbiosis. Plant Physiol 125:69–72

Long SR, Egelhoff T, Fisher RF, Jacobs TW, Mulligan JT (1985) Fine structure studies of *R. meliloti nodDABC* genes. In: Evans HJ, Bottomley PJ, Newton WE (eds) Nitrogen fixation research progress. Martinus Nijhoff Publishers, Boston, pp 87–94

MacLean AM, Finan TM, Sadowsky MJ (2007) Genomes of the symbiotic nitrogen-fixing bacteria of legumes. Plant Physiol 144:615–622

Madsen EB, Madsen LH, Radutoiu S, Olbryt M, Rakwalska M, Szczyglowski K, Sato S, Kaneko T, Tabata S, Sandal N, Stougaard J (2003) A receptor kinase gene of the LysM type is involved in legume perception of rhizobial signals. Nature 425:637–640

Martinez E, Romero D, Palacios R (1990) The *Rhizobium* genome. Crit Rev Plant Sci 9:59–93

Martinez-Romero E, Caballero-Mellado J (1996) *Rhizobium* phylogenies and bacterial genetic diversity. Crit Rev Plant Sci 15:113–140

Mergaert P, van Montagu M, Prome JC, Holsters M (1993) Three unusual modifications, a D-arabinosyl, an N-methyl, and a carbamoyl group, are present on the Nod factors of *Azorhizobium caulinodans* strain ORS571. Proc Natl Acad Sci USA 90:1551–1555

Merrick MJ (1992) Regulation of nitrogen fixation genes in free-living and symbiotic bacteria. In: Stacey G, Burris B, Evans HJ (eds) Biological nitrogen fixation. Chapman and Hall, New York, pp 835–876

Mosier AR, Duxbury JM, Freney JR, Heinemeyer O, Minami K (1996) Nitrous oxide emissions from agricultural fields: assessment, measurement and mitigation. Plant Soil 181:95–108

Moulin L, Munive A, Dreyfus B, Boivin-Masson C (2001) Nodulation of legumes by members of the β-subclass of Proteobacteria. Nature 411:948–950

Mulligan JT, Long SR (1985) Induction of *Rhizobium meliloti nodC* expression by plant exudate requires *nodD*. Proc Natl Acad Sci USA 82:6609–6613

Niner BM, Hirsch AM (1998) How many *Rhizobium* genes, in addition to *nod, nif/fix*, and *exo*, are needed for nodule development and function. Symbiosis 24:51–102

Pérez-Mendoza D, Domínguez-Ferreras A, Muñoz S, Soto MJ, Olivares J, Brom S, Girard L, Herrera-Cervera JA, Sanjuán J (2004) Identification of functional *mob* regions in *Rhizobium etli*: evidence for self-transmissibility of the symbiotic plasmid pRetCFN42d. J Bacteriol 186:5753–5761

Price NP, Relić B, Talmont F, Lewin A, Promé D, Pueppke SG, Maillet F, Dénarié J, Promé JC, Broughton WJ (1992) Broad-host-range *Rhizobium* species strain NGR234 secretes a family of carbamoylated, and fucosylated, nodulation signals that are O-acetylated or sulphated. Mol Microbiol 6:3575–3584

Pueppke SG (1996) The genetic and biochemical basis for nodulation of legumes by rhizobia. Crit Rev Biotechnol 16:1–51

Pueppke SG, Broughton WJ (1999) *Rhizobium* sp. strain NGR234 and *R. fredii* USDA257 share exceptionally broad, nested host ranges. Mol Plant Microbe Interact 12:293–318

Rao JR, Fenton M, Jarvis BDW (1994) Symbiotic plasmid transfer in *Rhizobium leguminosarum* biovar *trifolii* and competition between the inoculant strain ICMP2163 and transconjugant soil bacteria. Soil Biol Biochem 26:339–351

Reeve W, Chain P, O'Hara G, Ardley J, Nandesena K, Bräu L, Tiwari R, Malfatti S, Kiss H, Lapidus A, Copeland A, Nolan M, Land M, Hauser L, Chang YJ, Ivanova N, Mavromatis K, Markowitz V, Kyrpides N, Gollagher M, Yates R, Dilworth M, Howieson J (2010) Complete genome sequence of the *Medicago* microsymbiont *Ensifer* (*Sinorhizobium*) *medicae* strain WSM419. Stand Genomic Sci 28:77–86

Rivas R, Velázquez E, Willems A, Vizcaíno N, Subba-Rao NS, Mateos PF, Gillis M, Dazzo FB, Martínez-Molina E (2002) A new species of *Devosia* that forms a unique nitrogen-fixing root nodule symbiosis with the aquatic legume *Neptunia natans*. Appl Environ Microbiol 68:5217–5222

Roche P, Debellé F, Maillet F, Lerouge P, Faucher C, Truchet G, Dénarié J, Promé JC (1991) Molecular basis of symbiotic host specificity in *Rhizobium meliloti*: *nodH* and *nodPQ* genes encode the sulfation of lipooligosaccharide signals. Cell 67:1131–1143

Rostas K, Kondorosi E, Horvath B, Simoncsits A, Kondorosi A (1986) Conservation of extended promoter regions of nodulation genes in Rhizobium. Proc Natl Acad Sci USA 83:1757–1761

Ruvkun GB, Ausubel FM (1980) Interspecies homology of nitrogenase genes. Proc Natl Acad Sci USA 77:191–195

Sadowsky MJ (2005) Soil stress factors influencing symbiotic nitrogen fixation. In: Wernerand D, Newton WE (eds) Nitrogen fixation in agriculture, forestry, ecology and the environment. Springer, Dordrecht, pp 89–112

Sadowsky MJ, Cregan PB, Gottfert M, Sharma A, Gerhold D, Rodriguez-Quinones F, Keyser HH, Hennecke H, Stacey G (1991) The *Bradyrhizobium japonicum nolA* gene and its involvement in the genotypespecific nodulation of soybeans. Proc Natl Acad Sci USA 88:637–641

Sagan M, Gresshoff PM (1996) Developmental mapping of nodulation events in pea (*Pisum sativum* L.) using supernodulating plant genotypes and bacterial variability reveals both plant and *Rhizobium* control of nodulation regulation. Plant Sci 117:167–179

Sanjuan J, Carlson RW, Spaink HP, Bhat UR, Barbour WM, Glushka J, Stacey G (1992) A 2-O-methylfucose moiety is present in the lipo-oligosaccharide nodulation signal of *Bradyrhizobium japonicum*. Proc Natl Acad Sci USA 89:8789–8793

Schlaman HRM, Phillips DA, Kondorosi E (1998) Genetic organization and transcriptional regulation of rhizobial nodulation genes. In: Spaink HP, Kondorosi A, Hooykaas PJJ (eds) The Rhizobiaceae. Kluwer Academic Publishers, Dordrecht, pp 351–386

Schmeisser C, Liesegang H, Krysciak D, Bakkou N, Le Quéré A, Wollherr A, Heinemeyer I, Morgenstern B, Pommerening-Röser A, Flores M, Palacios R, Brenner S, Gottschalk G, Schmitz RA, Broughton WJ, Perret X, Strittmatter AW, Streit WR (2009) *Rhizobium* sp. strain NGR234 possesses a remarkable number of secretion systems. Appl Environ Microbiol 75:4035–4045

Schultze M, Quiclet-Sire B, Kondorosi E, Virelizer H, Glushka JN, Endre G, Géro SD, Kondorosi A (1992) *Rhizobium meliloti* produces a family of sulphated lipooligosaccharides exhibiting different degrees of plant host specificity. Proc Natl Acad Sci USA 89:192–196

Sen D, Weaver RW (1988) Nitrogenase acetylene activities of isolated peanut and cowpea bacteroids at optimal oxygen availability and comparison with whole nodule activities. J Exp Botany 35:785–789

Shearman CA, Rossen L, Johnston AWB, Downie JA (1986) The *Rhizobium leguminosarum* nodulation gene *nodF* encodes a polypeptide similar to acyl carrier protein and is regulated by *nodD* plus a factor in pea root exudate. EMBO J 5:647–652

Singleton PW, Stockinger KR (1983) Compensation against ineffective nodulation in soybean (*Glycine max*). Crop Sci 23:69–72
Spaink HP (1995) The molecular basis of infection and nodulation by rhizobia—the ins and outs of sympathogenesis. Annu Rev Phytopathol 33:345–368
Spaink HP, Wijffelman CA, Pees E, Okker RJH, Lugtenberg BJJ (1987) *Rhizobium* nodulation gene *nodD* as a determinant of host specificity. Nature 328:337–340
Spaink HP, Sheeley DM, van Brussel AA, Glushka J, York WS, Tak T, Geiger O, Kennedy EP, Reinhold VN, Lugtenberg BJ (1991) A novel highly unsaturated fatty acid moiety of lipooligosaccharide signals determines host specificity of *Rhizobium*. Nature 354:124–130
Stacey G, Libault M, Brechenmacher L, Wan J, May GD (2006) Genetics and functional genomics of legume nodulation. Curr Opin Plant Biol 9:110–121
Sugawara M, Okazaki S, Nukui N, Ezura H, Mitsui H, Minamisawa K (2006) Rhizobitoxine modulates plant-microbe interactions by ethylene inhibition. Biotechnol Adv 24:382–388
Sullivan JT, Ronson CW (1998) Evolution of rhizobia by acquisition of a 500-kb symbiosis island that integrates into a phe-tRNA gene. Proc Natl Acad Sci USA 95:5145–5149
Sy A, Giraud E, Jourand P, Garcia N, Willems A, de Lajudie P, Prin Y, Neyra M, Gillis M, Boivin-Masson C, Dreyfus B (2001) Methylotrophic *Methylobacterium* bacteria nodulate and fix nitrogen in symbiosis with legumes. J Bacteriol 183:214–220
Triplett EW, Sadowsky MJ (1992) Genetics of competition for nodulation of legumes. Annu Rev Microbiol 46:399–428
Truchet G, Roche P, Lerouge P, Vasse J, Camut S (1991) Sulphated lipo-oligosaccharide signals of *Rhizobium meliloti* elicit root nodule organogenesis in alfalfa. Nature 351:670–673
Trujillo ME, Willems A, Abril A, Planchuelo AM, Rivas R, Ludeña D, Mateos PF, Martínez-Molina E, Velázquez E (2005) Nodulation of *Lupinus albus* by strains of *Ochrobactrum lupini* sp. nov. Appl Environ Microbiol 71:1318–1327
Udvardi MK, Tabata S, Parniske M, Stougaard J (2005) *Lotus japonicus*: legume research in the fast lane. Trends Plant Sci 10:222–228
van Brussel AA, Recourt K, Pees E, Spaink HP, Tak T, Wijffelman CA, Kijne JW, Lugtenberg BJ (1990) A biovar specific signal of *Rhizobium leguminosarum* bv. *viceae* induces increased nodulation gene-inducing activity in root exudate of *Vicia sativa* subsp. *nigra*. J Bacteriol 172:5394–5401
van Rhijn P, Vanderleyden J (1995) The *Rhizobium*-plant symbiosis. Microbiol Rev 59:124–142
Vandamme P, Goris J, Chen WM, de Vos P, Willems A (2002) *Burkholderia tuberum* sp. nov. and *Burkholderia phymatum* sp. nov., nodulate the roots of tropical legumes. Syst Appl Microbiol 25:507–512
Young JPW, Haukka KE (1996) Diversity and phylogeny of rhizobia. New Phytol 133:87–94
Young ND, Udvardi M (2009) Translating *Medicago truncatula* genomics to crop legumes. Curr Opin Plant Biol 12:93–201
Young JP, Crossman LC, Johnston AW, Thomson NR, Ghazoui ZF, Hull KH, Wexler M, Curson AR, Todd JD, Poole PS, Mauchline TH, East AK, Quail MA, Churcher C, Arrowsmith C, Cherevach I, Chillingworth T, Clarke K, Cronin A, Davis P, Fraser A, Hance Z, Hauser H, Jagels K, Moule S, Mungall K, Norbertczak H, Rabbinowitsch E, Sanders M, Simmonds M, Whitehead S, Parkhill J (2006) The genome of *Rhizobium leguminosarum* has recognizable core and accessory components. Genome Biol 7:R34
Zurdo-Piñeiro JL, Rivas R, Trujillo ME, Vizcaíno N, Carrasco JA, Chamber M, Palomares A, Mateos PF, Martínez-Molina E, Velázquez E (2007) *Ochrobactrum cytisi* sp. nov., isolated from nodules of *Cytisus scoparius* in Spain. Int J Syst Evol Microbiol 57:784–788

Chapter 6
Plant Growth Promotion by Rhizosphere Bacteria Through Direct Effects

Yael Helman, Saul Burdman, and Yaacov Okon

Abstract The area of the soil influenced by plant roots is named rhizosphere. Among the microorganisms inhabiting the rhizosphere, several are plant growth promoting rhizobacteria (PGPR). Among PGPR species, some promote plant growth through direct effects on the plant, while other rhizosphere bacterial species benefit plant growth through reduction of damage caused by plant pathogens. This chapter deals with the aforementioned first class of PGPR. We focus mainly on bacteria belonging to the *Azospirillum* genus, as these bacteria have been deeply investigated in terms of plant growth promotion mechanisms. Moreover, inoculants (products that contain bacterial cells in a suitable carrier for agricultural use) with *Azospirillum* strains have been developed and are being used on a commercial scale. Other PGPR with potential to promote crop yields include species from the *Herbaspirillum*, *Gluconacetobacter*, *Burkholderia*, *Pseudomonas*, and *Paenibacillus* genera. Selected species from these genera, with plant growth promotion potential, are also described in this chapter, with emphasis on the properties of these bacteria that are associated with survival, fitness, and plant growth promotion mechanisms.

6.1 Introduction

The rhizosphere is the area of the soil that is influenced by the plant roots. It is rich in microorganisms, with their composition differing from the rest of the soil due to the so named rhizosphere effect. Among the microorganisms inhabiting the rhizosphere – including viruses, bacteria, protozoa, and fungi – several species promote root and plant growth (Hartmann et al. 2008; Spaepen et al. 2009). Other species are either neutral or deleterious to plant growth (Van Loon 2007).

Y. Helman • S. Burdman • Y. Okon (✉)
Department of Plant Pathology and Microbiology, The R. H. Smith Faculty of Agriculture, Food and Environment. The Hebrew University of Jerusalem, Rehovot 76100, Israel
e-mail: okon@agri.huji.ac.il

E. Rosenberg and U. Gophna (eds.), *Beneficial Microorganisms in Multicellular Life Forms*,
DOI 10.1007/978-3-642-21680-0_6,

Among the PGPR species, there are those that directly promote plant growth by different ways: production and secretion of plant growth substances, supply of readily available micro and macroelements, biological nitrogen fixation (diazotrophs), and solubilization of phosphorus, among others (Spaepen et al. 2009). Other rhizosphere bacterial species benefit plant growth through indirect effects, which are mainly associated with reduction of damage caused by plant pathogens. These are the so named biological control agents that can act through direct effects on the pathogen or by induction of systemic resistance in the plant (Van Loon 2007; Weller 2007).

This chapter deals with the aforementioned, first class of beneficial rhizobacteria, namely PGPR that promote plant growth through direct effects on the plant. Here, we present details about the biology of selected PGPR and properties related to their habitat, survival, and influence in the rhizosphere. We focus on one of the best studied PGPR genera, *Azospirillum* spp., for which bacterial inoculants for use in agriculture have been developed and commercialized. More detailed information about the rhizosphere habitat, bacterial endophytes, indirect plant growth promotion, and symbionts is discussed in Chaps. 5, 7, and 8 of this volume.

6.2 From Basic Research to Application in the Field: The *Azospirillum brasilense* Case

6.2.1 Properties of Azospirillum

The *Azospirillum* genus belongs to the alpha-proteobacteria class and comprises free-living, nitrogen fixing, vibrio- or spirillum-shaped rods that exert beneficial effects on plant growth and yield of many agronomical important crops (Okon 1994; Baldani et al. 2005). Azospirilla are able to fix nitrogen in association with plants but extensive measurements of nitrogen fixation in greenhouse and field experiments indicated that nitrogen fixation does not play a major role in plant growth promotion in most systems evaluated so far (Spaepen et al. 2009). On the other hand, azospirilla are able to produce and excrete plant growth regulators (phytohormones) such as auxins (indole-3-acetic acid, IAA), cytokinins, and gibberellins, as well as nitric oxide, which likely are key components of plant growth promotion effects (Creus et al. 2005; Spaepen et al. 2007, 2009; Molina-Favero et al. 2008).

The genomes of *A. brasilense* Sp245 and *A. lipoferum* CRT1 have been sequenced, but the full annotations of these genomes have not been published yet (I. Zhulin and F. Wisniewsky-Dye; personal communication, 2010). It is apparent that the ability of azospirilla to successfully establish beneficial relationships with plants is, at least partially, the result of horizontal gene transfer, which aquatic bacteria such as *Aquaspirillum*, that are closely related to *Azospirillum*, underwent a transition to soil and rhizosphere environments by acquiring necessary genes from other soil bacteria. The acquired gene clusters belong to several functional categories, including transcription, transport, carbohydrate metabolism, and signal transduction (I. Zhulin, unpublished data).

The genomes of *Azospirillum* strains differ in size – from 4.8 to 9.7 Mb – and some strains possess numerous plasmids (Wisniewski-Dyé and Vial 2008). For example, *A. brasilense* Sp7 contains five large plasmids, three with molecular masses of 46, 90, and 115 MDa, respectively, and two with molecular masses higher than 300 MDa (Wisniewski-Dyé and Vial 2008). The Sp7 90-MDa, termed pRhico, has been sequenced (Vanbleu et al. 2004), it is widespread among *A. brasilense* strains, and contains genes involved in motility (formation of polar and lateral flagella), growth in minimal medium, assembly and export of cell surface polysaccharides such as exopolysaccharides (EPS) and lipopolysaccharides (LPS), and interaction with plant roots (Vanbleu et al. 2004). EPS and LPS, as well as cell surface proteins such as outer membrane proteins and flagellin, have been shown to be involved in attachment, adherence, and colonization of the root surface by azospirilla (Burdman et al. 2000).

In bacteria, poly-beta-hydroxyalkanoates (PHAs) are intracellular energy and carbon storage compounds that can be mobilized and used when carbon is a limiting resource. Intracellular accumulation of PHAs enhances survival of bacteria under environmental stress conditions in water and soil (Castro-Sowinski et al. 2010). In PHA-producing *A. brasilense*, PHAs are major determinant for overcoming periods of carbon and energy starvation. Production of PHAs is of critical importance for improving shelf-life, efficiency, and reliability of commercial inoculants (Kadouri et al. 2005).

6.2.2 *Mechanisms of Direct Plant Growth Promotion by* **Azospirillum brasilense**

One of the most pronounced effects of inoculation with azospirilla on root morphology is the proliferation of root hairs as observed in several grasses, cereals, and legumes under controlled conditions in the greenhouse as well as in the field (Fig. 6.1). Inoculation with *Azospirillum* or with other diazotrophs, can also promote the elongation of primary roots and increase the number and length of lateral roots (Okon and Kapulnik 1986). These effects are generally dependent on the inoculum concentration (Fig. 6.1) and are consistent with exogenous IAA levels, indicating that the morphological and physiological effects on the roots upon inoculation with azospirilla are mainly due to the production and secretion of IAA by the bacterium (Dobbelaere and Okon 2007; Spaepen et al. 2007, 2009).

It was also observed that secretion of the diffusible molecule nitric oxide by azospirilla is required for the enhancement of lateral root and adventitious root formation (Molina-Favero et al. 2008). Increasing evidence indicates that nitric oxide is a key signaling molecule that is involved in a wide range of functions in plants (Creus et al. 2005; Molina-Favero et al. 2008). It has been demonstrated that nitric oxide plays an important role in auxin-regulated signaling cascades in plants, influencing root growth and development (Pagnussat et al. 2003). Nitric oxide was

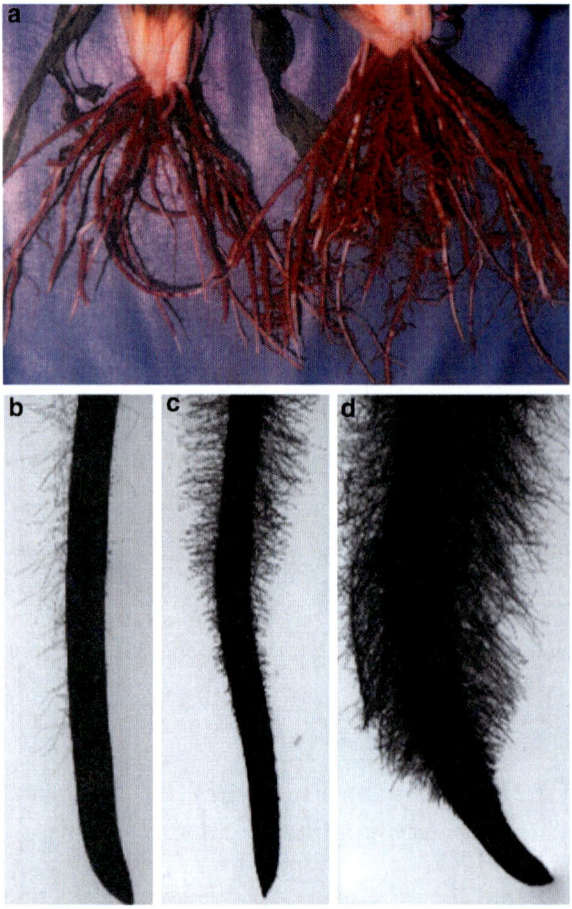

Fig. 6.1 Positive effects of inoculation with *Azospirillum brasilense* on roots. (**a**) Effects of field inoculation of maize with a commercial peat inoculant carrying *A. brasilense* strain Cd [Soygro (Pty) Ltd., Potchefstroom, South Africa]: inoculated (*right*) versus non-inoculated (*left*) roots. (**b–d**) Effect of seed inoculation with *A. brasilense* strain Sp245 on root morphology and root hair formation of 1-week-old-wheat seedlings: (**b**) non-inoculated; (**c** and **d**) inoculated with 10^7 and 10^9 CFU $seed^{-1}$ (Spaepen et al. 2008)

shown to be produced by *A. brasilense* Sp245 under aerobic conditions, mainly due to the activity of periplasmic nitrate reductase (Nap). A *nap*$^-$ *A. brasilense* mutant produces only 5% of the nitric oxide produced by the wild type, and is not able to promote root and plant development as compared to the wild type (Molina-Favero et al. 2008).

Inoculation with *A. brasilense* was shown to increase the respiration rate of root tips, and lead to an increase in the specific activity of several enzymes from root extracts (Dobbelaere and Okon 2007). The described effects on root growth and activity result in enhanced mineral and water uptake from soil by the inoculated roots. This was repeatedly demonstrated under greenhouse and field conditions in various important crops such as maize and wheat. Therefore, inoculation with *A. brasilense* can clearly benefit crops growing under suboptimal mineral and water soil conditions (Dobbelaere and Okon 2007). The positive effects of inoculation with *A. brasilense* on root morphology and activity in greenhouse and field conditions consistently correlate with improved performance of the aerial parts of the plant, which is reflected in improved leaf surface area and mineral content, timing of flowering, anthesis, and yield (Dobbelaere and Okon 2007).

Biological nitrogen fixation in *Azospirillum*-inoculated plants (wheat, maize, and other grasses) has been deeply investigated by various techniques including the acetylene reduction assay (ARA), the ^{15}N dilution technique, ^{15}N fixation and Kjeldhal N-content measurements. In most grain and forage crops grown under various conditions, the contribution of nitrogen fixation by *Azospirillum* has been estimated to be no more than 10 kg N ha^{-1} year$-^{1}$, while for instance, fertilization of maize in modern agriculture is in the order of 200–300 kg N ha^{-1}. Therefore, it appears that the contribution of biological nitrogen fixation by *A. brasilense* is too low to contribute significant amounts of nitrogen to field crops, and that nitrogen fixation by azospirilla does not play a major role in plant growth promotion (Okon and Labandera-Gonzalez 1994).

6.2.3 *Co-inoculation of Legumes with Symbiotic Rhizobia and* Azospirillum brasilense

Bacteria of the genus *Rhizobium* interact with leguminous plants in a host-specific manner through formation of nitrogen-fixing root nodules (Long 1989). Dual inoculation of several legumes with rhizobia and azospirilla, as well as with few other PGPR has been shown to significantly increase nodulation, nitrogen fixation, accumulation of macro and microelements, and biomass as compared to inoculation with *Rhizobium* alone (Sarig et al. 1986; Burdman et al. 1998; Rodelas et al. 1996, 1999).

Early events of nodule formation by rhizobia require expression of bacterial nodulation (*nod*) genes – including *nodABC* – which are induced by plant flavonoids secreted by the roots (Cooper 2007). Inoculation of both common bean and alfalfa seedlings with *A. brasilense* resulted in an increased production of plant root flavonoids and an enhanced capacity to induce *Rhizobium nod* gene expression as compared to non-inoculated controls (Burdman et al. 1996; Volpin et al. 1996). The presence of azospirilla in the rhizosphere was reported to activate the hydrolysis of conjugated phytohormones and flavonoids in the root tissue, thus leading to the release of compounds in their more active forms (Dobbelaere and Okon 2007).

Recent detailed studies support the above observations associated with co-inoculation effects. The effects of *A. brasilense* strain Cd on plant growth and nodulation, as well as on flavonoid and lipochitooligosaccharide nodulation factor (LCO Nod factor) production, were assessed in a *Rhizobium*-common bean hydroponics growth system (Dardanelli et al. 2008). *A. brasilense* promoted root branching in bean seedling roots and increased secretion of *nod* gene-inducing flavonoid compounds, as detected by high-performance liquid chromatography (HPLC). *A. brasilense* also promoted a longer, more persistent exudation of flavonoids by common bean roots relative to controls. A general positive effect of *Azospirillum-Rhizobium* co-inoculation on the expression of *nod* genes by *Rhizobium tropici* CIAT899 and *Rhizobium etli* ISP42, and on Nod factor production, was observed in the presence of root exudates of co-inoculated plants as compared to inoculation with *Rhizobium* alone (Dardanelli et al. 2008). Nodulation and nitrogen fixation are inhibited by high levels of salt, with the consequent reduction in symbiotic effectiveness (Zahran 1999). The negative effects on *nod* gene expression and on Nod factor production by salt stress were relieved in common bean plants co-inoculated with rhizobia and azospirilla (Dardanelli et al. 2008). In agreement with these findings, *A. brasilense* was shown to significantly reduce the negative effects on growth of chickpeas caused by irrigation with saline water (Hamaoui et al. 2001).

A. brasilense Sp7 was recently compared with several mutants affected in production of IAA and nitric oxide for their effects on the symbiosis between *Vicia sativa* spp. *nigra* (vetch) and *Rhizobium leguminosarum* bv. *viciae* (Star et al. 2011). Results from this study confirmed that IAA and nitric oxide produced by *A. brasilense* also play an important role in co-inoculation of legumes with rhizobia and azospirilla, and confirmed that IAA production by *A. brasilense* is a key component for enhancement of secretion of *nod* gene-inducing flavonoids by legume roots (Star et al. 2011).

6.2.4 Agronomical Impacts: Utilization of Commercial Inoculants

Bacterial inoculants for agricultural use are commercial formulations containing PGPR, which can be applied to the seeds or to the soil at planting. During production, transportation, and storage of inoculants, bacteria should survive under several stress factors such as acidity, desiccation, chemical pesticides, and non-optimal temperatures (Rebah et al. 2007).

Inoculation of soybeans with *Bradyrhizobium japonicum*, mainly via liquid inoculants, has developed into a widely utilized commercial practice, with about 40 million hectares of soybeans being currently inoculated in South America (Izaguirre-Mayoral et al. 2007). In the past 12 years, all over the world, but mainly in Latin America, commercial inoculant companies like Nitragin, Nitrasoil, and Rizobacter in Argentina and ASIA-Biofábrica Siglo XXI in Mexico, as well as agricultural research institutions – INTA in Argentina, EMBRAPA in Brasil, RENARE-MGAP in Uruguay, and INIFAP in Mexico among others – have developed and tested in the field commercial inoculants containing free-living PGPR. These products are generally nominated as biofertilizers, with *Azospirillum* being the most widely used in commercial scale. However, other potential PGPR, including *Pseudomonas, Gluconacetobacter*, and *Herbaspirillum* strains, are also being assessed.

Extensive field inoculation experiments with *A. brasilense* carried out in Israel in the 1980s (Okon et al. 1988), clearly showed an average significant increase in crop yield of maize, wheat, sorghum, and other forage grasses. In these cases, the seeds were inoculated with freshly prepared peat inoculants (Table 6.1). More recently (1998–2011), there have been about 300 reports of field inoculation experiments utilizing mainly *A. brasilense* liquid- or peat-based commercial inoculants, under varied climatic and soil conditions, mainly in Argentina, Uruguay, Brazil, and Mexico (Fuentes-Ramirez and Caballero-Mellado 2005; Cassan and Garcia de

Table 6.1 Summary of field inoculation experiments with *Azospirillum brasilense* Cd (ATCC 29729) in Israel between 1979 and 1984

Crop	Number of field experiments[a]	Locations	Cultivars	Average increase in yield over controls (%)[b]
Maize	13	7	6	16.2
Wheat	10	6	5	8.1
Sorghum	5	2	3	17.0
Setaria italica	2	2	1	43.0
Panicum miliaceum	1	1	1	13.2

[a]All experiments were carried out with freshly prepared peat-based inoculants
[b]Overall success rate (statistical significance of yield promotion by *A. brasilense*) was ~70% (Okon et al. 1988)

Salamone 2008; Hungria et al. 2010), but also in other parts of the world (Okon and Labandera-Gonzalez 1994; Dobbelaere et al. 2001). Commercial inoculants have been applied in hundreds of thousands hectares, for instance, over 200,000 ha year^{-1} in Mexico since 2006, and about one million hectares in Brazil (S.C. Araujo, personal communication, 2010).

In all the above early and recent field inoculation experiments, consistent increases in crop yield, in the range of 5–30%, have been reported for maize, wheat, sorghum, rice, and legumes (nodulated by rhizobia following inoculation with *A. brasilense* inoculants (Table 6.2). Statistically significant increases in crop yield have been obtained for intermediate or adequate levels of N, P, K, microelements, and water, when the inoculant quality numbers (10^9 CFU per g or ml) of living bacteria and their physiological status has been maintained in the inoculant (Spaepen et al. 2009). These inoculant products have been increasingly approved and registered by government agencies, inoculant companies associations, and official agricultural research institutions (Cassan and Garcia de Salamone 2008). Thus, based on extensive field experimentation with proper agronomic design and accurate statistical evaluation of crop yield parameters, it is well accepted that the inoculation practice has the potential to promote crop yield in fields with adequate moisture and fertilized with 80–150 kg N ha^{-1}. Moreover, this practice can reduce by 20–50% the application of nitrogen fertilizers (that are generally in the range of 200–300 kg N ha^{-1}) while ensuring similar crop yields as in fully fertilized soils (Dobbelaere and Okon 2007; Cassan and Garcia de Salamone 2008; Spaepen et al. 2009).

6.3 Other Rhizosphere Bacterial Genus and Species with Potential of Direct Plant Growth Promotion

6.3.1 Herbaspirillum

Within the beta-proteobacteria, the genus *Herbaspirillum* comprises several diazotrophic species, some of which exhibit the potential of root surface, endophytic and systemic colonization of members of the *Graminaceae* family, although other

Table 6.2 Summary of commercial field inoculation experiments with inoculants prepared with various *Azospirillum* strains in Latin America during the last decade

Country/crop	Number of field experiments	Average increase in yield over controls (%)[a]
Mexico[b]		
Maize	17	34
Sorghum	20	23
Wheat	7	22
Barley	13	42
Argentina[c]		
Maize	110	5
Wheat	273	41
Uruguay[d]		
Maize	4	23
Sorghum	4	8
Rice	4	6
Brazil[e]		
Maize	9	8–30
Wheat	8	13–18

[a]Overall success rate (statistical significance of yield promotion by *A. brasilense*) was ~70%
[b]Experiments performed during 1999–2000 with commercial peat-based inoculants of *A. brasilense* strains CFN-535 and UAP-154 (Ministry of Agriculture Research Institute-INIFAP, Centro de Ciencias Genomicas, Universidad Nacional Autonoma de Mexico- UNAM); performed at various locations in the states of Campeche, Hidalgo, Oaxaca, Quintana Roo, Michoacan, Puebla, and Tabasco (Dobbelaere et al. 2001)
[c]Experiments performed in various locations during 2002–2007 with commercial liquid inoculants of *A. brasilense* strain Az39 (Nitragin, Argentina) (Diaz-zorita et al. 2009)
[d]Experiments performed during 2005–2010 with commercial liquid inoculants of *A. brasilense* strain Az39 (Lage y Cia., Uruguay). Results reported by ASINAGRO-Uruguay (contractor for field experiments)
[e]Experiments performed during 3 years for maize, 2 years for wheat in two locations, with liquid and peat-based inoculants of six *A. brasilense* and two *A. lipoferum* strains (Hungria et al. 2010)

plants can also serve as hosts (Reis et al. 2007; Rothballer et al. 2009). Endophytic colonization by *Herbaspirillum* strains is not very well defined. After surface sterilization of plant roots, *Herbaspirillum* cells could be observed and isolated from "within the plant." In addition, herbaspirilla were observed microscopically in intercellular spaces and in the vascular tissue of roots in a similar fashion as for bacterial pathogens, but without causing apparent damage to the plant (Reis et al. 2007). On the contrary, plant growth promotion by strains of this genus has been observed. *H. seropedica*, the most studied species in the *Herbaspirillum* genus, has been observed colonizing roots, intercellular spaces, and even within intact root cells of rice, wheat, and maize (Rothballer et al. 2009). The genome of *H. seropedica* strain Z78 has been sequenced. Interestingly, a cluster of type-III protein secretion system genes, possible reminiscent of a pathogenicity island, and potentially involved in *Herbaspirillum*-plant interactions, was identified (F. A. Pedrosa, GENOPAR Consortium).

Other plant-associated species are *H. frisingense*, isolated from the C4-fiber plant *Miscanthus spp.* and *Pennisetum purpureum* (Rothballer et al. 2008),

H. lusitanum, isolated from root nodules of common bean and *H. hiltneri*, isolated from surface sterilized wheat roots (Rothballer et al. 2008). *H. rubrisulalbicans* is capable of invading the intercellular spaces of the root and has been shown to be slightly pathogenic to sugarcane (Reis et al. 2007). There is detailed microscopic evidence supporting that most of the described *Herbaspirillum* species are true endophytes, but it is not clear as to which extent they proliferate in the plant. More quantitative measurements and evaluations of the extent of endophytic colonization and its impact on plant growth are needed (James and Olivares 1998). Endophytic colonization is considered to be a very important trait for PGPR, because it is hypothesized that endophytes could have a much more intimate and stable interaction with their plant hosts.

The different *Herbaspirillum* species possess most of the traits that are thought to contribute to plant growth promotion. They are capable of fixing atmospheric nitrogen, produce IAA, and some strains possess aminocyclopropane carboxylate (ACC) deaminase that hydrolyzes 1-aminocyclopropane-1-carboxylate (ACC), the precursor of the plant hormone ethylene. In several plant-microbe systems involving ACC deaminase-producing PGPR, it has been demonstrated that a decrease in ethylene levels in the root vicinity via bacterial ACC deaminase activity may contribute to plant growth stimulation (Glick et al. 2007). In addition, herbaspirilla are capable of producing bacterial autoinducer signaling molecules of the N-acylhomoserinelactone (AHL) type, which have been found to exert positive effects on diverse plants (Rothballer et al. 2008).

6.3.2 Gluconacetobacter

An additional nitrogen fixing, endophyte PGPR, which is capable of colonizing sugarcane and other plants is *Gluconacetobacter diazotrophicus* (alpha-proteobacteria) (Reis et al. 2007). The genome of one strain, PAL5, has been sequenced (J. I. Baldani, RioGene Consortium). *G. diazotrophicus* is well adapted to sugarcane tissues, as it shows optimal growth under 10% sucrose and a pH of 5.5. It has been demonstrated that this bacterium is well adapted to fix nitrogen within the sugarcane tissue providing fixed nitrogen to the plant. The quantities of nitrogen supplied to sugarcane by *G. diazotrophicus* nitrogen fixation are still to be determined, but apparently some sugarcane cultivars can obtain about 50% of the required nitrogen this way. Since there are other diazotrophs colonizing the surface and the internal tissues of sugarcane roots, it is difficult to assess the specific contribution of *G. diazotrophicus* (Sevilla et al. 2001).

Nitrogen fixation and IAA production abilities make this bacterium a potential PGPR to be used especially for sugarcane, but also other plant species could be assessed. In this regard, *G. diazotrophicus* has also been isolated from *Pennisetum purpureum*, sweet potato, coffee, pineapple, tea, mango, and banana. Other characterized species in this genus are *G. johannae* and *G. azotocaptans* (Reis et al. 2007).

6.3.3 Burkholderia

Nitrogen fixing, plant-associated *Burkholderia* species represent a great potential for agrobiotechnological applications (Estrada de los Santos et al. 2001; Caballero-Mellado et al. 2007) Some species are also capable of degrading aromatic compounds and have the potential to be used in bioremediation (Caballero-Mellado et al. 2007). *Burkholderia* belongs to the beta-proteobacteria, and taxonomical analyses of 60 species within this genus have revealed two major clusters. The first cluster includes human pathogens such as the *Burkholderia cepacia* complex. The second and more recently established cluster (including 25 species described so far) comprises related environmental, non-pathogenic species, mostly associated with plants (Caballero-Mellado et al. 2007).

Some plant-associated burkholderias are able to colonize the rhizosphere or the root intercellular spaces of several plants and promote plant growth. Increases in plant nutrient availability via biological nitrogen fixation, IAA production, phosphate solubilization, ACC deaminase, and siderophore production were demonstrated under controlled greenhouse conditions for several species, including *B. unamae*, *B. tropica*, and *B. silvatlantica* (Caballero-Mellado et al. 2007; Suarez-Moreno et al. 2010). Interestingly, some *Burkholderia* species like *B. tuberum* and *B. mimosarum* have the ability to form symbiotic interactions with plants and form nitrogen-fixing nodules similar to those produced by rhizobia in legumes (Elliott et al. 2009). Despite the PGPR potential of plant-associated burkholderias, their closeness to the *B. cepacia* group has made the development of commercial, burkholderia-based inoculants controversial so far.

6.3.4 Pseudomonas

The bacterial genus *Pseudomonas* (gama-proteobacteria) comprises many bacterial species with agricultural importance. While some species like *P. syringae* are plant pathogenic bacteria of important crops, others are involved in disease suppression (Weller 2007) and/or direct plant growth-promotion (Glick et al. 2007). Extensive research on direct PGPR by pseudomonads has been carried out with *Pseudomonas putida*, mostly under controlled laboratory and greenhouse conditions. A nitrogen-fixing, ACC deaminase-producing strain of *P. putida* promotes root initiation and elongation, likely by lowering the levels of ethylene in plants (Glick et al. 2007).

Other reports on direct PGPR effects by pseudomonads include *P. aurantiaca* strain SR1 in wheat, maize, soybean, alfalfa, and carob tree, both in the greenhouse and in the field (Rosas et al. 2009). Other PGPR within this genus is *Pseudomonas stutzeri* A1501, which was isolated from rice paddy soils and has been widely used as a crop inoculant in China. *P. stutzeri* can colonize the root surface and is capable of fixing nitrogen under microaerobic conditions. The genome of strain A1501 has been sequenced (Yan et al. 2008).

6.3.5 Paenibacillus

The genus *Paenibacillus* comprises over 30 species of facultative anaerobes and endospore-forming, periflagellated, heterotrophic, low G + C Gram-positive bacilli (Lal and Tabacchioni 2009). *Paenibacillus polymyxa* (formerly known as *Bacillus polymyxa*) is of biotechnological potential, mostly due to its biological control properties. It inhabits different niches such as soils, roots, and rhizospheres of various important crop plants. In the rhizosphere, *P. polymyxa* is involved in biological nitrogen fixation, which is considered to correlate with its PGPR activity as well as soil phosphorous solubilization. In some cases like in the wheat rhizosphere, it has been demonstrated that *P. polymyxa* produces IAA and cytokinin-like compounds (Lal and Tabacchioni 2009). The production of very resistant endospores by this bacterium could be an asset for inoculant formulations.

Recently, *P. vortex*, a pattern forming bacterium that was originally isolated from colonies of *Bacillus subtilis*, was also proposed as a potential PGPR (Sirota-Madi et al. 2010). Genome analysis as well as greenhouse experiments revealed that, although this bacterium does not exhibit nitrogen fixation activity, it may be able to induce plant growth, probably through phosphate solubilization and IAA production. Additionally, the extensive set of defense- and offense-related genes identified in its genome suggests that *P. vortex* can successfully inhabit the highly competitive environment of the rhizosphere as well as serve as an efficient plant beneficial microorganism.

6.4 Unresolved Questions and Future Research

PGPR commercial products are increasingly being used in agriculture, but there is a continuous need for investigation in order to understand more in depth the molecular bases of PGPR-plant interactions. As reviewed in this chapter, in the last three decades there have been significant advances in elucidation and understanding of mechanisms involved in plant growth promotion. However, the commercial products have been developed and used based mainly on empirical observations in the field. For a more comprehensive development and utilization of commercial inoculants, there are several issues that require further research:

1. Follow up of the dynamics of bacterial colonization of plants. In this regard, an important question that must be solved for most systems is whether PGPR are continuously colonizing (and promoting growth of) the developing roots or alternatively, the major effects are at the early stages of root development.
2. Assessment of optimal bacterial concentration. Bacterial numbers in the root tissue and/or in the root surface are very important. It seems that when the beneficial effects are through biological nitrogen fixation, phosphorous

solubilization, or disease control, for a significant impact in plant growth promotion, there is a need of bacterial colonization at relatively high numbers (10^6–10^7 CFU per g or cm of plant tissue). On the other hand, it is likely that lower colonization numbers (10^3–10^5 CFU per g or cm plant tissue) are necessary if the main mechanism of plant growth promotion is derived from the production of plant growth substances such as auxins. This is an issue that should be further verified for the different PGPR-plant systems.

3. The relevance of endophytic colonization. There is a need to deeply investigate the possible advantages of endophytic colonization in comparison to colonization of the rhizosphere and the root surface.
4. Signal exchange between the bacterium and the plant. In recent years, the use of high-throughput techniques contributed with significant advances in the elucidation of signal molecule exchange between bacterial pathogens/symbionts and their plant hosts. Similar advances are needed in the area of PGPR–plant interactions. More research is needed to understand which genes are turned on or turned off at the various stages of the interaction, by both the bacteria and the plants. This knowledge could contribute with new ideas as to which traits could be enhanced or reduced for more efficient plant growth promotion. The genomes of several PGPR and of plants of agronomic importance are being sequenced and annotated, enabling high-throughput research to assess the expression of multiple bacterial and plant genes, as well as understanding their regulation during the interaction.
5. PGPR survival properties. Investigating the traits that contribute to bacterial survival under adverse conditions during inoculant production, storage, inoculation, and colonization of seeds and plants is also very important. For example, it is important to better understand the roles of cell storage materials (e.g., PHAs, glycogen, polyphosphates, and others) and cell surface components (e.g., EPS, LPS, and surface proteins) in enhanced resistance of bacteria to diverse stress conditions like high salinity, desiccation, osmotic pressure, suboptimal temperature, and more. It is also necessary to carry out studies to assess inoculum performances at different times during storage.
6. Quality control of PGPR inoculants. This is a very critical issue that should be assessed and demands collaborative efforts among the producing companies, research institutions, and government agencies. Since these products are composed of living organisms, it is very important to confirm the bacterial strain at all stages of the production pipeline and to ensure that high bacterial numbers are maintained until the product reaches the farmer's hands.

This chapter is dedicated to the memory of Robert H. Burris and Jesus Caballero-Mellado, for their extensive contribution to the research of diazotrophic plant growth promoting rhizobacteria.

References

Baldani JI, Krieg NR, Baldani VLD, Hartmann A, Döbereiner J (2005) Genus II. *Azospirillum*. In: Brenner DJ, Krieg NR, Staley JT (eds) Bergey's manual of systematic bacteriology, vol II, 2nd edn. Springer, New York, pp 7–26

Burdman S, Volpin H, Kigel J, Kapulnik Y, Okon Y (1996) Promotion of *nod* gene inducers and nodulation in common bean (*Phaseolus vulgaris*) roots inoculated with *Azospirillum brasilense* Cd. Appl Environ Microbiol 62:3030–3033

Burdman S, Vedder D, German M, Itzigsohn R, Kigel J, Jurkevitch E, Okon Y (1998) Legume crop yield promotion by inoculation with *Azospirillum*. In: Elmerich C, Kondorosi A, Newton WE (eds) Biological nitrogen fixation for the 21st century. Kluwer Academic Publishers, Dordrecht, pp 609–612

Burdman S, Jurkevitch E, Okon Y (2000) Surface characteristics of *Azospirillum brasilense* in relation to cell aggregation and attachment to plant roots. Crit Rev Microbiol 26:91–110

Caballero-Mellado J, Onofre-Lemus J, Estrada De Los Santos P, Martinez-Aguilar L (2007) The tomato rhizosphere, an environment rich in nitrogen-fixing *Burkholderia* species with capabilities of interest in agriculture and bioremediation. Appl Environ Microbiol 73:5308–5319

Cassan FD, Garcia de Salamone I (2008) *Azospirillum* sp.: cell physiology, plant interactions and agronomic research in Argentina. Asociacion Argentina de Microbiologia

Castro-Sowinski S, Burdman S, Matan O, Okon Y (2010) Functions of PHA in the environment. In: Chen GQ (ed) Plastics from bacteria: natural functions and applications. Springer, Berlin, pp 39–61

Cooper JE (2007) Early interactions between legumes and rhizobia: disclosing complexity in a molecular dialogue. J Appl Microbiol 103:1355–1365

Creus CM, Graziano M, Casanovas EM, Pereyra MA, Simontacchi M, Puntarulo S, Barassi CA, Lamattina L (2005) Nitric oxide is involved in the *Azospirillum brasilense*-induced lateral root formation in tomato. Planta 221:297–303

Dardanelli MS, Fernandez de Cordoba FJ, Espuny MR, Rodriguez Carvajal MA, Soria Diaz ME, Gil Serrano A, Okon Y, Megias M (2008) Effect of *Azospirillum brasilense* coinoculated with *Rhizobium* on *Phaseolus vulgaris* flavonoids and Nod factor production under salt stress. Soil Biol Biochem 40:2713–2721

Dobbelaere S, Okon Y (2007) The plant growth promoting effects and plant responses. In: Elmerich C, Newton WE (eds) Associative and endophytic nitrogen-fixing bacteria and cyanobacterial associations. Springer, Heidelberg, pp 145–170

Dobbelaere S, Croonenborghs A, Thys A, Ptacek D, Vanderleyden J, Dutto P, Labandera-Gonzalez C, Caballero-Mellado J, Francisco Aguirre J, Kapulnik Y, Brener S, Burdman S, Kadouri D, Sarig S, Okon Y (2001) Responses of agronomically important crops to inoculation with *Azospirillum*. Aust J Plant Physiol 28:871–879

Diaz-Zorita M, Fernandez-Canigia MV (2009) Field performance of a liquid formulation of *Azospirillum brasilense* on dryland wheat productivity. Eur. J. Soil Biol. 45:3–11

Elliott GN, Chou JH, Chen WM, Bloemberg GV, Bontemps C, Martínez-Romero E, Velázquez E, Young JPW, Sprent JI, James EK (2009) *Burkholderia* spp. are the most competitive symbionts of *Mimosa*, particularly under N-limited conditions. Environ Microbiol 11:762–778

Estrada De Los Santos P, Bustillos-Cristales R, Caballero-Mellado J (2001) *Burkholderia*, a genus rich in plant-associated nitrogen fixers with wide environmental and geographic distribution. Appl Environ Microbiol 67:2790–2798

Fuentes-Ramirez LE, Caballero-Mellado J (2005) Bacterial biofertilizers. In: Sadiqui ZA (ed) PGPR: biological control and biofertilization. Springer, Dordrecht, pp 143–172

Glick BR, Cheng Z, Czarny J, Duan J (2007) Promotion of plant growth by ACC deaminase-producing soil bacteria. Eur J Plant Pathol 119:329–339

Hamaoui B, Abbadi JM, Burdman S, Rashid A, Sarig S, Okon Y (2001) Effects of inoculation with *Azospirillum brasilense* on chickpeas (*Cicer arietinum* L.) and fava beans (*Vicia faba* L.) under different growth conditions. Agronomie 21:553–560

Hartmann A, Schmid M, van Tuinen D, Berg G (2008) Plant-driven selection of microbes. Plant Soil 321:235–257

Hungria M, Campo RJ, Souza EM, Pedrosa FA (2010) Inoculation with selected strains of *Azospirillum brasilense* and *A. lipoferum* improves yields of maize and wheat in Brazil. Plant Soil 331:413–425

Izaguirre-Mayoral ML, Labandera C, Sanjuan S (2007) Biofertilizantes en Iberoamerica: Una visión técnica, científica y empresarial. Imprenta Denad Internacional SA, Montevideo

James EK, Olivares FL (1998) Infection and colonisation of sugarcane and other graminaceous plants by endophytic diazotrophs. Crit Rev Plant Sci 17:77–119

Kadouri D, Castro-Sowinski S, Jurkevitch E, Okon Y (2005) Ecological and agricultural significance of bacterial polyhydroxyalkanoates. Crit Rev Microbiol 31:55–67

Lal M, Tabacchioni S (2009) Ecology and biotechnological potential of *Paenibacillus polymyxa*: a minireview. Indian J Microbiol 49:2–10

Long SR (1989) *Rhizobium*-legume nodulation: life together in the underground. Cell 56:203–214

Molina-Favero C, Creus CM, Simontacchi M, Puntarulo S, Lamattina L (2008) Aerobic nitric oxide production by *Azospirillum brasilense* Sp245 and its influence on root architecture in tomato. Mol Plant-Microbe Interact 21:1001–1009

Okon Y (1994) *Azospirillum*/Plant associations. CRC Press, Boca Raton

Okon Y, Kapulnik Y (1986) Development and function of *Azospirillum*-inoculated roots. Plant Soil 90:3–16

Okon Y, Labandera-Gonzalez CA (1994) Agronomic applications of *Azospirillum*: an evaluation of 20 years world-wide field inoculation. Soil Biol Biochem 26:1591–1601

Okon Y, Kapulnik Y, Sarig S (1988) Field inoculation studies with *Azospirillum* in Israel. In: Suba Rao NS (ed) Biological nitrogen fixation recent developments. Oxford and IBH Publishing Co., New Delhi, pp 175–195

Pagnussat GC, Lanteri ML, Lamattina L (2003) Nitric oxide and cyclic GMP are messengers in the indole acetic acid-induced adventitious rooting process. Plant Physiol 132:1241–1248

Rebah FB, Prévost D, Yezza A, Tyagi RD (2007) Agro-industrial waste materials and wastewater sludge for rhizobial inoculant production: a review. Bioresour Technol 98:3535–3546

Reis V, Lee S, Kennedy C (2007) Biological nitrogen fixation in sugar cane. In: Elmerich C, Newton WE (eds) Associative and endophytic nitrogen-fixing bacteria and cyanobacterial associations. Springer, Heidelberg, pp 213–232

Rodelas B, González-López J, Salmerón V, Pozo C, Martínez-Toledo MV (1996) Enhancement of nodulation, N_2-fixation and growth of faba bean (*Vicia faba* L.) by combined inoculation with *Rhizobium leguminosarum* bv. *viceae* and *Azospirillum brasilense*. Symbiosis 21:175–186

Rodelas B, González-López J, Martínez-Toledo MV, Pozo C, Salmerón V (1999) Influence of *Rhizobium*/*Azotobacter* and *Rhizobium*/*Azospirillum* combined inoculation on mineral composition of faba bean (*Vicia faba* L.). Biol Fertil Soils 29:165–169

Rosas SB, Avanzini G, Carlier E, Pasluosta C, Pastor N, Rovera M (2009) Root colonization and growth promotion of wheat and maize by *Pseudomonas aurantiaca* SR1. Soil Biol Biochem 41:1802–1806

Rothballer M, Eckert B, Schmid M, Fekete A, SchloterM LA, Pollmann S, Hartmann A (2008) Endophytic root colonization of gramineous plants by *Herbaspirillum frisingense*. FEMS Microb Ecol 66:85–95

Rothballer M, Schmid M, Hartmann A (2009) Diazotrophic bacterial endophytes in Gramineae and other plants. In: Pawlowski K (ed) Prokaryotic symbionts in plants. Springer, Berlin, pp 273–302

Sarig S, Kapulnik Y, Okon Y (1986) Effect of *Azospirillum* inoculation on nitrogen fixation and growth of several winter legumes. Plant Soil 90:335–342

Sevilla M, Burris RH, Gunapala N, Kennedy C (2001) Comparison of benefit of sugarcane plant growth and $^{15}N_2$ incorporation following inoculation of sterile plants with *Acetobacter diazotrophicus* wild-type and Nif$^-$ mutant strains. Mol Plant-Microbe Interact 14:358–366

Sirota-Madi A, Olender Z, Helman Y, Ingham C, Brainis I, Roth D, Hagi E, Brodsky L, Leshkowitz E, Galatenko V, Nikolaev V, Mugasimangalam RC, Bransburg-Zabary S, Gutnick DL, Lancet D, Ben-Jacob E (2010) Genome sequence of the pattern forming *Paenibacillus vortex* bacterium reveals potential for thriving in complex environments. BMC Genomics 11:710–724

Spaepen S, Vanderleyden J, Remans R (2007) Indole-3-acetic acid in microbial and microorganism-plant signaling. FEMS Microbiol Rev 31:425–448

Spaepen S, Dobbelaere S, Croonenborghs A, Vanderleyden J (2008) Effects of *Azospirillum brasilense* indole-3-acetic acid production on inoculated wheat plants. Plant Soil 312:15–23

Spaepen S, Vanderleyden J, Okon Y (2009) Plant growth-promoting actions of rhizobacteria. Adv Bot Res 51:283–320

Star L, Matan O, Dardanelli MS, Kapulnik Y, Burdman S, Okon Y (2011) The *Vicia sativa* spp. *nigra*-*Rhizobium leguminosarum* bv. *viciae* symbiotic interaction is improved by *Azospirillum brasilense*. Plant Soil. doi:10.1007/s11104-010-0713-7 (in press)

Suarez-Moreno ZR, Devescovi J, Myers M, Hallack L, Mendonca-Previato L, Caballero-Mellado J, Venturi V (2010) Commonalities and differences in N-acyl homoserine lactone quorum sensing regulation in the species cluster of beneficial plant associated *Burkholderia*. Appl Environ Microbiol 76:4302–4317

Van Loon LC (2007) Plant responses to plant growth-promoting rhizobacteria. Eur J Plant Pathol 119:243–254

Vanbleu E, Marchal K, Lambrecht M, Mathys J, Vanderleyden J (2004) Annotation of the pRhico plasmid of *Azospirillum brasilense* reveals its role in determining the outer surface composition. FEMS Microbiol Lett 232:165–172

Volpin H, Burdman S, Castro-Sowinski S, Kapulnik Y, Okon Y (1996) Inoculation with *Azospirillum* increased exudation of rhizobial *nod*-gene inducers by alfalfa roots. Mol Plant-Microbe Interact 9:388–394

Weller DM (2007) *Pseudomonas* biocontrol agents of soilborne pathogens: looking back over 30 years. Phytopathology 97:250–256

Wisniewski-Dyé F, Vial L (2008) Phase and antigenic variation mediated by genome modifications. Antonie van Leeuw J Microb 94:493–515

Yan Y, Yang J, Dou Y, Chen M, Ping S, Peng J, Lu W, Zhang W, Yao Z, Li H (2008) Nitrogen fixation island and rhizosphere competence traits in the genome of root-associated *Pseudomonas stutzeri* A1501. Proc Natl Acad Sci USA 105:7564–7569

Zahran HH (1999) *Rhizobium*-legume symbiosis and nitrogen fixation under severe conditions and in an arid climate. Microbiol Mol Biolo Rev 63:968–989

Chapter 7
Rhizosphere Microorganisms

Dror Minz and Maya Ofek

Abstract Plant roots are the major source of available carbon in the soil. Therefore, soil bacteria maintain adaptive traits which enable them to exploit this highly competitive niche, yet microbial associations are inherent to plant adaptation to the highly heterogeneous soil environment. The rhizosphere has been, for many years, the focus of intense basic and applied research, aimed at modulating and gaining control over this environment to promote plant health and development. Such studies have revealed the high complexity of plant-associated microbial communities and the many factors that influence root-associated bacterial community composition, including the plant species and growth stage, as well as the soil physical, chemical, and biological characteristics. The high availability of carbon promotes copiotrophic/zymogenic life style in rhizosphere microorganisms. High diversity is maintained due to the prevalence of multiple micro-niches and utilization of different competition strategies. This may be fundamental to some of the most important plant growth and health promoting effect of root-microbe associations.

7.1 Introduction

Plant roots' association with microorganisms is a major hub of basic and applied research for over a century now. The term rhizosphere relates to the volume of soil affected by the presence and activity of plant roots (Brimecombe et al. 2001).

D. Minz • M. Ofek
Institute of Soil, Water and Environmental Sciences, Agricultural Research Organization of Israel, P.O. Box 6, Bet Dagan 50250, Israel
e-mail: minz@volcani.agri.gov.il

E. Rosenberg and U. Gophna (eds.), *Beneficial Microorganisms in Multicellular Life Forms*, 105
DOI 10.1007/978-3-642-21680-0_7, © Springer-Verlag Berlin Heidelberg 2011

In general, root depositions are considered the major driving force behind plant-microbe interactions in this environment (Pinton et al. 2001; Nelson 2004). Microbial activity (including that of pathogens) is stimulated by these root depositions (Nelson 2004; Rovira et al. 1983; Haichar et al. 2008), and in turn alters the rhizosphere (Brimecombe et al. 2001). Consequently, highly complex ecological interactions are formed, which affect plant health and development.

Plants have evolved in a microbial world, and therefore, plant-microbe interactions may be inherent to the plant adaptation to its environment. Indeed, bacteria and fungi can help the plant cope with both biotic and a-biotic stresses. Those stresses include soil-borne as well as foliar pathogens (Hoitink et al. 1997; Noble and Coventry 2005; Lugtenberg and Kamilova 2009), deficiency of inorganic nutrients such as phosphorus, iron, and nitrogen (Gyaneshwar et al. 2002; Richardson et al. 2009), salinity (Mayak et al. 2004; Zhang et al. 2008; Marulanda et al. 2010), and drought (Mayak et al. 2004; Figueiredo et al. 2008; Kohler et al. 2008, 2009; Liddycoat et al. 2009). The high degree of co-adaptation between plants and soil microbes is manifested by the high diversity of root-associated and endophytic species (Manter et al. 2010; Uroz et al. 2010) and the concomitant high frequency of plant growth-promotion related traits in soil and rhizosphere bacteria (Brito Alvarez et al. 1995; Cattelan et al. 1999; Berg et al. 2002, 2006; Ahmad et al. 2008; Garbeva et al. 2008; Zachow et al. 2008; Sato et al. 2009; Fürnkranz et al. 2009; Çakmakçi et al. 2010).

Rhizosphere microbial activities include solubilization of inorganic compounds, degradation and mineralization of organic compounds, and secretion of biologically active substances such as phyto-hormones, chelators, and antibiotics (Kapulnik and Okon 2002). Control and channeling of these attributes has much potential for agricultural purposes. Therefore, much effort has been dedicated to studying the interactions between roots and plant growth-promoting and biocontrol species (Lugtenberg and Kamilova 2009; Compant et al. 2010; Dutta and Podile 2010). However, it is acknowledged that such control can only be achieved by thorough understanding of the complex plant-soil-microbiome ecosystem as a whole.

Today, the destructive environmental impacts of long-term use of agrochemicals as means for plant nutrition and disease control are fully recognized (Matson et al. 1997; Power 2010). In consequence, development of sustainable agricultural practices is a pressing necessity (Lynch 2007; Royal Society 2009). Agrosystems are both providers and consumers of ecosystem services (Zhang et al. 2007). Many of these services, including nutrient cycling, disease control and preservation of soil fertility and structure are tightly linked to the microbial part of the equation (Sen 2003; Marschner and Rengel 2007). In addition to the traditional agricultural purposes, microbiological interventions in the rhizosphere can facilitate more recently advocated objectives such as phytoremediation (Wenzel 2009) and re-forestation (Gomes et al. 2010). Therefore, optimization of ecosystem performance may rely, in part, on our understanding and educated and effective manipulation of the rhizosphere.

7.2 The Rhizosphere

When Lorenz Hiltner first used the term rhizosphere, he referred to it as the soil compartment where the activity of plant roots creates opportunities for microorganisms to operate and interact (Hartmann et al. 2008). These are related to two main factors. First, roots release into their surrounding soil substantial amounts of organic matter (Hütsch et al. 2002; Kuzyakov and Schneckenberger 2004). Accordingly, measured concentrations of available organic carbon in the soil are inversely related to the distance from the root surface (Cheng et al. 1996; Séguin et al. 2004). Second, the roots retain a continuous film of water around them throughout their life, even when the soil is relatively dry (Carminati et al. 2010). Diffusion pathways control microbial activity and distribution in the soil (Koch 1990; Or et al. 2007). Therefore, this water film is essential for biochemical processes to occur. What's more, it also provides the necessary matrix for plant-microbe and microbe-microbe communication and interaction. In the following sections these two subjects will be elaborated on.

7.2.1 Roots, Rhizodepositions and the Rhizosphere Effect

Plant roots are linear units that can be divided into compartments differing in the degree of development and differentiation, as well as their function and physiological and biochemical characteristics. Furthermore, plant root systems exhibit high physiological and biochemical plasticity (Waisel and Eshel 2002), which is manifested by changes in root properties and activities (Neumann and Römheld 2002). Around each root, a rhizosphere is formed as roots change the chemical, physical, and biological properties of the soil in their immediate vicinity. The radial dimensions of the rhizosphere may range several millimeters in diameter for soluble nutrients (such as nitrate) or volatiles, but is much more restricted (<1 mm) for sparingly soluble minerals (such as P and Fe) (Neumann and Römheld 2002). Root compounds released into the soil may directly facilitate the plants acquisition of mineral nutrient. These include excreted and secreted compounds (carbon dioxide, bicarbonates, ethylene, protons, electrons, etc.) that affect the soil pH and redox potentials. Other secreted compounds, such as phyto-siderophores and enzymes, target specific nutrients and directly increase their availability to the plant. Rates of release of these compounds are highly affected by nutrient limitations.

Organic carbon is the major component of root-released compounds. A considerable percentage of the assimilated carbon, allocated to the roots is released into the soil. Organic rhizodepositions include lysates of sloughed-off cells, dead and senescent tissue, and root exudates released from intact root cells by either secretion/excretion or diffusion. Sloughed-off cells and tissues are the main constituent of root depositions and are thought to be the major carbon source for rhizosphere microorganisms, while root exudates (secreted or diffused) may constitute only a small fraction of root depositions (Uren 2001). However, root exudates can have direct and immediate function in rhizosphere processes (Neumann and Römheld 2002). High molecular weight mucilage secreted by exocytosis from the root tip, function in lubrication and water retention at the root surface (Carminati et al. 2010). Low molecular weight compounds deposited include sugars, carboxylates, amino acids, phenolics, flavonoids, hormones, and others. Many factors affect the quantity and composition of released organic carbon. First, different plant species release different types and amounts of organic carbon (Hütsch et al. 2002; Jones et al. 2009). In addition, many environmental factors contribute high variance in released organic carbon, including light intensity, temperature, mechanical impedance, soil solution pH, soil water content, and nutrient limitations (Neumann and Römheld 2002).

Organic compounds released by roots have short half-life in the soil: few hours to several days (Ryne et al. 2001; Jones et al. 2009), due to their rapid decomposition by microorganisms. The rhizosphere effect is manifested by the considerable difference between the composition of bulk soil, as compared with rhizosphere bacterial and fungal communities (Smit et al. 1999; Smalla et al. 2001; Kowalchuk et al. 2002). The rhizosphere is considered "biased" in favor of specific associated microbial populations based on modification of the root exudates profile. Indeed, rhizosphere of different plant species planted in the same soil harbor bacterial communities unique in their composition (Grayston et al. 1998; Marschner et al. 2001b; Kowalchuk et al. 2002). Using ^{13}C-labling, it was shown that this specificity of association was directly related to organic carbon released by different plant species (Haichar et al. 2008).

7.2.2 Plant-Microbe and Microbe-Microbe Interactions in the Rhizosphere

Microorganisms, as well as plants, are able to sense and modulate each other's presence, activity, and growth. Usually, these processes involve the production and secretion of diffusible molecules. Therefore, the temporal and spatial persistence of an aqueous film around roots provides the necessary matrix for such interactions to occur, even at considerable distances. In consequence, the probability for interaction in the rhizosphere, particularly at the root surface, is increased as compared to the bulk soil. Hence, cooperation, competition, and even transfer of mobile genetic elements may be facilitated (Barea et al. 2005; van Elsas et al. 2003).

Recruitment of the most important plant symbionts is mediated by plant-produced diffusates: Flavonoids produced by roots of legume plant species, regulate the initiation of symbiosis between those plants and N_2-fixing *Rhizobium* spp. (van Rhijn and Vanderleyden 1995); Sesquiterpenes and strigolactons produced by roots are involved in regulation of mycorrhizal-root association (Akiyama et al. 2005; Besserer et al. 2006).

It was shown that acyl-homoserine lactones (AHLs), responsible for quorum sensing in gram-negative bacteria, can move as far as 70 μm in the rhizosphere of soil-grown tomatoes and affect native soil populations (Steidel et al. 2001). Furthermore, plants can modulate bacterial population behavior by secretion of substances mimicking AHLs (Teplitski et al. 2000; Faure et al. 2009). At the same time, AHLs produced by bacteria were shown to affect plant physiology and root development (Schuhegger et al. 2006; Ortíz-Castro et al. 2008).

The ability to produce key plant growth-regulating hormones, such as auxines, cytokinins, gibberellins, ethylene, and abscisic acids, is common among soil bacteria and fungi (Arshad and Frankenberger 1991; Dodd et al. 2010). Those are involved in both plant growth-promoting and pathogenic processes. Plants produce a wide array of secondary metabolites and short peptides that attack both fungi and bacteria (Osbourn 1996; Dixon 2001; Castro and Fontes 2005). Those, and the many more produced by resident bacteria and fungi may exclude pathogens and other affected populations from the rhizosphere (Jousset et al. 2008; Raaijmakers et al. 2009).

7.2.3 Mycorrhiza

Mycorrhizal interactions are common associations between fungal community and the roots of most terrestrial plants (Smith and Read 1997). These associations have a great influence on plant nutrition and water availability, thus affecting plant growth, health, and ecology. It was further demonstrated that diversity of arbuscular mycorrhizal fungi in the soil can even determine plant biodiversity (van der Heijden et al. 1998a, b, 2003). Mycorrhiza occurs in several different forms, of which two are the most studied: ectomycorrhiza and endomycorrhiza. In the first, the fungal hyphae remains outside of plant cells, while in the second, part of the fungal hyphae intracellularly colonizes the host plant root cells.

Arbuscular mycorrhizal association is probably the most widespread terrestrial symbiosis. It is formed by 70–90% of land plant species, including most agriculturally relevant plants, with fungi that belong to the Glomeromycota phylum. In this relationship, the plant provides its fungus with soluble fixed carbon, while the arbuscular mycorrhizal fungi (AMF) have an enormous ability to absorb mineral nutrients from the soil and enhance their uptake by the plant (Jayachandran and Shetty 2003; Smith and Read 1997). In addition, AMF modification of natural plant microbial community composition (Marschner et al. 2001a; Marschner and Baumann 2003) improves plant defense against pathogens (Rabie 1998; Habte et al. 1999), drought, and pollution (Soares and Siqueira 2008; Ruiz-Lozano et al. 2001; Augé 2001; Vivas et al. 2003; Busquets et al. 2010) and improves soil structure and quality (Rillig and Mummey 2006; Miller and Jastrow 2000), thus alleviates environmental biotic and abiotic stresses on plants and accelerates their health and establishment (Ceccarelli et al. 2010; Caravaca et al. 2003).

7.3 Rhizosphere Microbial Ecology

7.3.1 Soil Versus Root Effects

The principal determinant of rhizosphere community composition is that of the soil. Seed-borne species may persist, but the seed surface and the young emerging roots are rapidly colonized by soil bacteria (Normander and Prosser 2000; Green et al. 2006). Therefore, it is not surprising that for a single plant species, the composition of rhizosphere microbial communities will differ when grown in different soils (Grayston et al. 1998; Marschner et al. 2001b; Kowalchuk et al. 2002; Ridder-Duine et al. 2005; Berg et al. 2006; Garbeva et al. 2008). The specific chemical/physical characteristics of a particular soil may influence the establishment of bacteria in the rhizosphere directly or indirectly- through changes in root growth and activity. In order to assess the contribution of the soil type to root-bacterial interactions; the native microbial component must be controlled for- a daunting task. One way to explore this issue is to use sterilized soils in order to examine the establishment of inoculated species. When examined in 23 sterilized soils, the soil type had negligible effect on wheat and tomato rhizosphere establishment of the plant growth-promoting bacterium *Azospirillum brasilense* (Bashan et al. 1995). Conversely, association of *Pseudomonas* spp. with flax (*Linium usitatissinum* L.) roots was affected by soil type (Latour et al. 1999). Furthermore, accumulation of native soil bacteria and particularly of *Pseudomonas* in the rhizosphere of wheat was significantly influenced by compaction of the soil (Watt et al. 2003). This effect was indirect and related to the rate of root elongation under the different mechanical impedance regimes. Competition with native microorganisms has been shown to adversely affect rhizosphere colonization by introduced strains, and particularly of plant growth-promoting bacteria (Jousset et al. 2008; Martinez-Toledo et al. 1988).

The influence of the root is highest in its immediate vicinity, with respect to microbial community composition (Kowalchuk et al. 2002; Inbar et al. 2005; Green et al. 2006; Weisskopf et al. 2008; Ofek et al. 2009). However, different bacterial lineages are not equally affected by root proximity. For example, cucumber seed- and root- colonization by Oxalobacteraceae was highly sensitive to the "root effect". However, *Chryseobacterium* (Bacteroidetes) populations were unaffected by root proximity, but were highly sensitive to modification of the potting mix (Green et al. 2007). The structure of the rhizosphere and root bacterial communities is also affected by root proximity and is characterized by reduced diversity and increased dominance (Marilley et al. 1998; Marilley and Aragno 1999; Weisskopf et al. 2008; Ofek et al. 2009). These outcomes are considered typical of carbon source-dependent selective stimulation. Root activity stimulates the growth of microbial species which are otherwise dormant in the soil (Weisskopf et al. 2008). Therefore, composition and diversity of plant species may be an important factor in the maintenance of soil microbial biodiversity. However, examination of the coupling between above-ground plant communities and soil microbial community diversities provided contradicting results (Kowalchuk et al. 2002; Garbeva et al. 2006; Chung et al. 2007; Kielak et al. 2008). This may be the result of the extremely high diversity of soil communities coupled with relatively low resolution level of bacterial community composition determination methodologies used. In that respect, the use of "next generation sequencing" approach, which allows for comprehensive sampling depth (Uroz et al. 2010) may give better estimation of such linkage.

7.3.2 Succession Along the Root Axis and During the Plant Life Cycle

A single root can be divided into several distinct zones which differ in the degree of differentiation and activity. Roughly, the root can be divided into root cap/tip, elongation zone, root hair zone, and mature/ramified root segments. Different studies have shown that distinct microbial assemblages form at these different root locations (Liljeroth et al. 1991; Kurakov and Kostina 2001; Marschner et al. 2001a; Folman et al. 2001, 2003; Clayton et al. 2005; Ofek 2011). Furthermore, the deviation in microbial community composition between these zones may increase under nutrient-limiting conditions, such as iron deficiency (Yang and Crowley 2000). These differences represent a process of succession determined by the differentiation of the plant root cells and the shifts in their properties and activity. In accordance, different bacterial taxa show preference in colonization of different root locations. Folman et al. (2001) have found that occurrence of slow-growing bacteria has increased with distance from the root tip. In contrast, abundance of native soil *Pseudomonas* was higher at the root tip zone (Watt et al. 2003, 2006; Gamalero et al. 2004). Many authors linked bacterial colonization to "hubs" of root exudation such as root tips and root hair zones. Generally, the exudation profile at the specific root location is thought to control colonization (Walker et al. 2003; Compant et al. 2010). However, some authors stressed the importance of root surface properties and tissue architecture on colonization of live (Dandurand et al. 1997) or decomposing (Lindedam et al. 2009) roots.

As the plant root system develops, new soil horizons are occupied by roots. The variance between soil microbial communities at these different soil compartments (Blume et al. 2002; Griffiths et al. 2003; Will et al. 2010) is manifested in the respective root associations (Chiarini et al. 2000; Rodríguez-Zaragoza et al. 2008). Furthermore, roots respond to local soil conditions and nutrient availability by changes in their activities, which are manifested in respective changes in their rhizosphere (Neumann and Römheld 2002). One example is cluster roots of many Proteaceae species and *Lupinus* (Fabaceae). These specialized roots are formed in response to extreme phosphorus deficiency, and increase the mobilization of P (and also of Fe, Mn, Zn, and Al) in the rhizosphere by extraordinarily high secretion of organic chelators (citrate, malate, and phenolics), as well as enzymes (phosphatase) (Neumann and Martinoia 2002). The activity of cluster roots was shown to have a strong effect on the composition of rhizosphere and root associated bacteria (Marschner et al. 2002; Weisskopf et al. 2005). Furthermore, it was indicated that bacteria, particularly the plant hormone IAA-producing species, may be involved in the initiation of cluster root formation (Skene 2000). Finally, rates and composition of root exudates have been shown to change with plant age and between developmental stages (Brimecombe et al. 2001). Typically, the relative amount of carbon released into the rhizosphere decreases with plant age, along with a decrease in associated bacterial abundance (Liljeroth and Bååth 1988). These are also accompanied by shifts in rhizosphere bacterial community composition (Marschner et al. 2001a; Green et al. 2006; Houlden et al. 2008).

7.4 Modifying Rhizosphere Bacterial Communities

The ability to effectively modify rhizosphere bacterial communities in order to support plant growth and health is essential for mass-productive sustainable agriculture. The most explored modifications are compost amendments (Hoitink and Boehm 1999; Inbar et al. 2005; Noble and Coventry 2005; Green et al. 2006, 2007; Hargreaves et al. 2008; Adesemoye et al. 2009; Ofek et al. 2009; Bonanomi et al. 2010) and application of different microbial preparations (Emmert and Handelsman 1999; Paulitz and Belanger 2001; Whipps 2001; Lugtenberg and Kamilova 2009; Compant et al. 2010). Those are widely distinct applications, as the first adds to the amended medium not only a rich and diverse consortium of biological agents, but also organic matter and nutrients. It was confirmed that the efficacy of such treatments involves the response of the soil, the plant, and the rhizosphere microbial communities (Hoitink et al. 1997; Zhang et al. 1998; Bonanomi et al. 2010; Minz et al. 2010). Therefore, one main objective of compost amendment or of inoculation with specific microbial strains is to modify the rhizosphere conditions, particularly *via* manipulation of the associated microbial community composition (Lynch 2002).

The response of soil and rhizosphere bacterial communities to different anthropogenic and other disturbances has been discussed in terms of resilience (Lynch 2002; Baumgarte and Tebbe 2005; Ager et al. 2010; Fuchs 2010). Indeed, the proximal root environment can be highly resistant to invasion (Inbar et al. 2005). Generally, introduction of new microorganisms produces only restricted spatial and temporal effects on soil, rhizosphere, and root microbial communities (Lottmann et al. 2000; Nautiyal et al. 2002; Castro-Sowinski et al. 2007). Thus, the plant growth-promoting effect of such treatments may be related to microbial events occurring during early stages of plant development. Such early effects were pointed out for inoculants of different bacterial species (Götz et al. 2006; Roesti et al. 2006) and for compost amendment (Tiquia et al. 2002; Inbar et al. 2005; Green et al. 2006).

Consequences of compost amendment or of single species inoculation often include shifts in the plant roots hormonal balance or a plant systemic response, namely induced systemic resistance (van Loon et al. 1998; Persello-Cartieaux et al. 2003; Compant et al. 2005). Thus, direct or indirect effects of the introduced microorganisms may result in similar modification of the root habitat. If so, bacterial assemblages of roots in phyto-effectively modified soils may share qualitative and quantitative characteristics, different from those exhibited by roots in un-modified soils. However, most community-level studies were focused on shifts in bacterial community compositions in response to soil modification and have not assessed other community-level parameters.

Previous studies have shown that specific microorganisms and composts can promote plant growth and biologically control plant diseases caused by soil-borne as well as foliar pathogens (Hoitink et al. 1997; Noble and Coventry 2005; Lugtenberg and Kamilova 2009). Even so, the efficacy of such applications has not yet reached satisfactory results and often, contradicting results are reported (Lazarovitz 2001; Vallad and Goodman 2004). These reports bring to light our lack of understanding of population and community-level determinants which are controlled during the assemblage process of microbial communities of seeds and roots.

7.5 Future Prospects

Recent advances in molecular biology and genomic technologies provide new tools for research in microbial ecology. These include mass sequencing that allows in-depth description of community composition and metagenomics. Combined with reverse transcription, mass sequencing allows studying the transcriptome that can reveal some of the community metabolic functioning. Other experimental strategies, such as environmental community proteomics and metabolomics, the use of stable isotope probing, NanoSIMS, and exciting new emerging methods and tools, are constantly being applied in environmental microbiology and microbial ecology. We anticipate that the application of such tools in the field on rhizosphere and root microbiology will, in the near future, result in an enormous increase in our understanding of this highly complex and essential environment. Such understanding, based on the principles of system biology, will provide the means to modify, control, and tailor the rhizosphere.

References

Adesemoye AO, Torbert HA, Klopper JW (2009) Plant growth-promoting rhizobacteria allow reduced application rates of chemical fertilizers. Microb Ecol 58:921–929

Ager D, Evans S, Li H, Killey AK, van der Gast CJ (2010) Anthropogenic disturbance affects the structure of bacterial communities. Environ Microbiol 12:670–678

Ahmad F, Ahmad I, Khan MS (2008) Screening of free-living rhizospheric bacteria for their multiple plant growth promoting activities. Microbiol Res 163:173–181

Akiyama K, Matsuzaki KI, Hayashi H (2005) Plant sesquiterpenes induce hyphal branching in arbuscular mycorrhizal fungi. Nature 435:824–827

Arshad M, Frankenberger WT (1991) Microbial production of plant hormones. Plant Soil 133:1–8

Augé RM (2001) Water relations, drought and vesicular-arbuscular mycorrhizal symbiosis. Mycorrhiza 11:3–42

Barea JM, Pozo MJ, Azcón R, Azcón-Aguilar C (2005) Microbial co-operation in the rhizosphere. J Exp Bot 56:1761–1778

Bashan Y, Puentes ME, Rodriguez-Mendoza MZ, Toledo G, Holguin G, ferrera-Cerrato R, Pedrin S (1995) Survival of *Azospirillum brasilense* in bulk soil and rhizosphere of 23 soil types. Appl Environ Microbiol 61:1938–1945

Baumgarte S, Tebbe CC (2005) Field studies on the environmental fate of the Cry1Ab Bt-toxin produced by transgenic maize MON810 and its effect on bacterial communities in the maize rhizosphere. Mol Ecol 14:2539–2551

Berg G, Roskot N, Steidle A, Ebrel L, Zock A, Smalla K (2002) Plant-dependent genotypic and phenotypic diversity of antagonistic rhizobacteria isolated from different Verticillium host plants. Appl Environ Microbiol 68:3328–3338

Berg G, Opelt K, Zachow CH, Lottmann J, Gotz M, Costa R, Smalla K (2006) The rhizosphere effect on bacteria antagonistic toward the pathogenic fungus *Verticillium* differs depending on plant species and site. FEMS Microbiol Ecol 56:250–261

Besserer A, Puech-Pagès V, Kiefer P, Gomez-Roland V, Jauneau A, Roy S, Portais JC, Roux C, Bécard G, Séjalon-Delmas N (2006) Strigolactones stimulate arbuscular mycorrhizal fungi by activating mitochondria. PLoS Biol 4:e226

Blume E, Bischoff M, Reichert JM, Moorman T, Konopka A, Turco RF (2002) Surface and subsurface microbial biomass, community structure and metabolic activity as a function of soil depth and season. Appl Soil Ecol 20:171–181

Bonanomi G, Antignani V, Capodilupo M, Scala F (2010) Identifying the characteristics of organic soil amendments that suppress soilborne plant diseases. Soil Biol Biochem 42:136–144

Brimecombe MJ, De Leij FA, Lynch JM (2001) The effect of root exudates on rhizosphere microbial populations. In: Pinton R, Varanini Z, Nannipieri P (eds) The rhizosphere. Marcel Dekker, New York, pp 95–140

Brito Alvarez MA, Gagne S, Antoun H (1995) Effect of compost on rhizosphere microflora of the tomato and on the incidence of plant growth-promoting rhizobacteria. Appl Environ Microbiol 61:194–199

Busquets M, Calvet C, Camprubí A, Estaún V (2010) Differential effects of two species of arbuscular mycorrhiza on the growth and water relations of Spartium *junceum* and *Anthyllis cytisoides*. Symbiosis 52:95–101

Çakmakçi R, Dönmez MF, Ertürk Y, Erat M, Haznedar A, Sekban R (2010) Diversity and metabolic potential of culturable bacteria from the rhizosphere of Turkish tea grown in acidic soils. Plant Soil 332:299–318

Caravaca F, Barea JM, Palenzuela J, Figueroa D, Alguacil MM, Roldán A (2003) Establishment of shrub species in a degraded semiarid site alter inoculation with native or allochtonous arbuscular mycorrhizal fungi. Appl Soil Ecol 22:103–111

Carminati A, Moradi AB, Vetterlein D, Vontobel P, Lehrmann E, Weller U, Vogel HJ, Oswald SE (2010) Dynamics of soil water content in the rhizosphere. Plant Soil 332:163–176

Castro MS, Fontes W (2005) Plant defence and antimicrobial peptides. Protein Pept Lett 12:11–16

Castro-Sowinski S, Herschkovitz Y, Okon Y, Jurkevitch E (2007) Effects of inoculation with plant growth-promoting rhizobacteria on resident rhizosphere microorganisms. FEMS Microbiol Lett 276:1–11

Cattelan AJ, Hartel PG, Fuhrmann JJ (1999) Screening for plant growth-promoting Rhizobacteria to promote soybean growth. Soil Sci Soc Am J 63:1670–1680

Ceccarelli N, Curadi M, Martelloni L, Sbrana C, Picciarelli P, Giovannetti M (2010) Mycorrhizal colonization impacts on phenolic content and antioxidant properties of artichoke leaves and flower heads two years after field transplant. Plant Soil 335:311–323

Cheng W, Zhang Q, Coleman DC, Carroll CR, Hoffman CA (1996) Is available carbon limiting microbial respiration in the rhizosphere? Soil Biol Biochem 28:1283–1288

Chiarini L, Giovannelli V, Bevivin A, Dalmastri C, Tabacchioni S (2000) Different portions of the maize root system host *Burkholderia cepacia* populations with different degrees of genetic polymorphism. Environ Microbiol 2:111–118
Chung H, Zak D, Reich PB, Ellsworth DS (2007) Plant species richness, elevated CO_2, and atmospheric nitrogen deposition alter soil microbial community composition and function. Glob Change Biol 13:980–989
Clayton SJ, Clegg CD, Murray PJ, Gregory PJ (2005) Determination of the impact of continuous defoliation of *Lolium perenne* and *Trifolium repens* on bacterial and fungal community structure in the rhizosphere soil. Biol Fertil Soils 41:109–115
Compant S, Duffy B, Nowak J, Clément C, Barka EA (2005) Use of plant growth promoting bacteria for biocontrol of plant diseases: principles, mechanisms of action, and future prospects. Appl Environ Microbiol 71:4951–4959
Compant S, Clément C, Sesstisch A (2010) Plant growth-promoting bacteria in the rhizo- and endosphere of plants: their role, colonization, mechanisms involved and prospects for utilization. Soil Biol Biochem 42:669–678
Dandurand LM, Schotzko DJ, Knudsen GR (1997) Spatial patterns of rhizoplane populations of *Pseudomonas fluorecens*. Appl Environ Microbiol 63:3211–3217
Dixon RA (2001) Natural products and plant disease resistance. Nature 411:843–847
Dodd IC, Zinovkina NY, Safronova VI, Belimov AA (2010) Rhizobacterial mediation of plant hormone status. Ann Appl Biol 157:361–379
Dutta S, Podile AR (2010) Plant growth promoting rhizobacteria (PGPR): the bugs to debug the root zone. Crit Rev Microbiol 36:232–244
Emmert EAB, Handelsman J (1999) Biocontrol of plant disease: a (Gram-) positive perspective. FEMS Microbiol Lett 171:1–9
Faure D, Vereecke D, Leveau JHJ (2009) Molecular communication in the rhizosphere. Plant Soil 321:279–303
Figueiredo MVB, Burity HA, Martinez CR, Chanway CP (2008) Alleviation of drought stress in common bean (*Phaseolus vulgaris* L.) by co-inoculation with *Paenibacillus polymyxa* and *Rhizobium tropici*. Appl Soil Ecol 40:182–188
Folman LB, Postma J, van Veen JA (2001) Ecophysiological characterization of rhizosphere bacterial communities at different root locations and plant developmental stages of cucumber grown on rockwool. Microb Ecol 42:586–597
Folman LB, Postma J, van Veen JA (2003) Inability to find consistent bacterial biocontrol agents of *Pythium aphanidermatum* in cucumber using screens based on ecophysiological traits. Microb Ecol 45:72–87
Fuchs JG (2010) Interactions between beneficial and farmful microorganisms: from the composting process to compost application. In: Insam H, Franke-Whittle I, Goberna M (eds) Microbes at work: from wastes to resources. Springer, Berlin/Heidelberg, pp 213–230
Fürnkranz M, Müller H, Berg G (2009) Characterization of plant growth promoting bacteria from crops in Bolivia. J Plant Dis Protect 116:149–155
Gamalero E, Lingua G, Capri FG, Fusconi A, Berta G, Lemanceau P (2004) Colonization pattern of primary tomato roots by *Pseudomonas fluorescens* A6RI characterized by dilution plating, flow cytometry, fluorescence, confocal and scanning electron microscopy. FEMS Microbiol Ecol 48:79–87
Garbeva P, Potsma J, van Veen JA, van Elsas JD (2006) Effect of above-ground plant species on soil microbial community structure and its impact on suppression of *Rhizoctonia solani* AG3. Environ Microbiol 8:233–246
Garbeva P, van Elsas JD, van Veen JA (2008) Rhizosphere microbial community and its response to plant species and soil history. Plant Soil 302:19–32
Gomes NCM, Cleary DFR, Pinto FN, Egas C, Almeida A, Cunha A, Mendoça-Hagler LCS, Smalla K (2010) Taking root: enduring effect of rhizosphere bacterial colonization in mangroves. PLoS One 5:e14065

Götz M, Gomes NC, Dratwinski A, Costa R, Berg G, Peixoto R, Mendoça-Hagler L, Smalla K (2006) Survival of *gfp*-tagged antagonistic bacteria in the rhizosphere of tomato plants and their effects on the indigenous bacterial community. FEMS Microbiol Ecol 56:207–218
Grayston SJ, Wang SQ, Campbell CD, Edwards AC (1998) Selective influence of plant species on microbial diversity in the rhizosphere. Soil Biol Biochem 30:369–378
Green SJ, Inbar E, Michel FC, Hadar Y, Minz D (2006) Succession of bacterial communities during early plant development: transition from seed to root and effect of compost amendment. Appl Environ Microbiol 72:3975–3983
Green SJ, Michel FC, Hadar Y, Minz D (2007) Contrasting patterns of seed and root colonization by bacteria from the genus *Chrysobacterium* and from the family Oxalobacteraceae. ISME J 1:291–299
Griffiths RI, Whiteley AS, O'Donnell AG, Bailey MJ (2003) Influence of depth and sampling time on bacterial community structure in an upland grassland soil. FEMS Microbiol Ecol 43:35–43
Gyaneshwar P, Naresh Kumar G, Parekh LJ, Poole PS (2002) Role of soil microorganisms in improving P nutrition of plants. Plant Soil 245:83–93
Habte M, Zhang YC, Schmitt DP (1999) Effectiveness of *Glomus* species in protecting white clover against nematode damage. Can J Bot 77:135–139
Haichar FZ, Marol C, Berge O, Rangel-Castro JI, Prosser JI, Balesdent J et al (2008) Plant host habitat and root exudates shape soil bacterial community structure. ISME J 2:1221–1230
Hargreaves JC, Adl MS, Warman PR (2008) A review of the use of composted municipal solid waste in agriculture. Agric Ecosyst Environ 123:1–14
Hartmann A, Rothballer M, Schmid M (2008) Lorenz Hiltner, a pioneer in rhizosphere microbial ecology and soil bacteriology research. Plant Soil 312:7–14
Hoitink HAJ, Boehm MJ (1999) Biocontrol within the context of soil microbial communities: a substrate-dependent phenomenon. Annu Rev Phytopathol 37:427–446
Hoitink HAJ, Stone AG, Han DY (1997) Suppression of plant diseases by composts. Hort Sci 32:184–187
Houlden A, Timmis-Wilson M, Day MJ, Bailey MJ (2008) Influence of plant developmental stage on microbial community structure and activity in the rhizosphere of three field crops. FEMS Microbiol Ecol 65:193–201
Hütsch BW, Augustin J, Merbach W (2002) Plant rhizodeposition – an important source of carbon turnover in soils. J Plant Nutr Soil Sci 165:397–407
Inbar E, Green SJ, Hadar Y, Minz D (2005) Competing factors of compost concentration and proximity to root affect the distribution of *Streptomyces*. Microb Ecol 50:73–81
Jayachandran K, Shetty KG (2003) Growth response and phosphorus uptake by arbuscular mycorrhizae of wet prairie sawgrass. Aquat Bot 76:281–290
Jones DL, Nguyen C, Finlay RD (2009) Carbon flow in the rhizosphere: carbon trading at the soil-root interface. Plant Soil 321:5–33
Jousset A, Scheu S, Bonkowski M (2008) Secondary metabolite production facilitates establishment of rhizobacteria by reducing both protozoan predation and the competitive effects of indigenous bacteria. Funct Ecol 22:714–719
Kapulnik Y, Okon Y (2002) Plant growth-promotion by rhizospheric bacteria. In: Waisel Y, Eshel A, Kafkafi U (eds) Plant roots: the hidden half, 3rd edn. Marcel Dekker, New York, pp 869–886
Kielak A, Pijl AS, van Veen JA, Kowalchuk GA (2008) Differences in vegetation composition and plant species identity lead to only minor changes in soil-borne microbial communities in a former arable field. FEMS Microbiol Ecol 63:372–382
Koch AL (1990) Diffusion- the crucial process in any aspects of the biology of bacteria. Adv Microb Ecol 11:37–70
Kohler J, Hernández JA, Caravaca F, Roldán A (2008) Plant-growth-promoting rhizobacteria and arbuscular mycorrhizal fungi modify alleviation biochemical mechanisms in water-stressed plants. Funct Plant Biol 35:141–151

Kohler J, Hernández JA, Caravaca F, Roldán A (2009) Induction of antioxidant enzymes is involved in the greater effectiveness of a PGPR versus AM fungi with respect to increasing the tolerance of lettuce to severe salt stress. Environ Exp Bot 65:245–252
Kowalchuk GA, Buma DS, de Boer W, Klinkhamer PGL, van Veen JA (2002) Effects of above-ground plant species composition and diversity on the diversity of soil-borne microorganisms. Anton Leeuw Int J G 81:509–520
Kurakov AV, Kostina NV (2001) Spatial peculiarities in the colonization of the plant rhizoplane by microscopic fungi. Microbiology 70:165–174
Kuzyakov Y, Schneckenberger K (2004) Review of estimation of plant rhizodeposition and their contribution to soil organic matter formation. Arch Agron Soil Sci 50:115–132
Latour X, Philippot L, Corberand T, Lemanceau P (1999) The establishment of an introduced community of fluorescent pseudomonads in the soil and in the rhizosphere is affected by the soil type. FEMS Microbiol Ecol 30:163–170
Lazarovitz G (2001) Management of soil-borne plant pathogens with organic soil amendments: a disease control strategy salvaged from the past. Can J Plant Pathol 23:1–7
Liddycoat SM, Greenberg BM, Wolyn DJ (2009) The effect of plant growth-promoting rhizobacteria on asparagus seedlings and germinating seeds subjected to water stress under green house conditions. Can J Microbiol 55:388–394
Liljeroth E, Bååth E (1988) Bacteria and fungi on roots of different barley varieties (*Hordeum vulgare* L.). Biol Fertil Soil 7:53–57
Liljeroth E, Burgers SLGE, van Veen JA (1991) Changes in bacterial populations along roots of wheat (*Triticum aestivum* L.) seedlings. Biol Fertil Soil 10:276–280
Lindedam J, Magid J, Poulsen P, Luxhøi J (2009) Tissue architecture and soil fertility controls on decomposer communities and decomposition of roots. Soil Biol Biochem 41:1040–1049
Lottmann J, Heuer H, de Vries J, Mahn A, Düring K, Wackernagel W, Smalla K, Berg G (2000) Establishment of introduced antagonistic bacteria in the rhizosphere of transgenic potatoes and their effect on the bacterial community. FEMS Microbiol Ecol 33:41–49
Lugtenberg B, Kamilova F (2009) Plant-growth-promoting rhizobacteria. Annu Rev Microbiol 63:541–556
Lynch JM (2002) Resilience of the rhizosphere to anthropogenic disturbance. Biodegradation 13:21–27
Lynch JP (2007) Roots of the second green revolution. Aust J Bot 55:493–512
Manter DK, Delgado JA, Holm DG, Stong RA (2010) Pyrosequencing reveals a highly diverse and cultivar-specific bacterial endophyte community in potato roots. Microb Ecol 60:157–166
Marilley L, Aragno M (1999) Phylogenetic diversity of bacterial communities differing in degree of proximity of *Lolium perenne* and *Trifolium repens* roots. Appl Soil Ecol 13:127–136
Marilley L, Vogt G, Blanc M, Aragno M (1998) Bacterial diversity in bulk soil and rhizosphere fractions of *Lolium perenne* and *Trifolium repens* as reveled by PCR restriction analysis of 16S rDNA. Plant Soil 198:219–224
Marschner P, Baumann K (2003) Changes in bacterial community structure induced by mycorrhizal colonisation in split-root maize. Plant and Soil 251:279–289
Marschner P, Rengel Z (2007) Contribution of rhizosphere interactions to soil biological fertility. In: Abbott LK, Murphy DV (eds) soil biological fertility. Springer, Netherlands, pp 81–98
Marschner P, Crowley DE, Lieberei R (2001a) Arbuscular mycorrhizal infection changes the bacterial 16S rDNA community composition in the rhizosphere of maize. Mycorrhiza 11:297–302
Marschner P, Yang CH, Lieberie R, Crowley DE (2001b) Soil and plant specific effect on bacterial community composition in the rhizosphere. Soil Biol Biochem 33:1437–1445
Marschner P, Neumann G, Kania A, Weisskopf L, Lieberei R (2002) Spatial and temporal dynamics of the microbial community structure in the rhizosphere of cluster roots of white lupin (*Lupinus albus* L.). Plant Soil 246:167–174

Martinez-Toledo MV, Gonzalez-Lopez J, Moreno RJ, Ramos-Cormenzana A (1988) Effect of inoculation with Azotobacter chroococcum on nitrogenase activity of Zea mays roots grown in agricultural soils under aseptic and non-sterile conditions. Biol Fertil Soil 6:170–173

Marulanda A, Azcón R, Chaumount F, Ruiz-Lozano JM, Aroca R (2010) Regulation of plasma membrane aquaporins by inoculation with a *Bacillus megaterium* strain in maize (*Zea mayse* L.) plants under unstressed and salt-stressed conditions. Planta 232:533–543

Matson PA, Parton WJ, Power AG, Swift MJ (1997) Agricultural intensification and ecosystem properties. Science 277:504–509

Mayak S, Tirosh T, Glick BR (2004) Plant growth-promoting bacteria confer resistance in tomato plants to salt stress. Plant Physiol Biochem 42:565–572

Miller RM, Jastrow JD (2000) Mycorrhizal fungi influence soil structure. In: Kapulnik Y (ed) Arbuscular mycorrhizas: physiology and function. Kluwer, Dordrecht, pp 3–18

Minz D, Green SJ, Ofek M, Hadar Y (2010) Compost microbial populations and interactions with plants. In: Insam H, Franke-Whittle I, Goberna M (eds) Microbes at work: from wastes to resources. Springer, Berlin/Hieldelberg

Nautiyal CS, Johri JK, Singh HB (2002) Survival of the rhizosphere-competent biocontrol strain *Pseudomonas fluorescence* NBRI2650 in the soil and phytosphere. Can J Microbiol 48:588–601

Nelson EB (2004) Microbial dynamics and interactions in the spermosphere. Annu Rev Phytopathol 42:271–309

Neumann G, Martinoia E (2002) Cluster roots – an underground adaptation for survival in extreme environments. Trends Plant Sci 7:162–167

Neumann G, Römheld V (2002) Root-induced changes in the availability of nutrients in the rhizosphere. In: Waisel Y, Eshel A, Kafkafi U (eds) Plant roots: the hidden half, 3rd edn. Marcel Dekker, New York, pp 617–650

Noble R, Coventry E (2005) Suppression of soil-borne plant diseases with composts: a review. Biocontrol Sci Technol 15:3–20

Normander B, Prosser JI (2000) Bacterial origin and community composition in the barley phytosphere as a function of habitat and pre-sowing conditions. Appl Environ Microbiol 66:4372–4377

Ofek M (2011) Root bacterial community assemblage: basic knowledge gained from application-oriented studies. PhD thesis, The Hebrew University of Jerusalem, Jerusalem

Ofek M, Hadar Y, Minz D (2009) Effects of compost amendment vs. single strain inoculation on root bacterial communities of young cucumber seedlings. Appl Environ Microbiol 75:6441–6450

Or D, Semts BF, Wraith JM, Dechesne A, Friedman SP (2007) Physical constraints affecting bacterial habitats and activity in unsaturated porus media – a review. Adv Water Resour 30:1505–1527

Ortíz-Castro R, Martínez-Trujillo M, López-Bucio J (2008) N-acyl-L-homoserine lactones: a class of bacterial quorum-sensing signals alter post-embryonic root development in Arabidopsis thaliana. Plant Cell Environ 31:1497–1509

Osbourn AE (1996) Preformed antimicrobial compounds and plant defense against fungal attack. Plant Cell 8:1821–1831

Paulitz TC, Belanger RR (2001) Biological control in greenhouse systems. Annu Rev Phytopathol 39:103–133

Persello-Cartieaux F, Nussaume L, Robaglia C (2003) Tales from the underground: molecular plant-rhizobacteria interactions. Plant Cell Environ 26:189–199

Pinton R, Varanini Z, Nannipieri P (2001) The rhizosphere as a site of biochemical interactions among soil components, plants and microorganisms. In: Pinton R, Varanini Z, Nannipieri P (eds) The rhizosphere. Marcel Dekker, New York, pp 1–18

Power AG (2010) Ecosystem services and agriculture: tradeoffs and synergies. Phil Trans R Soc 365:2959–2971

Raaijmakers JM, Paulitz TC, Steinberg C, Alabouvett C, Moënne-Loccoz Y (2009) The rhizosphere: a playground and pattlefield for soilborne pathogens and beneficial microorganisms. Plant Soil 321:341–361
Rabie GH (1998) Induction of fungal disease resistance in *Vicia faba* by dual inoculation with *Rhizobium leguminosarum* and vesicular-arbuscular mycorrhizal fungi. Mycopathologia 141:159–166
Richardson AE, Barea JM, McNeill AM, Prigent-Combaret C (2009) Acquisition of phosphorus and nitrogen in the rhizosphere and plant growth promotion by microorganisms. Plant Soil 321:305–339
Ridder-Duine AS, Kowalchuk GA, Klein Gunnewiek PJA, Smant W, van Veen JA, de Boer W (2005) Rhizosphere bacterial community composition in natural stands of *Carex arenaria* (sand sedge) is determined by bulk soil community composition. Soil Biol Biochem 37:349–357
Rillig MC, Mummey DL (2006) Mycorrhizas and soil structure. New Phytol 171:41–53
Rodríguez-Zaragoza S, González-Ruíz T, González-Lozano E, Lozada-Rojas A, Mayzlish-Gati E, Steinberger Y (2008) Vertical distribution of microbial communities under the canopy of two legume bushes in the Tehuacán desert, Mexico. Eur J Soil Ecol 44:373–380
Roesti D, Gaur R, Johri BN, Imfeld G, Sharma S, Kawaljeet K, Aragno M (2006) Plant growth stage, fertilizer management and bio-inoculation of arbuscular mycorrhizal fungi and plant growth promoting rhizobacteria affect the rhizobacterial community structure in rain-fed wheat fields. Soil Biol Biochem 38:1111–1120
Rovira AD, Bowen GD, Foster RC (1983) The significance of rhizosphere microflora and mycorrhizas in plant nutrition. In: Lauchli A, Bielski RL (eds) Encyclopedia of plant physiology: inorganic plant nutrition. Springer, Berlin/Heidelberg/New York, pp 61–93
Ruiz-Lozano JM, Collados C, Barea JM, Azcón R (2001) Arbuscular mycorrhizal symbiosis can alleviate drought-induced nodule senescence in soybean plants. New Phytol 151:493–502
Ryne PR, Delhaize E, Jones DL (2001) Function and mechanisms of organic anion exudation from plant roots. Annu Rev Plant Physiol Plant Mol Biol 52:527–560
Sato a, Watanabe T, Unno Y, Purnomo E, Osaki M, Shinano T (2009) Analysis of diversity of diazotrophic bacteria associated with the rhizosphere of a tropical arbor, *Melastoma malabathricum* L. Microb Environ 24:81–87
Schuhegger R, Ihring A, Gantner S, Bahnweg G, Knappe C, Vogg G et al (2006) Induction of systemic resistance in tomato by N-acyl-L-homoserine lactone-producing rhizosphere bacteria. Plant Cell Environ 29:909–918
Séguin V, Gagnon C, Courchesne F (2004) Changes in water extractable metals, pH and organic matter concentrations at the soil-root interface of forested soils. Plant Soil 260:1–17
Sen R (2003) The root–microbe–soil interface: new tools for sustainable plant production. New Phytol 157:391–398
Skene KR (2000) Pattern formation in cluster roots: some developmental and evolutionary considerations. Ann Bot 85:901–908
Smalla K, Wieland G, Buchner A, Zock A, Parzy J, Kaiser S, Rostok N, Heuer H, Berg G (2001) Bulk and rhizosphere soil bacterial communities studied by denaturing gradient gel electrophoresis: plant-dependent enrichment and seasonal shifts revealed. Appl Environ Microbiol 67:4742–4751
Smit E, Leeflang P, Glandorf B, van Elsas JD, Wernars K (1999) Analysis of fungal diversity in the wheat rhizosphere by sequencing of cloned PCR-amplified genes encoding 18S rRNA and temperature gradient gel electrophoresis. Appl Environ Microbiol 65:2614–2621
Smith SE, Read DJ (1997) Mycorrhizal symbiosis, 2nd edn. Academic, London
Soares CRFS, Siqueira JO (2008) Mycorrhiza and phosphate protection of tropical grass species against heavy metal toxicity in multi-contaminated soil. Biol Fertil Soils 44:833–841
Society R (2009) Reaping the benefits: science and the sustainable intensification of global agriculture. Marcel Dekker, London, 72

Steidel A, Sigl K, Schuhegger R, Ihring A, Schmid M, Gantner S, Stoffels M, Riedel K, Givskov M, Hartmann A, Langebartels C, Eberl L (2001) Visualization of N-acylhomoserine lactone-mediated cell-cell communication between bacteria colonizing the tomato rhizosphere. Appl Environ Microbiol 67:5761–5770

Teplitski M, Robinson JB, Bauer WD (2000) Plants secrete substances that mimic bacterial N-acyl homoserine lactone signal activities and affect population density-dependent behaviors in associated bacteria. Mol Plant Microbe Interact 13:637–648

Tiquia SM, Lloyd J, Herms DA, Hoitink HAJ, Michel FC (2002) Effects of mulching and fertilization on soil nutrients, microbial activity and rhizosphere bacterial community structure determined by analysis of TRFLPs of PCR-amplified 16S rRNA genes. Appl Soil Ecol 21:31–48

Uren NC (2001) Types, amounts, and possible functions of compounds released into the rhizosphere by soil-grown plants. In: Pinton R, Varanini Z, Nannipieri P (eds) The rhizosphere. Marcel Dekker, New York, pp 19–40

Uroz S, Buée M, Murat C, Fery-Klett P, Martin F (2010) Pyrosequencing reveals a contrasted bacterial diversity between oak rhizosphere and surrounding soil. Environ Microbiol 2:281–288

Vallad GE, Goodman RM (2004) Systemic acquired resistance and induced systemic resistance in conventional agriculture. Crop Sci 44:1920–1934

Van der Heijden MGA, Boller T, Wiemken A, Sanders IR (1998a) Different arbuscular mycorrhizal fungal species are potential determinants of plant community structure. Ecology 79:2083–2091

Van der Heijden MGA, Klironomos JN, Ursic M, Moutoglis P, Streitwolf-Engels R, Boller T, Wiemken A, Sanders IR (1998b) Mycorrhizal fungal diversity determines plant biodiversity, ecosystem variability and productivity. Nature 396:69–72

Van der Heijden MGA, Wiemken A, Sanders IR (2003) Different arbuscular mycorrizal fungi alter coexistence and resource distribution between co-occurring plant. New Phytol 157:569–578

Van Elsas JD, Turner S, Bailey MJ (2003) Horizontal gene transfer in the phytosphere. New Phytol 157:525–537

Van Loon LC, Bakker PAHM, Pieterse CMJ (1998) Systemic resistance induced by rhizosphere bacteria. Annu Rev Phytopathol 36:453–483

Van Rhijn P, Vanderleyden J (1995) The *Rhizobium*-plant symbiosis. Microbiol Rev 59:124–142

Vivas A, Vörös I, Biró B, Campos E, Barea JM, Azcón R (2003) Symbiotic efficiency of autochthonous arbuscular mycorrhizal fungus (*G. mosseae*) and *Brevibacillus* sp. isolated from cadmium polluted soil under increasing cadmium levels. Environ Pollut 126:179–189

Waisel Y, Eshel A (2002) Functional diversity of various constituents of a single root system. In: Waisel Y, Eshel A, Kafkafi U (eds) Plant roots: the hidden half, 3rd edn. Marcel Dekker, New York, pp 157–174

Walker TS, Bais HP, Grotewold E, Vivanco JM (2003) Root exudation and rhizosphere biology. Plant Physiol 132:44–51

Watt M, McCully ME, Kirkegaard JA (2003) Soil strength and rate of root elongation alter the accumulation of *Pseudomonas* spp. and other bacteria in the rhizosphere of wheat. Funct Plant Biol 30:483–491

Watt M, Hugenholtz P, White R, Vinall K (2006) Numbers and locations of native bacteria on field-grown wheat roots quantified by fluorescence *in situ* hybridization (FISH). Environ Microbiol 8:871–884

Weisskopf L, Fromin N, Tomasi N, Aragno M, Martinoia E (2005) Secretion activity of white lupin's cluster roots influences bacterial abundance, function and community structure. Plant Soil 268:181–194

Weisskopf L, Le Bayon RC, Kohler F, Page V, Jossi M, Gobat JM, Martinoia E, Aragno M (2008) Spatio-temporal dynamics of bacterial communities associated with two plant species differing in organic acid secretion: a one-year microcosm study on lupin and wheat. Soil Biol Biochem 40:1772–1780

Wenzel WW (2009) Rhizosphere processes and management in plant-assisted bioremediation (phytoremediation) of soils. Plant Soil 321:385–408

Whipps JM (2001) Microbial interactions and biocontrol in the rhizosphere. J Exp Bot 52:487–511

Will C, Thürmer A, Wollherr A, Nacke H, Herold N, Schrumpf M, Gutknecht J, Wubet T, Buscot F, Daniel R (2010) Horizon-specific bacterial community composition of German grassland soils, as revealed by pyrosequencing-based analysis of 16S rRNA genes. Appl Environ Microbiol 76:6751–6759

Yang CH, Crowley DE (2000) Rhizosphere microbial community structure in relation to root location and plant iron nutritional status. Appl Environ Microbiol 66:345–351

Zachow C, Tilcher R, Berg G (2008) Sugar beet-associated bacterial and fungal communities show a high indigenous antagonistic potential against plant pathogens. Microb Ecol 55:119–129

Zhang W, Han DY, Dick WA, Davis KR, Hiotink HAJ (1998) Compost and compost water extract-induced systemic acquired resistance in cucumber and *Arabidopsis*. Phytopathology 88:445–450

Zhang W, Rickett TH, Kremen C, Carney K, Swinton SM (2007) Ecosystem services and dis-services to agriculture. Ecol Econ 64:253–260

Zhang H, Kim MS, Dowd SE, Shi H, Paré PW (2008) Soil bacteria confer plant salt tolerance by tissue-specific regulation of the sodium transporter *HKT1*. Mol plant Microbe interact 21:737–744

Chapter 8
Microbial Protection Against Plant Disease

Eddie Cytryn and Max Kolton

Abstract The growing demand for food due to the expansion of world population necessitates development of practices that will enable high-yield, disease-free agriculture, with minimal use of chemical pesticides. Biocontrol-associated microorganisms can inhibit disease by either directly antagonizing phytopathogens or by stimulating a state of induced resistance in plants. Certain agronomic practices such as amendment of compost and biochar to soil curb plant disease by stimulating abundance and activity of indigenous soil and root-associated biocontrol agents (BCAs) resulting in so called "suppressive soils." Over the past half century, a multitude of attempts have been made to implement both BCAs and agronomic agricultural practices as plant protection strategies. Nonetheless, the use of these strategies in large-scale commercial agriculture is still very limited due to restricted efficacy and consistency. The following chapter focuses on well-documented BCAs and their modes of action, delineates microbial-associated biocontrol-inducing agronomic practices, and discusses future directions required for applying biocontrol in high-yield sustainable agriculture.

E. Cytryn (✉)
Institute of Soil, Water and Environmental Sciences, The Volcani Center, Agricultural Research Organization, P.O. Box 6, Bet Dagan 50250, Israel
e-mail: eddie@volcani.agri.gov.il

M. Kolton
Institute of Soil, Water and Environmental Sciences, The Volcani Center, Agricultural Research Organization, P.O. Box 6, Bet Dagan 50250, Israel

Department of Plant Pathology and Weed Research, The Volcani Center, Agricultural Research Organization, P.O. Box 6, Bet Dagan 50250, Israel

Institute of Plant Sciences and Genetics in Agriculture, The Robert H. Smith Faculty of Agriculture, Food and Environment, The Hebrew University of Jerusalem, P.O. Box 12, Rehovot, 76100, Israel

E. Rosenberg and U. Gophna (eds.), *Beneficial Microorganisms in Multicellular Life Forms*, 123
DOI 10.1007/978-3-642-21680-0_8, © Springer-Verlag Berlin Heidelberg 2011

8.1 Introduction

Plant disease can be caused by a multitude of phylogenetically diverse organisms, which include insects, nematodes, oomycetes, protozoa, fungi, bacteria, viruses, and parasitic plants. It is estimated that approximately 10% of global food production is lost from phytopathogen-related plant disease, which results in annual losses of billions of dollars a year and poses a major threat to food safety, especially in underdeveloped countries (Strange 2005).

Although plants harbor natural resistance mechanisms that combat pathogens, pathogens have evolved specialized strategies to suppress plant defense responses and induce disease susceptibility in resistant hosts (Abramovitch and Martin 2004). Intensive agriculture strengthens the selective advantage of these resistant pathogens especially in monoculture systems (Zhu et al. 2000), necessitating increased use of organophosphate, organochlorine, and carbamate pesticides. Although pesticides are often effective, they can have extremely detrimental environmental and epidemiological impacts and their extended use often leads to development of pesticide-resistant phytopathogens (van den Bosch and Gilligan 2008).

Environmentally-sustainable agriculture suitable for meeting the food requirements of the twenty-first century necessitates development of practices that will enable intensive, high-yield disease-free agriculture, with minimal use of chemical pesticides (Godfray et al. 2010). Over the past 50 years, a multitude of studies have demonstrated the capacity of certain microorganisms to reduce plant disease. Biocontrol can be generated by direct application of microorganisms to the soil, rhizosphere or phylosphere, or through implementation of agronomic practices, such as organic and inorganic carbon amendments (compost, biomass or biochar) or planting strategies, that curb plant disease by stimulating abundance and activity of beneficial soil and root-associated microorganisms (Gerhardson 2010).

The following chapter focuses on well-documented biocontrol-associated microorganisms and their modes of action, delineates microbial-associated biocontrol-inducing agronomic practices, and discusses future directions for applying biocontrol in high-yield sustainable agriculture.

8.2 Biocontrol Mechanisms

Microorganisms are an intricate part of the plant root ecosystem. A small fraction of rhizosphere fungi and bacteria are plant pathogens, but many others are associated with enhanced plant productivity (Graham and Vance 2000; Adesemoye and Kloepper 2009), plant growth promotion (Glick et al. 2007), and reduced susceptibility of plants to pathogens. Microorganisms that antagonize plant pathogen activity, termed biocontrol agents (BCAs), protect plants from disease either through direct interaction with phytopathogens or by elicitation of induced systemic resistance (ISR) in plant hosts. These mechanisms are highly complex and have been exhaustively studied from the perspective of both plant and biocontrol agent (Baker 1968; Compant et al. 2005; Elad 2003; Shoresh et al. 2010; Weller 1988).

The fluorescent pseudomonads and certain bacilli strains are undoubtedly the most studied biocontrol bacteria (Choudhary and Johri 2009; Haas and Défago 2005; Weller 2007). Members of these genera are associated with both direct antagonism towards pathogens and ISR. Nonetheless, many other biocontrol-associated bacterial have been characterized including selected members of the *Agrobacterium, Alcaligens, Azotobacter, Cellumonas, Enterobacter, Erwinia, Flavobacterium, Rhizobium,* and *Bradyrhizobium, Serratia, Streptomyces*, and *Xanthomonas* genera (Weller 1988). Several biocontrol fungi (BCF) have also been found to control numerous foliar, root, and fruit pathogens and even invertebrates such as nematodes, through direct interaction with plant pathogens or ISR elicitation. These include selected strains of the *Trichoderma* genus and the recently described order *Sebacinales* (Borneman and Becker 2007; Elad 2003; Shoresh et al. 2010). In certain cases, strains closely related to phytopathogens serve as phytopathogen antagonists, as is the case with the BCA *Agrobacterium radiobacter* K84, which is phylogenetically similar to *A. tumefaciens,* the causative agent of the crown gall disease (Smith and Hindley 1978).

8.3 Direct Biocontrol Agent-Pathogen Interactions

Plant roots excrete an array of organic compounds which include amino acids, fatty acids, nucleotides, organic acids, phenolics, plant growth regulators, putrescine, sterols, sugars, and vitamins (Lugtenberg and Kamilova 2009). The root surface (rhizoplane) thereby represents an oasis for microbial colonization. The microbial community structure of the rhizoplane is often dependent on plant genotype, which dictates the composition and concentration of specific exudates that are excreted at the root surface (Manter et al. 2010). This was elegantly demonstrated by Badri et al. (2009) who showed that an *Arabidopsis thaliana* ABC transporter mutant (*abcg30*), which causes increased levels of phenolic exudates and decreased levels of sugars (relative to the wild type strain), resulted in a significant shift in microbial community structure with a relatively greater abundance of potentially beneficial bacteria.

BCA efficiency is often dependent on "rhizosphere competence," the capacity of microbes to adhere, replicate, and survive on the root surface (Ahmad and Baker 1987). Rhizosphere competence is dictated abiotic parameters as well as a wide array of microbial physiological mechanisms. Chemotaxis coupled to high motility is a key factor for successful root colonization by some BCAs (Lugtenberg and Dekkers 1999; Lugtenberg and Kamilova 2009; Williams 2007). Once microorganisms reach preferred locations on the root surface, adhesion, often associated with biofilm formation, is generally required. This involves the combined action of transport mechanisms such as type III secretion systems (Mavrodi et al. 2011), and adhesive matrix building components such as lipopolysaccharides (LPS), extrapolysaccharides (EPS) and type IV pili (Compant et al. 2005; Dörr et al. 1998; Rudrappa et al. 2008; Weller 2007).

Many BCAs inhibit plant disease through direct interactions with phytopathogens. Niche exclusion, due to colonization of root surfaces has been suggested as one potential mechanism of biocontrol. Microscopic analyses of roots generally show that regions of microbial colonization are sparse; indicating that niche competition is most likely restricted to "organic hotpots" that are rich in organic exudates or other nutrients (Weller 1988). Biocontrol strains often mop-up key nutrients in the rhizosphere and thus prevent proliferation of phytopathogens. One of the prime mechanisms is production of siderophores that chelate low molecular weight Fe^{3+}. Siderophores of some BCAs have extremely high iron (Fe) affinity and are can inhibit growth of plant pathogens with low Fe^{3+} affinity in the iron-limited rhizosphere (Compant et al. 2005; Tortora et al. 2011).

Many biocontrol strains produce extracellular enzymes that directly antagonize phytopathogens. For example, bacterial and fungal chitinases have been shown to degrade fungal cell walls and degrade chitinous components of nematode eggs (Cohen-Kupiec and Chet 1998; Neeraja et al. 2010); proteases produced by biocontrol bacteria and fungi have been shown to be involved in antifungal activity and nematode predation (Ko et al. 2009; Lopez-Llorca et al. 2010). Chitin monomers produced by BCA-mediated chitinase activity can also serve to elicit ISR in plants (see below), indicating that in certain cases BCAs can inhibit pathogens by more than one antagonistic mechanism.

Antibiotic production is strongly associated with biocontrol strains. One of the first examples of biocontrol via antibiosis was for the non-pathogenic strain *Agrobacterium radiobacter* K84, which suppresses the crown gall disease agent *A. tumefaciens* via the antibiotic agrocin 84 (Smith and Hindley 1978; Penyalver et al. 2009). Since then a variety of antibiotics have been identified in gram negative- (hydrogen cyanide, oomycin A, phenazine, and the antifungal agent 2,4-diacetylphloroglucinol), as well as in gram positive-BCAs (oligomycin A, kanosamine, zwittermicin A, and xanthobaccin) (Compant et al. 2005). Antibiotic production by BCAs is generally believed to occur following direct interaction with bacterial and fungal phytopathogens in the rhizosphere as a means of competitive exclusion. Nonetheless, plants also have the capacity to stimulate BCA antibiotic production. In a split-root experiment (where pathogen and BCA are physically separated), enhanced transcription of the *phlA* gene, which encodes for the 2,4-diacetylphloroglucinol in the BCA *Pseudomonas fluorescens* st. CHA0, was observed following inoculation of plant roots with the oomycete pathogen *Pythium ultimum* (Jousset et al. 2011). Biocontrol capacity often requires a threshold concentration of the BCAs in order to antagonize phytopathogens. This is supported by the fact that the production of biocontrol allelochemicals such as antibiotics have been found to be triggered by quorum-sensing associated homoserine lactone (HSL) molecules (Kay et al. 2005; Liu et al. 2007). For example, De Maeyer et al. (2011) recently isolated a large pool of fluorescent pseudomonads from red cocoyam roots. They found that all of the phenazine-producing isolates showing antagonistic activity towards the root rot pathogen *Pythium myriotylum* produced acyl-HSLs, and that non-antagonistic pseudomonad strains did not produce these quorum-sensing molecules.

8.4 Induction of Systemic Resistance in Plants

Plants have developed highly sophisticated defense systems that recognize pathogen molecules. These systems are primarily regulated by defense-signaling networks mediated by small-molecule hormones, most notably salicylic acid, ethylene, auxin (Indole-3-acetic acid), abscisic acid, and jasmonic acid (Pieterse et al. 2009).

Plants respond to pathogenic attack by local responses and by a hormone-mediated global plant resistance response termed systemic acquired resistance (SAR). SAR involves the production of pathogenesis-related (PR) proteins, which are mediated via a salicylic acid (SA)-dependent process. Plant membrane-localized pattern recognition receptors (PRRs) initiate immune responses following identification of complementary elicitors termed microbe-associated molecular patterns (MAMPs). MAMPs include bacterial flagellin, the elongation factor (EF)-Tu, lipopolysaccharides, peptidoglycans, and fungal cell wall components such as chitin (N-acetyl-chitooligosaccharideoligomers) (Saijo 2010). Recognition of MAMPs is central to the generation of an enhanced state of immunity against potentially infectious microbes, termed MAMP-triggered immunity (MTI), which is characterized by short- and long-term physiological responses which include reactive oxygen species (ROS) spiking, ethylene release, MAPK activation, reprogramming of transcriptome, and metabolome (e.g., production of phytoalexins and SA) and callose deposition (Boller and Felix 2009).

Unlike SAR, ISR is systemically generated in response to colonization of plant roots by non-pathogenic bacteria and fungi, most notably from the taxa *Pseudomonas*, *Bacillus,* and *Trichoderma* (Bakker et al. 2007; Kloepper et al. 2004; McSpadden Gardener 2007; Shoresh et al. 2010; Van Loon et al. 1998). This pathway is mediated by the phytohormones jasmonic acid (JA) and ethylene (ET) and the plant regulatory gene NPR1, and unlike SAR, it does not involve expression of PR proteins or other immune-related responses against the root-associated microorganism (Van Loon et al. 1998). ISR is associated with priming of the plant immune system, which results in enhanced defense upon subsequent infection with phytopathogenic elements (Conrath et al. 2002). In primed plants, defense responses are not directly induced by the priming agent, but are rapidly activated following identification of biotic or abiotic stress signals, which consequently results in an enhanced level of resistance against the phytopathogenic elements encountered (Van der Ent et al. 2009; van Hulten et al. 2006; van Wees et al. 2008). Van Hulten et al. (2006) demonstrated that the fitness costs of priming are lower than the cost of continual activation of plant defense mechanisms, and that the costs of priming are significantly lower than the benefits under conditions of disease pressure. ISR is induced by microbial elicitors that are similar to pathogenic MAMPs that trigger SAR. Well documented ISR elicitors include cell membrane-associated constituents such as flagellin, LPS (van loon, 2008) and lipoproteins (van Wees et al. 2008), antibiotics such as phenazines and 4-diacetylphloroglucinol (DAPG) (Bakker et al. 2007), extracellular enzymes and degradation products, volatile organic compounds such as N-alkylated benzylamine and 2,3-butanediol (Ongena et al. 2005), quorum sensing-associated HSLs (Schuhegger et al. 2006), and bacterial siderophores (van Loon et al. 2008). ISR-stimulating MAMPs appear to be host specific because applications of isolated components have differential resistance-inducing activities when applied to different plant species. Furthermore, there appears to be complementation of elicitors in at least some of the ISR-inducing microorganisms. For example, mutations in the ISR-eliciting components flagellin and LPS in certain beneficial pseudomonad species induced ISR at wild-type-like levels (Bakker et al. 2007). Although pathogen and ISR-inducing MAMPs appear to be recognized by the plant in a largely similar manner, contrary to pathogen association, interaction between beneficial microbes and plants does not involve plant-associated response against BCAs as it remains accommodated by the plant, suggesting a high degree of coordination and a continuous molecular dialog between the plant and the beneficial organism (Van Wees et al. 2008).

Recent studies have demonstrated that BCA-induced resistance may involve alternative pathways to the model NPR1-regulated JA and ethylene-activation pathway. For example, although several strains of *Bacillus* spp. elicit ISR through the traditional model pathway, in other strains ISR is dependent on SA and independent of JA and NPR1. In addition, while ISR by *Pseudomonas* spp. does not lead to accumulation of the defense gene PR1 in plants, in some cases, ISR by *Bacillus* spp. does (Choudhary and Johri 2009).

8.5 Environmentally Sustainable Agricultural Practices

A wide array of BCAs have been shown to inhibit plant disease in contained microcosm or greenhouse experiments, but their capacity to suppress phytopathogens in the field is often highly inconsistent and ineffective. This is believed to be primarily associated with lack of *ecological competence*, the ability of a specific microorganism to survive in a defined environment (Weller 1988). Ecological competence is determined by a wide array of abiotic and biotic conditions, including temperature, soil type, organic carbon levels, water potential, plant phenotype, and competition with the indigenous soil and root-associated microbiota (Weller 1988).

Some soils are known to be naturally suppressive to plant disease and in many instances this natural suppressiveness is biologically associated (Borneman and Becker 2007). Suppressive soils have been defined as "soils in which the pathogen does not establish or persist, establishes and causes disease for a while but thereafter the disease is less important, although the pathogen may persist in the soil" (Borneman and Becker 2007). Certain agronomic practices have been shown to induce suppressiveness in non-suppressive soils. The following section focuses on characterization of microbial-associated suppressive soils and prevalent biocontrol-inducing agronomic practices that have been shown to suppress plant pathogens by induction of BCAs. These practices are discussed below in the context of their BCA-inducing capacities.

8.5.1 Detection of Biocontrol Agents in Suppressive Soils

One of the most challenging aspects of understanding soil suppressiveness is the ability to identify BCAs that contribute to phytopathogen inhibition within the highly diverse soil microbiome (Borneman and Becker 2007). Two general approaches are commonly used: (A) Well-defined BCAs can be identified and quantified using classical culturing techniques, 16S rRNA analyses, and analytical chemical analyses; or (B) the microbial community composition of suppressive soils can be assessed concomitant to non-suppressive soils, using molecular fingerprinting techniques (i.e., DGGE, T-RFLP) or through sequencing-based technologies, which in some cases can enable inference of specific BCAs. The potential pitfalls of these methods are (A) that only a small fraction of soil microbes can be isolated; and (B) many BCAs are closely related to non-BCAs (sometimes even to phytopathogenic strains) and therefore, it is not always possible to infer biocontrol capacity based on phylogenetic analysis. One possible solution is to focus on specific genes or metabolites that are known to be associated with biocontrol. For example, Mazurier et al. (2009) assessed BCA abundance in soils suppressive to *Fusarium* wilt disease by comparing gene expression levels of the antibiotics 2,4-diacetylphloroglucinol (DAPG) (*phlDp*) and phenazine (*phzCp*) in suppressive versus conducive soils. As previously discussed, these antibiotics are produced by several fluorescent pseudomonads and are strongly associated with biocontrol. Expression levels of *phlDp* were similar in suppressive and conducive soils; but *phzCp* transcripts were only detected in the suppressive soil. This indicates that redox-active phenazines produced by fluorescent pseudomonads contribute to the natural soil suppressiveness to *Fusarium* wilt disease.

8.5.2 Compost Application

Application of certain composts to soil has been shown to confer resistance to plant pathogens. These "suppressive composts" can provide broad spectrum biological control of diseases caused by well-known soilborne plant pathogens (Hoitink and Boehm 1999). Compost-induced soil suppressiveness is believed to occur due to induction of BCAs which antagonize phytopathogens by way of antibiosis, parasitism, and ISR. For example, a suppressive compost mix fortified with selected BCAs induced ISR against fungal and bacterial pathogens in cucumber and *Arabidopsis*, respectively (Zhang et al. 1998). Autoclaving destroyed the ISR-inducing effect of the compost mix, and inoculation of the autoclaved mix with non-autoclaved compost mix or *Pantoea agglomerans* 278A restored the effect, suggesting the ISR-inducing activity of the compost mix was biological in nature (Zhang et al. 1998). Suppressive composts appear to contain organic substrates that are optimal for BCAs, enabling them to establish themselves in the rhizosphere, which in turn leads to sustained biological control (Hoitink and Boehm 1999). Several studies have demonstrated that the decomposition level of organic matter critically affects the composition of bacterial taxa as well as the populations and activities of BCAs (Boehm et al. 1993; Hoitink and Boehm 1999; Hoitink et al. 2006). This is evident by the fact that mature composts are generally more suppressive than immature composts. Furthermore, although the microbial community composition of compost-amended soils are substantially different from non-amended soils, microbial community analyses show that these are soil-associated communities, indicating that observed shifts stem from abiotic factors and not from influx of compost derived microbial communities (Green et al. 2006; Cytryn and Minz 2011).

Bacterial isolates and their metabolites exhibiting *in-vitro* biocontrol capacity are often used as indicators of compost suppressiveness. For example, compost amendment substantially increased the incidence of rhizosphere bacterial isolates exhibiting antagonism towards the fungal pathogens *Fusarium oxysporum* f. sp. radicis-lycopersici, *Pyrenochaeta lycopersici*, *Pythium ultimum*, and *Rhizoctonia solani* in tomato. Many of these isolates were characterized by the presence of biocontrol-associated factors such as siderophores, phosphate-solubilizing enzymes, and the plant growth-stimulating homolog indoleacetic acid (De Brito et al. 1995).

Pseudomonad abundance has been directly associated with suppressiveness of several composts (Boehm et al. 1993; Bradley and Punja 2010; Mazurier et al. 2009). For example, in suppressive compost-amended soils inoculated with pythium, the causative agent of damping-off, pseudomonads typically accounted for 25–45% of the total number of strains isolated from root tip segments of cucumber seedlings. In contrast, pseudomonads were not detected in root tip segments of seedlings grown in conducive decomposed peat mix (Boehm et al. 1993). Bradley and Punja (2010) demonstrated that composts containing the pseudomonas-associated antibiotic 2,4-diacetylphloroglucinol suppressed *Fusarium*-related disease in cucumber plants supporting the notion that this antibiotic is involved in biocontrol activity. Recently, a 16S rRNA taxonomic microarray was used to monitor rhizobacterial community composition in soils harboring cultivated tobacco concomitantly amended with conducive and suppressive composts sharing similar physicochemical properties. Fluorescent pseudomonads were significantly induced in the suppressive compost-amended soils, as were the *Azospirillum*, *Gluconacetobacter*, *Burkholderia*, *Comamonas*, and *Sphingomonadaceae* taxa, which are known to contain biocontrol strains (Kyselková et al. 2009).

8.5.3 Promotion of Biocontrol Agents Through Cropping Strategies

Certain wheat cultivars have been shown to enhance the suppressive capacity of agricultural soils in a number of plant-pathogen model systems and therefore have been used to enhance the suppressive capacity of soils prior to principal crop cultivation (Mazzola and Gu 2002; Meyer et al. 2010; Weinert et al. 2011). For example, Mazzola and Gu (2002) demonstrated that *Rhizoctonia solani*-associated disease was suppressed in apple orchard soils pre-treated with three successive 28-day cycles of specific wheat cultivars in the greenhouse prior to infestation. Pasteurization of soils after wheat cultivation and prior to pathogen introduction eliminated the disease-suppressive potential of the soil indicating that the disease suppression was associated with soil-derived microorganisms. Wheat cultivars that induced disease suppression enhanced populations of specific fluorescent pseudomonad genotypes with antagonistic activity toward *R. solani* in three different orchard soil types, but cultivars that did not elicit a disease-suppressive soil did not modify the antagonistic capacity of this bacterial community.

8.5.4 Biochar Application

Biochar is the solid coproduct of biomass pyrolysis, a technique used for second-generation biofuels (Lehmann 2007). Application of biochar to soil results in long-term sequestration of fixed carbon and it is associated with improved soil tilth, nutrient retention, and crop productivity (Glaser et al. 2002). Recently conducted experiments have shown that addition of biochar to soil also suppresses proliferation of fungal pathogens in pepper, tomato, and strawberry plants and enhances resistance to the broad mite pest (*Polyphagotarsonemus latus* Banks) in pepper (Elad et al. 2010; Meller Harel et al. 2011). The fact that the site of infection (leaves) is spatially separate from that of the amendment (roots) indicates that the biochar induces ISR in the inoculated plants. This phenomenon may, at least in part, be associated with biochar stimulation of soil and root-associated BCAs, with ISR-inducing capacities. Graber et al. (2010) used general and selective cultivation media to screen bulk soil and root-associated microbial communities from biochar amended and non-amended coconut-tuff potting mixtures. In biochar-amended pots, abundance of root-associated yeast, *Trichoderma* spp., and bacilli were approximately 3, 2, and 1 log units, respectively, while all three were not measurable in biochar-free pots. Significantly greater numbers of general bacteria, pseudomonads, and fungi were also observed in the bulk potting mixture in the biochar-amended treatments relative to the bulk potting mixtures of the control treatment. Phylogenetic characterization of selected bacterial isolates from the biochar-amended growing mixtures based on partial 16S rRNA gene analysis revealed that several of the biochar-associated isolates from the bulk soil and the rhizoplane were affiliated with previously documented BCAs.

Molecular microbial analyses were recently implemented to assess the effect of biochar amendments on bacterial community structure in both bulk soil and the rhizosphere, with the objective of identifying microorganisms potentially responsible for the observed biochar-induced plant resistance (Kolton et al. 2011). Fingerprinting methods (PCR-DGGE and T-RLFP) supported the notion that biochar amendment significantly alters bacterial community composition in both the bulk soil and on the pepper roots. We employed high throughput pyrosequencing of bacterial 16S rRNA gene amplicons to specifically focus on root-associated bacteria. Analyses indicated that the relative abundance of the *Bacteroidetes* phylum was significantly higher, and relative abundance of the *Proteobacteria* phylum was significantly lower in biochar-amended roots relative to the control roots (13.2 vs. 23.5% and 62.2 vs. 48.2%, for the *Proteobacteria* and *Bacteroidetes*, respectively). The increase in *Bacteroidetes* was primarily attributed to an increase in the *Flavobacterium* genus (4.2% of the total defined in control samples vs. 19.5% in the biochar-amended soils). Flavobacteria have been associated with biocontrol in general and specifically as elicitors of ISR (Alexander and Stewart 2001; Gunasinghe and Karunaratne 2009; Hebbar et al. 1991). It may be suggested that the gliding motility, highly diverse arsenal of extracellular macromolecule degrading enzymes, capacity to produce antibiotic substances and ubiquitous presence in the rhizoplane makes flavobacteria a potential candidate for elicitation of the observed biochar-induced plant resistance.

8.6 Exploitation of Beneficial Microorganisms for Enhancing Plant Resistance to Disease: Future Directions

The growing demand for food due to the expansion in world population , concomitant to the ever-increasing public and regulatory requirements to reduce pesticide use, necessitates development of superior biocontrol tools capable of containing plant disease. Although many commercial products based on isolated BCAs have been marketed over the years, the efficacy of these products is often limited and their employment is highly restricted in comparison to commercial fertilizer use. Agronomical practices that induce biocontrol are becoming more prevalent (especially in organic agriculture) however, their suppressive capacity is often inconsistent and the biocontrol mechanisms generated by these practices are, to a great extent, still a "black box".

Optimization of biocontrol capacity requires in-depth understanding of the microbial ecology of prominent BCAs in the soil and rhizosphere, particularly in the context of BCA-pathogen and BCA-plant interactions. Ecological competence of BCAs can be substantially enhanced by isolating indigenous strains from a specific environment and reintroducing BCA consortia following enrichment. Although pseudomonad and bacilli strains are undoubtedly prime BCAs, efforts should be made to isolate other biocontrol-associated microbes that may be more compatible with certain environments. The genomic era and rapid developments in genetic engineering can undoubtedly enhance the potential for developing superior BCAs. Once a greater understanding of ecology and physiology is acquired, it may be possible to develop super strains with enhanced biocontol capabilities catering to specific environments and ultimately "engineer" composts or soils with enhanced suppressive capacities.

References

Abramovitch RB, Martin GB (2004) Strategies used by bacterial pathogens to suppress plant defenses. Curr Opin Plant Biol 7:356–64

Adesemoye AO, Kloepper JW (2009) Plant-microbes interactions in enhanced fertilizer-use efficiency. Appl Microbiol Biotechnol 85:1–12

Ahmad JS, Baker R (1987) Rhizosphere competence of *Trichoderma harzianum*. Phytopathology 77:182–189

Alexander BJR, Stewart A (2001) Glasshouse screening for biological control agents of *Phytophthora cactorum* on apple (*Malus domestica*). N Z J Crop Hortic Sci 29:159–169

Badri DV, Quintana N, El Kassis EG, Kim HK, Choi YH, Sugiyama A, Verpoorte R, Martinoia E, Manter DK, Vivanco JM (2009) An ABC transporter mutation alters root exudation of phytochemicals that provoke an overhaul of natural soil microbiota. Plant Physiol 151:2006–2017

Baker R (1968) Mechanisms of biological control of soil-borne pathogens. Annu Rev Phytopathol 6:263–294

Bakker PA, Pieterse CM, van Loon LC (2007) Induced systemic resistance by fluorescent Pseudomonas spp. Phytopathology 97:239–243

Boehm MJ, Madden LV, Hoitink HAJ (1993) Effect of organic matter decomposition level on bacterial species diversity and compositin in relationship to Pythium damping-off severity. Appl Environ Microbiol 59:4171–4179

Boller T, Felix G (2009) A renaissance of elicitors: perception of microbe-associated molecular patterns and danger signals by pattern-recognition receptors. Annu Rev Plant Biol 60:379–406

Borneman J, Becker JO (2007) Identifying microorganisms involved in specific pathogen suppression in soil. Annu Rev Phytopathol 45:153–172

Bradley GG, Punja ZK (2010) Composts containing fluorescent pseudomonads suppress fusarium root and stem rot development on greenhouse cucumber. Can J Microbiol 56:896–905

Choudhary DK, Johri BN (2009) Interactions of *Bacillus* spp. and plants- with special reference to induced systemic resistance (ISR). Microbiol Res 164:493–513

Cohen-Kupiec R, Chet I (1998) The molecular biology of chitin digestion. Curr Opin Biotechnol 9:270–277

Compant S, Duffy B, Nowak J, Clément C, Barka EA (2005) Use of plant growth-promoting bacteria for biocontrol of plant diseases: principles, mechanisms of action, and future prospects. Appl Environ Microbiol 71:4951–4959

Conrath U, Pieterse CM, Mauch-Mani B (2002) Priming in plant-pathogen interactions. Trends Plant Sci 7:210–216

Cytryn E, Kautsky L, Ofek M, Mandelbaum RT, Minz D (2011) Short-term structure and functional changes in bacterial community composition following amendment with biosolids compost Appl Soil Ecol 48:160–167

de Brito AM, Gagne S, Antoun H (1995) Effect of compost on rhizosphere microflora of the tomato and on the incidence of plant growth-promoting rhizobacteria. Appl Environ Microbiol 61:194199

De Maeyer K, D'aes J, Hua GK, Perneel M, Vanhaecke L, Noppe H, Höfte M (2011) N-Acylhomoserine lactone quorum-sensing signalling in antagonistic phenazine-producing *Pseudomonas* isolates from the red cocoyam rhizosphere. Microbiology 157:459–72
Dörr J, Hurek T, Reinhold-Hurek B (1998) Type IV pili are involved in plant-microbe and fungus-microbe interactions. Mol Microbiol 30:7–17
Elad Y (2003) Biocontrol of foliar pathogens: mechanisms and application. Commun Agric Appl Biol Sci 68:17–24
Elad Y, Rav David D, Meller Harel Y, Borenshtein M, Ben Kalifa H, Silber A, Graber ER (2010) Induction of systemic resistance in plants by biochar, a soil-applied carbon sequestering agent. Phytopathology 100:913–921
Gerhardson B (2010) Biological substitutes for pesticides. Trends Biotechnol 20:338–343
Glaser B, Lehmann J, Zech W (2002) Ameliorating physical and chemical properties of highly weathered soils in the tropics with charcoal – a review. Biol Fertil Soils 35:219–230
Glick BR, Cheng Z, Czarny J, Duan J (2007) Promotion of plant growth by ACC deaminase-producing soil bacteria. E J Plant Path 119:329–339
Godfray HC, Beddington JR, Crute IR, Haddad L, Lawrence D, Muir JF, Pretty J, Robinson S, Thomas SM, Toulmin C (2010) Food security: the challenge of feeding 9 billion people. Science 12:812–818
Graber ER, Meller Harel Y, Kolton M, Cytryn E, Silber A, Rav David D, Tsechansky L, Borenshtein M, Elad I (2010) Biochar impact on development and productivity of pepper and tomato grown in fertigated soilless media. Plant Soil 337:481–496
Graham PH, Vance CP (2000) Nitrogen fixation in perspective: an overview of research and extension needs. Field Crop Res 65:93–106
Green SJ, Inbar E, Michel FC Jr, Hadar Y, Minz D (2006) Succession of bacterial communities during early plant development: transition from seed to root and effect of compost amendment. Appl Environ Microbiol 72:3975–3983
Gunasinghe WK, Karunaratne AM (2009) Interactions of *Colletotrichum musae* and *Lasiodiplodia theobromae* and their biocontrol by *Pantoea agglomerans* and *Flavobacterium* sp in expression of crown rot of "Embul" banana. Biocontrol 54:587–596
Haas D, Défago G (2005) Biological control of soil-borne pathogens by fluorescent pseudomonads. Nat Rev Microbiol 3:307–319
Hebbar P, Berge O, Heulin T, Singh SP (1991) Bacterial antagonists of sunflower (*Helianthus-Annuus* L) fungal pathogens. Plant Soil 133:131–140
Hoitink HAJ, Boehm MJ (1999) Biocontrol within the context of soil microbial communities: a substrate dependent phenomenon. Annu Rev Phytopathol 37:427–446
Hoitink HA, Madden LV, Dorrance AE (2006) Systemic resistance Induced by Trichoderma spp.: Interactions between the host, the pathogen, the biocontrol agent, and soil Organic matter quality. Phytopathology 96:186–189
Jousset A, Rochat L, Lanoue A, Bonkowski M, Keel C, Scheu S (2011) Plants respond to pathogen infection by enhancing the antifungal gene expression of root-associated bacteria. Mol Plant Microbe Interact 24:352–358
Kay E, Dubuis C, Haas D (2005) Three small RNAs jointly ensure secondary metabolism and biocontrol in *Pseudomonas fluorescens* CHA0. Proc Natl Acad Sci USA 102:17136–1741
Kloepper JW, Ryu CM, Zhang S (2004) Induced systemic resistance and promotion of plant growth by *Bacillus* spp. Phytopathology 94:1259–1266
Ko HS, Jin RD, Krishnan HB, Lee SB, Kim KY (2009) Biocontrol ability of *Lysobacter antibioticus* HS124 against Phytophthora blight is mediated by the production of 4-hydroxyphenylacetic acid and several lytic enzymes. Curr Microbiol 59:608–615
Kolton M, Meller-Harel Y, Pasternak Z, Graber ER, Tsechansky L, Rav David D, Silber A, Elad Y, Cytryn E (2011) Impact of biochar application to soil on the root-associated bacterial community structure of fully developed greenhouse pepper plants. Appl Envir Microbiol 77:4924–4930

Kyselková M, Kopecký J, Frapolli M, Défago G, Ságová-Marecková M, Grundmann GL, Moënne-Loccoz Y (2009) Comparison of rhizobacterial community composition in soil suppressive or conducive to tobacco black root rot disease. ISME J 3:1127–1138

Lehmann J (2007) Bio-energy in the black. Front Ecol Environ 5:381–387

Liu X, Bimerew M, Ma Y, Müller H, Ovadis M, Eberl L, Berg G, Chernin L (2007) Quorum-sensing signaling is required for production of the antibiotic pyrrolnitrin in a rhizospheric biocontrol strain of *Serratia plymuthica*. FEMS Microbiol Lett 270:299–305

Lopez-Llorca LV, Gómez-Vidal S, Monfort E, Larriba E, Casado-Vela J, Elortza F, Jansson HB, Salinas J, Martín-Nieto J (2010) Expression of serine proteases in egg-parasitic nematophagous fungi during barley root colonization. Fungal Genet Biol 47:342–351

Lugtenberg BJ, Dekkers (1999) What makes Pseudomonas bacteria rhizosphere competent? Environ Microbiol 1:9–13

Lugtenberg B, Kamilova F (2009) Plant-growth-promoting rhizobacteria. Annu Rev Microbiol 63:541–556

Manter DK, Delgado JA, Holm DG, Stong RA (2010) Pyrosequencing reveals a highly diverse and cultivar-specific bacterial endophyte community in potato roots. Microb Ecol 60:157–166

Mavrodi DV, Joe A, Mavrodi OV, Hassan KA, Weller DM, Paulsen IT, Loper JE, Alfano JR, Thomashow LS (2011) Structural and functional analysis of the type III secretion system from *Pseudomonas fluorescens* Q8r1–96. 193:177–189

Mazurier S, Corberand T, Lemanceau P, Raaijmakers JM (2009) Phenazine antibiotics produced by fluorescent pseudomonads contribute to natural soil suppressiveness to *Fusarium* wilt. ISME J 3:977–991

Mazzola M, Gu YH (2002) Wheat genotype-specific induction of soil microbial communities suppressive to disease incited by *Rhizoctonia solani* anastomosis group (AG)-5 and AG-8. Phytopathology 92:1300–1307

McSpadden Gardener BB (2007) Diversity and ecology of biocontrol *Pseudomonas* spp. in agricultural systems. Phytopathology 97:221–226

Meller-Harel Y, Pasternak Z, Graber ER, Tsechansky L, Rav David D, Silber A, Elad Y, Cytryn E (2011) Impact of biochar application to soil on the root-associated bacterial community structure of fully developed greenhouse pepper plants. Appl Envir Microbiol. 77:4924–4930

Meyer JB, Lutz MP, Frapolli M, Péchy-Tarr M, Rochat L, Keel C, Défago G, Maurhofer M (2010) Interplay between wheat cultivars, biocontrol pseudomonads, and soil. Appl Environ Microbiol 76:6196–6204

Neeraja C, Anil K, Purushotham P, Suma K, Sarma P, Moerschbacher BM, Podile AR (2010) Biotechnological approaches to develop bacterial chitinases as a bioshield against fungal diseases of plants. Crit Rev Biotechnol 30:231–241

Ongena M, Jourdan E, Schäfer M, Kech C, Budzikiewicz H, Luxen A, Thonart P (2005) Isolation of an N-alkylated benzylamine derivative from *Pseudomonas putida* BTP1 as elicitor of induced systemic resistance in bean. Mol Plant Microbe Interact 18:562–569

Penyalver R, Oger PM, Su S, Alvarez B, Salcedo CI, López MM, Farrand SK (2009) The S-adenosyl-L-homocysteine hydrolase gene *ahcY* of *Agrobacterium radiobacter* K84 is required for optimal growth, antibiotic production, and biocontrol of crown gall disease. Mol Plant Microbe Interact 22:713–724

Pieterse CMJ, Leon-Reyes A, Van der Ent S, Van Wees SCM (2009) Networking by small-molecule hormones in plant immunity. Nat Chem Biol 5:308–316

Rudrappa T, Biedrzycki ML, Bais HP (2008) Causes and consequences of plant-associated biofilms. FEMS Microbiol Ecol 64:153–166

Saijo Y (2010) ER quality control of immune receptors and regulators in plants. Cell Microbiol 12:716–724

Schuhegger R, Ihring A, Gantner S, Bahnweg G, Knappe C, Vogg G, Hutzler P, Schmid M, Van Breusegem F, Eberl L, Hartmann A, Langebartels C (2006) Induction of systemic resistance in tomato by N-acyl-L-homoserine lactone-producing rhizosphere bacteria. Plant Cell Environ 29:909–918

Shoresh M, Harman GE, Mastouri F (2010) Induced systemic resistance and plant responses to fungal biocontrol agents. Annu Rev Phytopathol 48:1–23

Smith VA, Hindley J (1978) Effect of agrocin 84 on attachment of *Agrobacterium tumefaciens* to cultured tobacco cells. Nature 276:498–500

Strange RN (2005) Plant disease: a threat to global food security. Annu Rev Phytopathol 43:83–116

Tortora ML, Díaz-Ricci JC, Pedraza RO (2011) *Azospirillum brasilense* siderophores with antifungal activity against *Colletotrichum acutatum*. Arch Microbiol. 193:275–86

van den Bosch F, Gilligan CA (2008) Models of fungicide resistance dynamics. Annu Rev Phytopathol 46:123–147

van der Ent S, Van Wees SC, Pieterse CM (2009) Jasmonate signaling in plant interactions with resistance-inducing beneficial microbes. Phytochemistry 70:1581–1588

van Hulten M, Pelser M, van Loon LC, Pieterse CM, Ton J (2006) Costs and benefits of priming for defense in *Arabidopsis*. Proc Natl Acad Sci USA 103:5602–5607

van Loon LC, Bakker PA, Pieterse CM (1998) Systemic resistance induced by rhizosphere bacteria. Annu Rev Phytopathol 36:453–483

van Loon LC, Bakker PA, van der Heijdt WH, Wendehenne D, Pugin A (2008) Early responses of tobacco suspension cells to rhizobacterial elicitors of induced systemic resistance. Mol Plant Microbe Interact 21:1609–1621

van Wees SC, Van der Ent S, Pieterse CM (2008) Plant immune responses triggered by beneficial microbes. Curr Opin Plant Biol 11:443–448

Weinert N, Piceno Y, Ding GC, Meincke R, Heuer H, Berg G, Schloter M, Andersen G, Smalla K (2011) PhyloChip hybridization uncovered an enormous bacterial diversity in the rhizosphere of different potato cultivars: many common and few cultivar-dependent taxa. FEMS Microbiol Ecol 75:497–506

Weller D (1988) Biological control of soilborne plant pathogens in the rhizosphere with bacteria. Annu Rev Phytopathol 26:379–407

Weller D (2007) *Pseudomonas* biocontrol agents of soilborne pathogens: looking back over 30 years. Phytopathology 97:250–256

Williams P (2007) Quorum sensing, communication and cross-kingdom signalling in the bacterial world. Microbiology 153:3923–3938

Zhang W, Han DY, Dick WA, Davis KR, Hoitink HAJ (1998) Compost and compost water extract-induced systemic acquired resistance in cucumber and Arabidopsis. Phytopathology 88:450–455

Zhu Y, Chen H, Fan J, Wang Y, Li Y, Chen J, Fan J, Yang S, Hu L, Leung H, Mew TW, Teng PS, Wang Z, Mundt CC (2000) Genetic diversity and disease control in rice. Nature 406:718–722

Part III
Coral–Microbe Symbioses

Chapter 9
Bacterial Symbionts of Corals and Symbiodinium

Kim B. Ritchie

Abstract Multipartite symbiosis in corals is an exciting area of research that is not well studied. Research to date indicates that bacterial associates of corals may protect the host by producing antibiotics and other beneficial compounds and nutrients, and are likely to play a role in the stability of the coral animal as a whole. These bacterial mutualists communicate with the host and host-associated microbes to regulate activities on the coral surface. The influence of bacteria in association with the coral algal endosymbiont (*Symbiodinium* spp.) and/or the coral host can be very specific and can involve biochemical interactions that affect the behavior of the algal symbiont. This defines a complex symbiosis between the coral, *Symbiodinium,* and associated bacteria.

9.1 Introduction

The vast majority of metabolic and biochemical diversity on earth resides in microbes, which make-up two domains of life (Bacteria and Archaea) and dominate the third (Eukarya). Multicellular eukaryotes, although more physically prominent, are limited in metabolic potential. It is, consequently, not surprising that many eukaryotes have become dependent on bacterial symbionts, thereby greatly expanding their ecological niche. Long-term symbiosis often results in genetic exchange between partners (Douglas 1994). Eukaryotes, themselves, are the products of a symbiosis between early cells and bacteria that is so intricate that they are no longer considered separate organisms (mitochondria and chloroplasts).

K.B. Ritchie (✉)
Mote Marine Laboratory, Center for Coral Reef Research, 1600 Ken Thompson Parkway, Sarasota, FL 34236, USA
e-mail: ritchie@mote.org

E. Rosenberg and U. Gophna (eds.), *Beneficial Microorganisms in Multicellular Life Forms,* 139
DOI 10.1007/978-3-642-21680-0_9, © Springer-Verlag Berlin Heidelberg 2011

Mutualism has been a prevailing force throughout the evolution of life, with many organisms coevolving interactions into beneficial, even obligate, partnerships. Multicellular organisms acquire processes from prokaryotes that include energy metabolism, defense, and development and these relationships can range from highly codependent "obligate" relationships to more casual or "facultative" symbioses (Dale and Moran 2006).

Most studies of marine microbial symbiosis are based on obvious interactions; from bioluminescent bacterial cultures maintained in squid and fish to photosynthetic dinoflagellates utilized by corals. The majority of microbial-scale symbioses remain obscure. Given that cnidarians are among the oldest metazoans known to form symbiotic relationships, it would be surprising *not* to find associations with bacteria that exhibit specific metabolic capabilities.

Recent research has focused on the nature of coral–*Symbiodinium* relationships, particularly what environmental conditions cause these associations to break down. The single-celled algae, commonly referred to as zooxanthellae, falls within the genus *Symbiodinium*. For hermatypic corals, *Symbiodinium*, the primary algal symbiont, is acquired either maternally by vertical inheritance or horizontally via transmission from the water column. Although the coral/algae symbiosis is an obligate mutualism, this relationship may be temporarily interrupted, or specific endosymbionts shuffled, when conditions are not favorable for either the host or symbiont (Baird et al. 2008). Corals also host a diverse array of other microbes within their tissues and secreted mucus layers (Knowlton and Rohwer 2003). These organisms include complex communities of bacteria, archaea, viruses, and other microbes (Rohwer et al. 2002; Kellogg 2004; Ritchie 2006; Ainsworth, et al. 2010), some of which are thought to be coral species-specific (Ritchie and Smith 1997; Rohwer et al. 2002; Shottner et al. 2009). We are just now beginning to understand the metabolic capabilities of these microbes, their function on the coral surface, and their potential benefit to the coral host or endosymbiotic *Symbiodinium* spp.

9.2 Bacteria Associated with Corals

Much less is known about the nature of coral–bacterial associations. Because the bacterial assemblages on corals are complex and dynamic, it's been challenging to identify trends in diversity on a large scale. A wide range of both culture-based and molecular methods have been used to understand the nature of microbial communities within the coral skeleton, tissues, and surfaces (reviewed in Rosenberg et al. 2007). Research in this vein illustrates that microbial communities are very diverse, complex, and, in some cases, partitioned specifically within coral host structures (Rohwer et al. 2001, 2002). Research also provides evidence that corals may favor specific populations of bacterial associates that are predicted to be mutualistic with the coral host (Ritchie and Smith 1997, 2004; Rohwer et al. 2002; Shottner et al. 2009). Fraune and Bosch (2010) demonstrate that the microbial community of the basal metazoan, *Hydra*, is predominantly dictated by the host.

Yet, little is known about the dynamics of coral–bacterial associations, when these associations are initiated, what precise functions each partner contributes and how specific these partnerships may be for the overall health of the coral/endosymbiont hosts. Symbiotic bacterial associations are likely to be important in coral biology but have been largely overlooked as microbial associates have been difficult to culture and study in appropriate model systems in the laboratory. Here, I examine the evidence that bacteria associated with corals may play an important role in symbioses of the coral holobiont.

9.2.1 Antibiotic-Producing Bacteria

We now know that mucus excreted by corals has antibiotic activity against a range of Gram-positive and Gram-negative bacteria, including coral pathogens and microbes present in the water column. This suggests that corals utilize a biochemical mechanism for disease resistance that may function as a primary defense against pathogens (Geffen and Rosenberg 2005; Ritchie 2006). Mucus collected from corals during a period of increased water temperatures does not show significant antibiotic activity, suggesting that the protective mechanism employed by certain corals may be lost when temperatures increase (Ritchie 2006). This observation provides a mechanism that may explain how increased temperatures lower coral resistance resulting in an increased susceptibility to diseases. However, antimicrobial assays with numerous Red Sea corals reveal high variability within coral species with regard to antibiotic production capabilities (Kelman et al. 2006).

The source of antibiotic activity is difficult to pinpoint as the coral holobiont is made up of numerous organisms, each with the potential to contribute metabolic byproducts. It has been shown that up to 30% of bacteria isolated from some coral species have antibiotic capabilities (Castillo et al. 2001; Ritchie 2006), many displaying broad-spectrum activity against a wide range of bacterial types, including coral pathogens (Ritchie 2006; Shnit-Orland and Kushmaro 2008; Nissimov et al. 2009). In situ, antibiotic production by associated bacteria is known to be a means of securing a niche by controlling microbial populations competing for the same resources (Neilson et al. 2000; Rao et al. 2005). These results collectively support the concept of a probiotic function for coral-associated bacteria. Moreover, it has been shown that there is a shift from antibiotic-producing bacteria to pathogen dominance when temperatures increase (Ritchie 2006). This finding suggests that a balance of potentially beneficial microbes is important in maintaining mutualistic partnerships in corals.

More evidence supporting a mutualistic relationship between the coral host and associated bacteria was demonstrated by using the coral mucus, and associated antibiotic activities, to *select* for antibiotic-producing bacteria and against other types of potentially opportunistic bacteria, or "visiting" bacteria that may be simply trapped in the mucus from water column sources (Ritchie 2006). The ability to isolate a higher number of antibacterial-producing coral isolates using this selection scheme suggests that these bacteria may be true mutualists that provide a service to the coral. Moreover, the ability to select *against* some coral mucus-associated bacteria provides evidence that the compounds present within the coral surface mucus are able to regulate the composition and activities of surface bacterial associates (Ritchie 2006). Since the coral system as a whole (coral, zooxanthellae, and associated bacteria) cannot be easily separated for experimentation, it is difficult to show whether these bacteria produce the total antibiotic effect found on the coral surface in situ, or whether there are multiple contributors to this innate immune defense. The specific source of the coral surface's antibacterial activity is unknown and could be one or a combination of allelopathic chemicals produced by the coral, bacteria, or endosymbiotic zooxanthellae. A study conducted by Marquis et al. 2005 showed that, out of eggs from 11 coral species tested, the only species that exhibited antibiotic activity was also the only coral species that incorporates *Symbiodinium* into the egg before being released. This finding suggests a potential allelopathic contribution of *Symbiodinium* and/or *Symbiodinium*-associated bacteria (see Sect. 9.3).

9.2.2 Bacteria and Early Life Stages of Corals

The precise initiation of the coral–bacterial association potentially depends on the coral reproductive strategy, as potential bacterial partners could be transmitted *vertically*, during reproduction, and/or *horizontally* via exposure to the water column during any stage of the host life cycle. Certainly, corals interact with bacteria in these early developmental stages. Apprill et al. (2009) showed in a study with the Hawaiian coral *Pocillopora meandrina* that the bacterial community composition in larvae appears to vary depending on the coral developmental stage and that many of these bacterial associates belong to the *Roseobacter* clade of α-proteobacteria (see Sect. 9.3). Fluorescence in situ hybridization with *Roseobacter*-specific probes provides evidence that bacteria are acquired horizontally early in development (as soon as 79 h after fertilization) with preferential uptake of members of the *Roseobacter* clade (Apprill et al. 2009). Using Fluorescence in situ hybridization with general bacterial probes, Sharp et al. (2010) showed that vertical transmission does not occur in a wide range of mass-spawning corals and the larvae do not acquire detectable numbers of bacteria until after larval settlement and metamorphosis. Although it is not yet determined what types of bacteria colonize corals during those early developmental stages, studies on interactions between coral larvae and bacteria have shown that bacteria can influence coral settlement success. *Pseudoalteromonas* sp. isolated from crustose coralline algae (CCA) was shown to induce coral larval metamorphosis and settlement (Negri et al. 2001), demonstrating that bacteria may play a role in development and coral recruitment (Negri et al. 2001; Webster et al. 2004; Ritson-Williams et al. 2010).

9.2.3 Quorum Sensing and Disease Defense

Single-celled bacteria use a sophisticated cell–cell communication system that involves chemical signals. This process, called quorum sensing (QS) is moderated by small diffusible chemical compounds called autoinducers, which, when accumulated to a certain threshold concentration within a diffusion-limited environment, result in synchronized group behaviors in bacteria. These behaviors include antibiotic production, bioluminescence, and pathogenic mechanisms, to list a few. QS allows bacterial populations to act in unison, thereby magnifying their ecological impact. There are at least a dozen known classes of autoinducer molecules (reviewed in Ng and Bassler 2009). Even though this cell–cell communication system may differ among species, QS drives important interactions between symbiotic bacterial communities and their hosts (Rassmussen and Givskov 2006; Dobretsov et al. 2009). QS in bacterial pathogens controls expression of virulence genes during interactions with the host by initiating a coordinated attack once bacterial cell numbers get to a critical mass (Dobretsov et al. 2009). Both eukaryotes and prokaryotes have evolved to recognize and counter QS in pathogens, there is also evidence that eukaryotic signal-mimics stimulate QS responses in bacteria (Teplitski et al. 2011). Other bacteria can counter attack by producing quorum *quenching* acylases or lactonases that break down signaling molecules (reviewed in Teplitski et al. 2011). In addition to the AHL-degrading enzymes, eukaryotes can inhibit or activate bacterial QS by producing compounds that mimic QS signals. For example, Rajamani et al. (2008) demonstrated that lumichrome (a derivative of the vitamin riboflavin) produced by a unicellular alga *Chlamydomonas reinhardtii* (as well as other prokaryotes and eukaryotes) can interact with the bacterial receptor for QS signals and elicit QS responses.

Coral extracts contain compounds capable of interfering with QS activities (Skindersoe et al. 2008; Alagely et al. 2011). The source of this activity is difficult to pinpoint and could originate from the coral, the dominant endosymbiont or any associated microbes. Alagely et al. (2011) recently showed that both coral and *Symbiodinium*-associated bacteria can alter swarming and biofilm formation in the coral pathogen *Serratia marcescens*. These phenotypes are typically controlled by QS, although inhibition of QS by these isolates remains to be demonstrated. There are few studies on the in situ roles of QS in corals but this process is likely to be used in both pathogenesis and mutualistic interactions (Teplitski and Ritchie 2009; Krediet et al. 2009a, b; Tait et al. 2010). While it is clear that at least some coral-associated commensals and pathogens produce QS signals under laboratory conditions (Tait 2010; Alagely et al. 2011), it is not clear whether these signals accumulate to the threshold concentrations in natural environments. QS may inhibit or activate pathogenesis, antibiotic production, exoenzyme production, and attachment by beneficial bacteria within coral tissues and on surfaces. It is feasible that *Symbiodinium* spp. also produce signaling molecules that control bacterial cell–cell communication, which would contribute to the specific complement of bacteria that is associated with corals. Perhaps bacterial species-specificity in corals is at least, in part, driven by the particular complement of *Symbiodinium* clades with a given coral species.

9.3 Symbiodinium–Bacteria Associations

Symbiodinium spp. are able to establish mutualisms with numerous hosts, including Cnidaria, Molluscs, Porifera, and other protists (Coffroth and Santos 2005). Due to genome complexity, the diversity of *Symbiodinium* spp. has been difficult to assess and is currently classified into large groups, called clades. Based on genome variation these clades appear to be evolutionarily distinct, have different host niches, and can be functionally diverse. Although much work has been directed toward *Symbiodinium*–coral mutualism, little is known about the nature of free-living *Symbiodinium*, including what bacterial mutualisms may be present before coral acquisition of *Symbiodinium,* in the case that the algal symbiont is not transmitted vertically.

Most dinoflagellates are difficult to grow in pure culture free from associated bacteria (Dixon and Syrett 1988). Antibiotic treatments, in combination with other physical methods, to remove contaminating bacteria have historically produced inconsistent results (Anderson and Kawachi 2005). One possibility for this phenomenon is that bacteria present in dinoflagellate cultures provide some necessary component(s) for the successful growth of the dinoflagellate outside of the host.

Bacteria present in cultures of *Symbiodinium* spp. (representing six clades) have been characterized via both culture-based and culture-independent methods (Ritchie et al. In Prep). Members of the Roseobacteriales group are specifically present in association with *Symbiodinium* cultures (Table 9.1). These results are striking in that *Symbiodinium* cultures were derived from different corals and coral reef invertebrates from different oceans (Table 9.2), suggesting that these bacteria are true mutualists of *Symbiodinium* spp. In addition, members of this group are shown to increase growth rates of *Symbiodinium* (Fig. 9.1). Another type of dinoflagellate (*Pfisteria*) was observed to be unable to grow and eventually died in axenic culture. However, addition of the co-occurring α-Proteobacteria strain restored normal growth (Alavi et al. 2001).

A wide variety of bacteria have been observed in close association with corals, but of particular interest are those belonging to the phylum α-proteobacteria (Apprill et al. 2009; Raina et al. 2009). The α-proteobacteria (particularly the Roseobacteriales) are abundant in the oceans, often comprising one-third of the bacterioplankton (Wagner-Dobler and Biebl 2006). This same group of bacteria is also closely associated with phytoplankton (Webster et al. 2004), including the

Table 9.1 Illustrates major bacterial groups associated with *Symbiodinium* cultures verified using culture-based methods, PCR-DGGE, T-RFLP analysis, and 16S rDNA clone library production

Bacteria	Symbiodinium clades
Roseobacter	A1, B1, C1, D1a, D2, E1, F2
CFB group[a]	A1, B1, C1, D2, F2
Marinobacter	A1, B1, E1, D1a, F2
Gram-positive bacteria	A1, C1, F2

[a]Cytophaga-Flavobacterium-Bacteroides (CFB)

Table 9.2 Zooxanthellae clade derivations

Culture number	Clade	Host species Common name (*Genus species*)	Location	Originally isolated by
61	A1	Jellyfish (*Cassiopeia xamachana*)	Caribbean/Florida	Trench (LeJeunesse)
KB8	A1	*Cassiopeia* sp.	Hawaii/Oahu	Kinzie (Santos)
147	B1	*Gorgonian (Pseudopterogorgia bipinnata*)	Caribbean/Jamaica	Trench (LeJeunesse)
MF11	B1	*Montastraea faveolata*	Caribbean/Fl or Panama	Coffroth (Santos)
152	C1	Corallimorph (*Rhodactis [Heteractis] lucida*)	Caribbean/Jamaica	Trench (LeJeunesse)
A001	D1a	Scleractinian (*Acropora sp.*)	NW Pacific/Okinawa	Hidaka (LeJeunesse)
401	D2	Forams	Red Sea	Trench (LeJeunesse)
383	E1	Actinarian (*Anthopleura elegantissima*)	East Pacific/California	Trench (LeJeunesse)
CCMP 421	E1	Free living *Symbiodinium* spp.	New Zealand/Wellington Harbor	Chang (Santos)
Mv	F1	Scleractinian (non rep) *Montipora capulata*	Hawaii/Oahu	Kinzie (LeJeunesse)
133	F2	Scleractinian (*Meandrina meandrites*)	Caribbean/Jamaica	Trench (LeJeunesse)

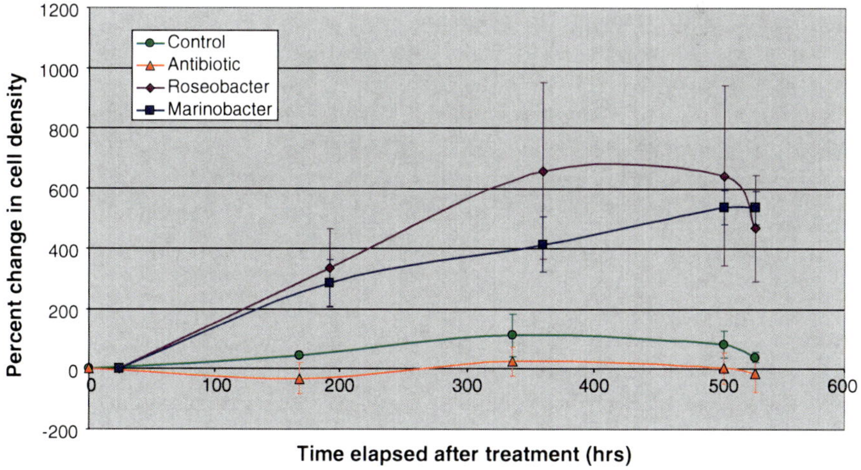

Fig. 9.1 Growth curve (Clade A1). In growth experiments, addition of *Symbiodinium*-associated bacteria, *Roseobacter* and *Marinobacter*, resulted in faster doubling time of *Symbiodinium* cultures as compared to control or antibiotic-treated (bacterial controlled) cells

dinoflagellate coral endosymbiont *Symbiodinium* (Ritchie et al. in prep.) and are known to be involved in dimethylsulfonioproprionate metabolism in the marine environment (González et al. 2003). Many of these bacteria, now classified as *Ruegeria* spp., were originally designated *Silicibacter* spp. (Yi et al. 2007).

Are these true mutualisms? Of interest is the observation that many obligate bacterial symbionts are so intimately codependent that the genome size is reduced in a process called degenerative minimalism. Some of these are incredibly small. For example, the aphid obligate symbiont *Buchnera* sp., is only 160 kb in size, having lost genes that are no longer necessary, including regulatory, cell envelope components, DNA repair, and genes involved in pathways that are redundant with the host (Nikoh et al. 2010). For this reason, both the host and the symbiont have become irreversibly dependent on each other. Also of interest is the observation that the mostly obligate coral algal endosymbiont, *Symbiodinium* sp., has one of the smaller genomes among dinoflagellates (LaJeunesse et al. 2005), suggesting that it could have lost necessary components, rendering it more likely to coevolve a semi-dependent relationship with a host organism.

The observed association between α-proteobacteria and dinoflagellates may be a true symbiosis with benefits for both the bacteria and the algal host. The bacteria may benefit by having a readily available source of organic compounds such as dimethylsulfoniopropionate (DMSP), a preferred source of reduced sulfur (Miller and Belas 2004; Raina et al. 2009). The algae may derive benefits from the bacterial production of antimicrobials such as tropodithietic acid (TDA; Geng and Belas 2010) and bioactive compounds, such as vitamin B-12 (Croft et al. 2005). A genomic comparison of the *Roseobacter* clade of α-Proteobacteria indicates that some type of surface-associated lifestyle is central to the ecology of all members of the group (Slightom and Buchan 2009).

Interestingly, members of the α-proteobacteria are resistant to antimicrobial activities of some sponge hosts (Kelman et al. 2009). Other Roseobacters have been shown to produce broad-spectrum antibiotics active against a wide range of Gram-positive and Gram-negative bacteria, including coral pathogens, supporting the idea of a protective function for these bacterial associates against invading pathogens (Nissimov et al. 2009). α-proteobacteria have also been found in association with many coral species. An important adaptive property of the α-proteobacteria is the presence of a bacterial system for diversity generation called Gene Transfer Agents (GTAs; Paul 2008). GTAs are defective bacteriophages that are able to randomly package bacterial host DNA and transfer DNA to other α-proteobacteria (Paul 2008). It has recently been shown that *Symbiodinium*-associated α-proteobacteria produce GTAs and are able to transfer genes to a range of bacteria in the marine environment (McDaniel et al. 2010). Furthermore, it was shown that gene transfer via this mechanism is much higher in the coral reef environment, suggesting an alternate mode of adaptation via swapping of potentially beneficial genes between marine bacteria (McDaniel et al. 2010) and possibly the coral holobiont.

Other mutualistic services provided by coral-associated microbes may include nitrogen-fixation by cyanobacteria and other bacteria located within the coral epithelium (Lesser et al. 2004; Olsen et al. 2009). The abundance of many nitrogen-fixing bacteria in corals is positively correlated with the abundance of endosymbiotic *Symbiodinium* suggesting a physiological partnership between *Symbiodinium* and associated bacteria (Olsen et al. 2009).

9.4 Unresolved Questions and Future Research

Attempts to characterize symbiotic bacteria in corals reveal a range of associations that may be beneficial to the coral, the endosymbiotic algae, and some associated bacteria. In many cases, the distinction between mutualism and pathogenesis may be blurred. Parameters that result in shifts associated with a specific bacterium may be beneficial under one set of conditions but harmful under another set. Are there specific "primary" symbionts (specialized within the host tissue and required by the host) in coral or *Symbiodinium* biology or are bacterial mutualisms at best "secondary" (facultative symbionts that do not reside exclusively in specialized host tissues and not strictly required for host survival)? In the case that most bacterial mutualisms in coral biology are secondary, is there functional redundancy in diverse secondary symbionts and how many different functional roles are required for the coral and or *Symbiodinium* host?

Some additional unresolved questions remain: (1) what is the nature of multi-partite partnerships and what are the precise services provided? (2) What is the evolutionary origin of multipartite mutualisms and at what level do microbes, viruses, and virus-like particles drive evolution that may influence coadaptation of bacteria, dinoflagellates, and corals? (3) What are the levels of mutualism within a coral holobiont (from obligate to facultative mutualisms)? (4) What are the roles of cell-to-cell communication between coral, *Symbiodinium*., and bacteria? (5) How, and at what stage, are bacterial symbionts acquired?

Model systems are fundamentally required to address specific interactions. Much work has been done on the basal metazoan *Hydra* to illustrate the value of a model systems approach (Bosch et al. 2009; See Chap. 23). Because *Hydra* is associated with a limited number of bacteria it has provided valuable insight into immunity and symbioses in simple animals. Cnidarian and dinoflagellate models could also be used to elucidate roles of bacteria in both coral and *Symbiodinium* biology. Ideally, these models would require cultured symbionts (bacterial and algal) and an easily maintained Cnidarian host (Weise et al. 2008). Our ability to culture many of these bacterial symbionts will aid in exploration of functions that are otherwise impossible to study due to the complex nature of the coral holobiont. Genome sequence data generation from model organisms will exponentially enhance our basic understanding of symbiotic associations at the molecular level. This includes reconstruction of host–symbiont phylogenies, analysis of genes important in specific interactions, comparative genomics, and advanced technologies.

Acknowledgments I gratefully acknowledge preliminary experimental input from Mote Marine Laboratory student interns Patrick Hutchins, Courtney Kiel, Stephanie Thornton, Roxanna Myers, Carmel Norman, Andrew Collingwood, and Jamie Schell. Todd LaJeunesse and Scott Santos kindly provided *Symbiodinium* culture information and strains. I thank Koty Sharp, Garriet Smith, John H. Paul, Max Teplitski, John Pringle, Eugene Rosenberg, and Joel Thurmond for their helpful discussions and/or input on this manuscript. Funding for this contribution was provided by Florida Protect Our Reefs Plate Grants and the Dart Foundation.

References

Ainsworth TD, Thurber RV, Gates RD (2010) The future of coral reefs: a microbial perspective. Trends Ecol Evol 25(4):233–240

Alagely A, Krediet CJ, Ritchie KB, Teplitski M (2011) Signaling-mediated cross-talk modulates swarming and biofilm formation in a coral pathogen *Serratia marcescens*. ISME J. doi:10.1038/ismej.2011.45

Alavi M, Miller T, Erlandson K, Schneider R, Belas R (2001) Bacterial community associated with Pfiesteria-like dinoflagellate cultures. Environ Microbiol 3(6):380–396

Anderson RA, Kawachi M (2005) Traditional microalgae isolation techniques. In: Anderson RA (ed) Algal culturing techniques. Elsevier, New York, pp 124–125

Apprill A, Marlow HQ, Martindale MQ, Rappe MS (2009) The onset of microbial associations in the coral *Pocillopora meandrina*. ISME J 3:685–699

Baird AH, Bhagooli R, Ralph PJ, Takahashi S (2008) Coral bleaching: the role of the host. Trends Ecol Evol 24(1):16–20

Bosch TCG, Augustin R, Anton-Erxleben F, Fraune S, Hemmrich G et al (2009) Uncovering the evolutionary history of innate immunity: the simple metazoan Hydra uses epithelial cells for host defence. Dev Comp Immunol 33:559–569

Castillo I, Lodeiros C, Nunez M, Campos I (2001) *In vitro* evaluation of antibacterial substances produced by bacteria isolated from different marine organisms. Rev Biol Trop 49 (3–4):1213–1222

Coffroth MA, Santos SR (2005) Genetic diversity of symbiotic dinoflagellates in the genus *Symbiodinium*. Protist 156:19–34

Croft MT, Lawrence AD, Raux-Deery E, Warren MJ, Smith AG (2005) Algae acquire vitamin B12 through a symbiotic relationship with bacteria. Nature 438:90–93

Dale C, Moran NA (2006) Molecular interactions between bacterial symbionts and their hosts. Cell 126(3):453–465

Dixon GK, Syrett PJ (1988) The growth of dinoflagellates in laboratory cultures. New Phytol 109:297–302

Dobretsov S, Teplitski M, Paul V (2009) Mini-review: quorum sensing in the marine environment and its relationship to biofouling. Biofouling 25:413–427

Douglas AE (1994) Symbiotic interactions. Oxford University Press, Oxford

Fraune S, Bosch TCG (2010) Why bacteria matter in animal development and evolution. Bioessays 32:571–580

Geffen Y, Rosenberg E (2005) Stress-induced rapid release of antibacterials by scleractinian corals. Mar Biol 146:931–935

Geng HF, Belas R (2010) Molecular mechanisms underlying roseobacter-phytoplankton symbioses. Curr Opin Biotechnol 21(3):332–338

González JM, Covert JS, Whitman WB, Henriksen JR, Mayer F, Scharf B, Schmitt R, Buchan A, Fuhrman JA, Kiene RP, Moran MA (2003) Silicibacter pomeroyi sp. nov. and Roseovarius nubinhibens sp. nov., dimethylsulfoniopropionate-demethylating bacteria from marine environments. Int J Syst Evol Microbiol 53:1261–1269

Kellogg CA (2004) Tropical Archaea: diversity associated with the surface microlayer of corals. Mar Ecol Prog Ser 273:81–88

Kelman D, Kashman K, Rosenberg E, Kushmaro A, Loya Y (2006) Antimicrobial activity of Red Sea corals. Mar Biol 149:357–363

Kelman D, Kashman K, Hill RT, Rosenberg E, Loya Y (2009) Chemical warfare in the sea: the search for antibiotics from Red Sea Corals and sponges. Pure Appl Chem 81(6):1113–1121

Knowlton N, Rohwer F (2003) Multispecies microbial mutualisms on coral reefs: the host as a habitat. Am Nat 162:S51–S62

Krediet C, Teplitski M, Ritchie KB (2009a) Catabolite control of enzyme induction and biofilm formation in a coral pathogen *Serratia marcescens* PDL100. Dis Aquat Org 87:57–66

Krediet CJ, Ritchie KB, Cohen M, Lipp E, Sutherland K, Teplitski M (2009b) Utilization of mucus from the coral *Acropora palmata* by environmental and pathogenic isolates of *Serratia marcescens*. Appl Environ Microbiol 75(12):3851–3858

LaJeunesse TC, Lambert G, Anderson RA, Coffroth MA, Galbraith DW (2005) *Symbiodinium* (Pyrrhophyta) genome sizes (DNA content) are smallest among dinoflagellates. J Phycol 41:880–886

Lesser MP, Mazel CH, Gorbunov MY, Falkowski PG (2004) Discovery of symbiotic nitrogen-fixing cyanobacteria in corals. Science 305:997–1000

Marquis CP, Baird AH, de Nys R, Holmstrom C, Koziumi N (2005) An evaluation of the antimicrobial properties of the eggs of 11 species of scleractinian corals. Coral Reefs 24:248–253

McDaniel LE, Young E, Delaney J, Ruhnau F, Ritchie KB, Paul JH (2010) High frequency of horizontal gene transfer in the oceans. Science 330:50

Miller TR, Belas R (2004) Dimethylsulfoniopropionate metabolism by Pfiesteria-associated Roseobacter spp. Appl Environ Microbiol 70(6):3383–3391

Negri AP, Webster NS, Hill RT, Heyward AJ (2001) Metamorphosis of broadcast spawning corals in response to bacteria isolated from crustose coraline algae. Mar Ecol Prog Ser 223:121–131

Neilson AT, Tolker-Neilsen K, Barken K, Molin S (2000) Role of commensal relationships on the spatial structure of a surface-attached microbial consortium. Environ Microbiol 2:59–68

Ng WL, Bassler BL (2009) Bacterial quorum-sensing network architectures. Annu Rev Genet 43:197–222

Nikoh N, McCutcheon JP, Kudo T, Miyagishima S, Moraln NA, Nakabachi A (2010) Bacterial genes in the aphid genome: absence of functional gene transfer from Buchnera to its host. PLoS Genetics 6(2):e1000827. doi:10.1371/journal.pgen.1000827

Nissimov J, Rosenberg E, Munn CB (2009) Antimicrobial properties of resident coral mucus bacteria of *Oculina patagonica*. FEMS Microbiol Lett 292:210–215

Olsen ND, Ainsworth TD, Gates RD, Takabayashi M (2009) Diazotrophic bacteria associated with Hawaiian Montipora corals: diversity and abundance in correlation with symbiotic dinoflagellates. J Exp Mar Biol Ecol 371:140–146

Paul JH (2008) Prophages in marine bacteria: dangerous molecular time bombs or the key to survival in the seas? ISME J 2:579–589

Raina JB, Tapiolas D, Willis BL, Bourne DG (2009) Coral-associated bacteria and their role in the biogeochemical cycling of sulfur. Appl Environ Microbiol 75(11):3492–3501

Rajamani S, Bauer WD, Robinson JB, Farrow JM 3rd, Pesci EC, Teplitski M, Gao M, Sayre RT, Phillips DA (2008) The vitamin riboflavin and its derivative lumichrome activate the LasR bacterial quorum-sensing receptor. Mol Plant Microbe Interact 21(9):1184–1192

Rao D, Webb JS, Kjelleberg S (2005) Competitive interactions in mixed-species biofilms containing the marine bacterium *Pseudoalteromonas tunicata*. Appl Environ Microbiol 71:1729–1736

Rassmussen TB, Givskov M (2006) Quorum sensing inhibitors: a bargain of effects. Microbiology 152:895–904

Ritchie K (2006) Regulation of marine microbes by coral mucus and mucus-associated bacteria. Mar Ecol Prog Ser 322:1–14

Ritchie KB Smith, GW (1997) Physiological comparisons of bacteria from various species of scleractinian corals. In: Proceedings of the 8th international symposium for reef studies, vol 1, Panama, pp 521–526
Ritchie KB, Smith GW (2004) Microbial communities of coral surface mucopolysaccharide layers. In: Rosenberg, Loya (eds) Coral health and disease. Springer, Berlin, pp 259–263
Ritchie KB, Kiel C, Thornton S, Collingwood A, Hutchins P, Schell J, Myers R, Norman C, Thurmond J (In Prep) Identification of bacterial groups associated with Symbiodinium cultures
Ritson-Williams, Raphael, Paul, Valerie J, Arnold SN, Steneck RS (2010) Larval settlement preferences and post-settlement survival of the threatened Caribbean corals *Acropora palmata* and *A.cervicornis*. Coral Reefs 29(1):71–81
Rohwer F, Breitbart M, Jara J, Azam F, Knowlton N (2001) Diversity of bacteria associated with the Caribbean coral *Montastraea franksii*. Coral Reefs 20:85–91
Rohwer F, Seguritan V, Azam F, Knowlton N (2002) Diversity and distribution of coral-associated bacteria. Mar Ecol Prog Ser 243:1–10
Rosenberg E, Koren O, Reshef L, Efrony R, Zilber-Rosenberg I (2007) The role of microbes in coral health, disease and evolution. Nat Rev Microbiol 5:355–362
Sharp KH, Ritchie KB, Schupp P, Ritson-Williams R, Paul VJ (2010) Bacterial acquisition by gametes and juveniles from several broadcast spawning coral species. PLoS ONE 5(5):e10898. doi:10.1371/journal.pone.0010898
Shnit-Orland M, Kushmaro A (2008) Coral mucus bacteria as a source for antibacterial activity. In: Proceedings of the 11th international coral reef symposium, Lauderdale, FL, pp 257–258
Shottner S, Hoffman F, Wild C, Rapp HT, Boetius A, Rametter A (2009) Inter- and intra-habitat bacterial diversity associated with cold-water corals. ISME J 3:756–759
Skindersoe ME, Alhede M, Phipps R, Yang L, Jensen PO, Rasmussen TB, Bjarnsholt T, Tolker-Nielsen T, Hoiby N, Givskov M (2008) Effects of antibiotics on quorum sensing in Psuedomonas aeruginosa. Antimicrob Agent Chemother 52(10):3648–3663
Slightom RN, Buchan A (2009) Surface colonization by marine Roseobacters: integrating genotype and phenotype. Appl Environ Microbiol 75(19):6027–6037
Tait K, Hutchison Z, Thompson FL, Munn CB (2010) Quorum sensing signal production and inhibition by coral-associated vibrios. Environ Microbiol Rep 2(1):145–150
Teplitski M, Ritchie KB (2009) How feasible is the biological control of coral disease? Trend Ecol Evol 24(7):378–385
Teplitski M, Mathesius U, Rumbaugh KP (2011) Perception and degradation of N-acyl homoserine lactone quorum sensing signals by mammalian and plant cells. Chem Rev 111 (1):100–116
Wagner-Dobler I, Biebl H (2006) Environmental biology of the marine Roseobacter lineage. Annu Rev Microbiol 60:255–280
Webster NS, Smith LD et al (2004) Metamorphosis of a scleractinian coral in response to microbial biofilms. Appl Environ Microbiol 70(2):1213–1221
Weise VM, Davy SK, Hoegh-Guldberg O, Rodriguez-Lanetty M, Pringle JR (2008) Cell biology in model systems as the key to understanding corals. Trend Ecol Evol 23(7):369–376
Yi H, Woon Lim Y, Chun J (2007) Taxonomic evaluation of the genera *Ruegeria* and *Silicibacter*: a proposal to transfer the genus *Silicibacter* Petursdottir and Kristjansson 1999 to the genus *Ruegeria* Uchino et al. 1999. Int J Syst Evol Microbiol 57:815–819

Chapter 10
Coral-Associated Heterotrophic Protists

L. Arotsker, E. Kramarsky-Winter, and A. Kushmaro

Abstract Protists are microscopic eukaryotic microorganisms that are ubiquitous, diverse, and major participants in oceanic food webs and in marine biogeochemical cycles. A survey of protist abundance in waters of coral reef environments was determined to be between $3.5{\cdot}10^3$ and $7.9{\cdot}10^3$ protists ml^{-1}. Recent studies showed that live corals harbor Stramenopile protists in and on their tissues. Analyses of large polyped coral species from the Gulf of Eilat (Northern Red Sea) and the Great Barrier Reef revealed numerous colonies with distinct white coatings covering their surface. Upon closer examination, this coating was found to be made up of numerous morphologically distinct microorganisms, containing a nucleus, mitochondria, and golgi complexes. These microorganisms were then characterized using molecular methods and identified as stramenopile protists belonging to the order Labyrinthulida (Kramarsky-Winter et al., 2006); family Thraustochytriidae. Thraustochytrids are a ubiquitous group of microorganisms found in association with marine invertebrates from sponges to echinoderms. One of the distinctive characters of this group is that almost all species develop ectoplasmic extensions from one or more points on the cell and form branched networks. These ectoplasmic nets provide mobility and contain hydrolytic enzymes that are surface-bound or are

L. Arotsker
Avram and Stella Goldstein-Goren Department of Biotechnology Engineering, Ben-Gurion University of the Negev, P.O. Box 653, Be'er-Sheva 84105, Israel

E. Kramarsky-Winter
Department of Zoology and the Porter Super Center for Ecological and Environmental Studies, Tel Aviv University, Ramat Aviv, Tel Aviv, Israel

A. Kushmaro (✉)
Avram and Stella Goldstein-Goren Department of Biotechnology Engineering, Ben-Gurion University of the Negev, P.O. Box 653, Be'er-Sheva 84105, Israel

National Institute of Biotechnology in the Negev, Ben-Gurion University of the Negev, P.O. Box 653, Be'er-Sheva 84105, Israel
e-mail: arielkus@bgu.ac.il

E. Rosenberg and U. Gophna (eds.), *Beneficial Microorganisms in Multicellular Life Forms*, 151
DOI 10.1007/978-3-642-21680-0_10,

secreted into the surrounding medium, helping in the digestion of organic material. In addition, members of this group of organisms are known to produce polyunsaturated fatty acids. Therefore, the presence of these microorganisms on the surface of massive and solitary coral species during bleaching events may explain why these corals survive bleaching better than branched species.

10.1 Heterotrophic Protists in the Marine Environment

Protists are microscopic eukaryotes that are ubiquitous, and diverse, and are major participants in oceanic food webs and in marine biogeochemical cycles (Sherr et al. 2007). Stramenopile protists are common protists found in the marine environment in sediment, and on and in marine organisms (Moss 1986; Raghukumar 2002). In a survey of four coral atolls in the Northern Line Islands (central Pacific), protist abundance in the sea water was determined using DAPI (Sigma Aldrich): 4′,6-Diamidino-2-phenylindole dihydrochloride staining with epifluorescence microscopy and metagenomic analysis of ~20 m^2 of the coral reef water (Dinsdale et al. 2008). Protist abundance doubled from Kingman and Palmyra atolls (3,575 $\pm$ 457.1 and 3,486 $\pm$ 275.4 protists ml^{-1}, respectively) to Tabuaeran and Kiritimati atolls (7,917 $\pm$ 2,037.1 and 7,124 $\pm$ 868.1 protists ml^{-1}, respectively) ($P < 0.001$), while protists/microbe ratio declined. On Kingman, 66% of the protists were strict heterotrophs (i.e., contained no chlorophyll) compared with 22% on Kiritimati. Increasing atoll size and oceanographically more oligotrophic water were directly correlated with significant increases in protists (Dinsdale et al. 2008). Another study on microeukaryotes in extreme aquatic environment of Mariager Fjord estimated a high richness (568 $\pm$ 114) of protist phylotypes (Zuendorf et al. 2006). Approximately 80% of the sequences analyzed in this study were identified as protist targets, majority of which seem to belong to strict or facultative anaerobe organisms, while 41% belong to Alveolates and 28% to Stramenopiles.

The Thraustochytrids are widely distributed marine stramenopile heterotrophs (saprophytic or occasionally parasitic) that are believed to have major impacts on the marine environment (Raghukumar 1987; Raghukumar et al. 1994, 2000; Honda et al. 1998; Kimura et al. 2001; Fan et al. 2002; Anderson et al. 2003; Kvingedal et al. 2006). Thraustochytrids have been demonstrated not only on the surface of marine organisms but also inside marine invertebrate tissues such as sponges and corals (Rinkevich 1999), echinoderms (Wagner–Merner et al. 1980; Thorsen 1999), and hydroids (Raghukumar 1988). Indeed, cell lines of the scleractinian coral *Porites lutea* (Frank et al. 1994), the sponges *Latrunculia magnifica* and *Negombata magnifica* (Ilan et al. 1996; Rinkevich et al. 1998), tunicates, and bivalves (Ellis and Bishop 1989) were all overgrown by thraustochytrid cells that were present undetected and intermixed with *in vitro* cell cultures. Moreover, of the best known marine protists are the photosynthetic symbiotic protists, found in all reef-building corals – the dinoflagellates of the genus *Symbiodinium*, also known as "zooxanthellae" (reviewed in Knowlton and Rohwer 2003).

10.2 Coral Associated Heterotrophic Protists

In recent years, corals have come to be considered as holobionts, an integration of a wide variety of symbiotic organisms, including the coral host and its associated protists, together with microbial and archaeal consortia. All of these organisms have unique and complementary roles in coral ecosystem (Rohwer et al. 2001, 2002; Rohwer and Kelly 2004; Zilber–Rosenberg and Rosenberg 2008). Of these associations, those with Thraustochytrids have recently come to light. Thraustochytrids are a family in the order Labyrinthulida, the Labyrinthulomycetes class (Raghukumar 2002; Leander et al. 2004), that are considered to be part of the Stramenopiles phylum (Patterson 1999). Early studies on Thraustochytrids in corals revealed presence of *Corallochytrium limacisporum* in scleractinian corals, of the genus *Acropora* sp. and *Porites* sp. (Raghukumar 1987; Raghukumar and Balasubramanian 1991). *C. limacisporum* cell counts in *Acropora secale* and *Porites* sp. were low, and an unidentified *Acropora* sp. revealed numerous thraustochytrid cells using immunofluorescence, 3.5–4.5 μm in diameter (Raghukumar and Balasubramanian 1991). In the same study, the authors successfully isolated thraustochytrids from *Pocillopora* sp. and several *Acropora* spp., including the *Acropora variabilis* and *A. arbuscula* from several locations in the Arabian Sea. These colonies harbored numerous thraustochytrids, including *C. limacisporum*, *Labyrinthuloides minula*, and *Thraustochytrium motivum*.

An additional association has been recently highlighted to occur in corals in the Gulf of Eilat. These studies showed distinct white coatings (Fig. 10.1) covering the surface of a number of large polyped scleractinian corals (Kramarsky–Winter et al. 2006; Harel et al. 2008; Siboni et al. 2010) occurring in shallow waters. Upon further investigation, this phenomenon was found to be geographically widespread (see Veron 2000; Johnston and Rohwer 2007; Siboni et al. 2010). Siboni et al. (2010)

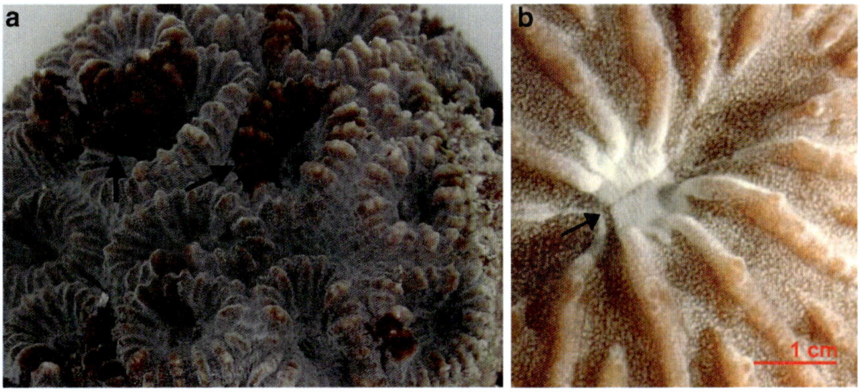

Fig. 10.1 Corals with aggregate coating: (**a**) *Favia* sp. coral from the GBR (Siboni et al. 2010); (**b**) *Fungia granulosa*, close up of mouth area with high concentration of aggregates (*arrow*) (Kramarsky–Winter et al. 2006)

showed that its prevalence on colonies of the coral *Favia* sp. in the Red Sea averaged 55% during 2005–2006, with an appearance that is inversely proportional to the ambient water temperatures. This coating was found to be comprised of aggregate-like microorganisms, sized between 5 and 30 μm and made up of numerous 1 μm coccoid bodies, patchily embedded in the coral mucus and tissue layers (Kramarsky–Winter et al. 2006). The coral mucus layer provides the coral host with protection as well as with the ability to ensnare particulate food (Schlichter and Brendelberger 1998; Goldberg 2002). It can also act as a growth medium for microorganisms supporting growth of numerous heterotrophic microorganisms. Transmission electron microscopy (TEM) micrographs of the mucus of these corals demonstrated numerous microorganisms, including colonies of microorganisms with a morphology that is typical of eukaryotic cell, comprising a nucleus, mitochondria, and golgi complexes (Kramarsky–Winter et al. 2006). Furthermore, molecular investigation of the make-up of the coral mucus flora indicated the presence of stramenopile protists. Indeed, 90% of the sequences from 18S rRNA clone libraries of the eukaryotic samples collected from the coral surface mucus were found to be stramenopiles of the family Thraustochytriidae (with a predominance of *Aplanochytrium* sp., *Thraustochytrium* sp., and *Labyrinthuloides* sp.), while other families, such as the Labyrinthulidae, Licmophoraceae, Naviculaceae, and Bacillariophyceae, were present at lower abundances (Siboni et al. 2010). TEM investigation showed that the protists found on *Favia* species from the GBR and from the Red Sea, and those from the fungiid coral *Fungia granulosa* from the Red Sea were morphologically similar (Kramarsky–Winter et al. 2006; Siboni et al. 2010). Although the method of acquisition of these symbionts is still largely unknown, Siboni et al. (2010) showed that in the Red Sea *Favia* sp., these protists are not found in the larvae and thus not transferred from the parent but rather are acquired from the environment once the coral larvae settle and metamorphose into polyps.

Interestingly, differences in density of stramenopiles were noted in the mucus coating on the coral surface in the different seasons (Siboni et al. 2010). This may have occurred due to increases in nutrient availability via vertical mixing occurring in the northern Red Sea (Erez et al. 1991). Thus, the environment and, perhaps the host, can play a role in the species and numbers of the symbionts on their surface. The stramenopiles are known to be capable of efficient saprophytic utilization of dissolved organic nutrients from the environment or from their hosts. Thus, those growing on the surface of aquatic plants (Austin 1988) or in the coral surface mucus most likely make use of nutrients supplied by the host. This may explain why such a large concentration of protists appear on the tissues surrounding the coral mouth opening, a site of nutrient exchange (Kramarsky–Winter et al. 2006). The acquisition of nutrients from their hosts or from the environment is possible through the use of a wide spectrum of enzymes, including some that degrade cellulose (Raghukumar et al. 1994; Sharma et al. 1994; Bremer and Talbot 1995; Raghukumar 2008). For example, *C. limacisporum*, coral-associated protist, is of considerable interest due to its capacity to grow on media containing inorganic nitrogen sources (Sumathi et al. 2006). Interestingly, this protist's evolutionary position is still unclear. It is considered to be a close relative to fungi due to the presence of a distinct enzyme, namely alpha-aminoadipate reductase (alpha-AAR), a fungal enzyme involved in the synthesis of lysine and ergosterol. In addition, the presence of the sterol C-14 reductase gene, a gene involved in the sterol pathway of animals and fungi, was also detected in the organism (Sumathi et al. 2006). This led to the assertion that these organisms comprise sister clades to both animals and fungi (Sumathi et al. 2006).

The coral-associated protist *C. limacisporum* belonging to the Corallochytrea class in the Mesomycetozoa subphylum of the Neomonada phylum (Cavalier-Smith 1997; Mendoza et al. 2002) was reported by Raghukumar (1987). Its cell life cycle traits are characterized by single, diad, or tetrad vegetative spherules with spores and an elongated limax amoebic stage. It possesses a cell wall, lacks flagella, and when found in the marine saprotrophic environment, it is a spherical, single-celled organism, 4.5–20.0 μm in diameter, undergoing several binary fissions to release numerous elongated daughter cells (up to 32 daughters per single cell) (Mendoza et al. 2002). Another case of coral-associated protist (*Fng1*) was documented in the mucus of the hermatypic coral *Fungia granulosa* from the Gulf of Eilat (Harel et al. 2008). In this study, *Fng1* (18S rRNA: AY870336; mitochondrial 16S rRNA: AY870337) was identified as belonging to the family Thraustochytridae (phylum Stramenopile, order Labyrinthulida). Its cell wall is characterized by an arrangement in a laminated morphology, which yields thin, flexible, and circular scales (Darley et al. 1973). Almost all species develop ectoplasmic extensions from one or more points on the cell and form a branched network that is generated by an organelle termed the bothrosome (or sagenogenetosome), located at the periphery of the cell (Porter 1972). Many of the species exhibit a gliding mobility associated with these ectoplasmic networks (Perkins 1973; Leander and Porter 2001; Raghukumar 2002; Leander et al. 2004). In addition, the ectoplasmic net contains hydrolytic enzymes that are surface-bound or are secreted into the surrounding medium, helping in the digestion of organic material (Coleman and Vestal 1987). The organisms' vegetative stages consist of single granular cells, which are globose to subglobose, measuring 4–20 μm in diameter (Raghukumar 2002). Most Thraustochytrids reproduce by means of zoospores, which possess a long anterior, tinsel flagellum along with a short, posterior, whiplash flagellum (Porter 1990; Raghukumar 2002). Morphological examination of this strain revealed a non-motile organism 35 μm in diameter, which is able to thrive on carbon-deprived media, and whose growth and morphology are inoculum dependent (Fig. 10.2). A similar protist was also isolated from the mucus of the coral *Favia* sp. (AY870338, AY870339, Harel et al. 2008). Microscopic scans revealed aggregated protist cells in the coral mucus layer and inside the coral tissue. In one of his experiments (Harel M, MSc thesis), he observed bleached coral (placed in dark for 9 months) in association with the protist aggregates. Although the symbiotic algae were gone, coral tissue seemed intact for a long time period, implying that the aggregates provide the coral with a carbon source (maybe by producing Polyunsaturated fatty acids – PUFA, or predation) while using it for shelter and nitrogen source (Harel et al. 2008). *Fng1* has the

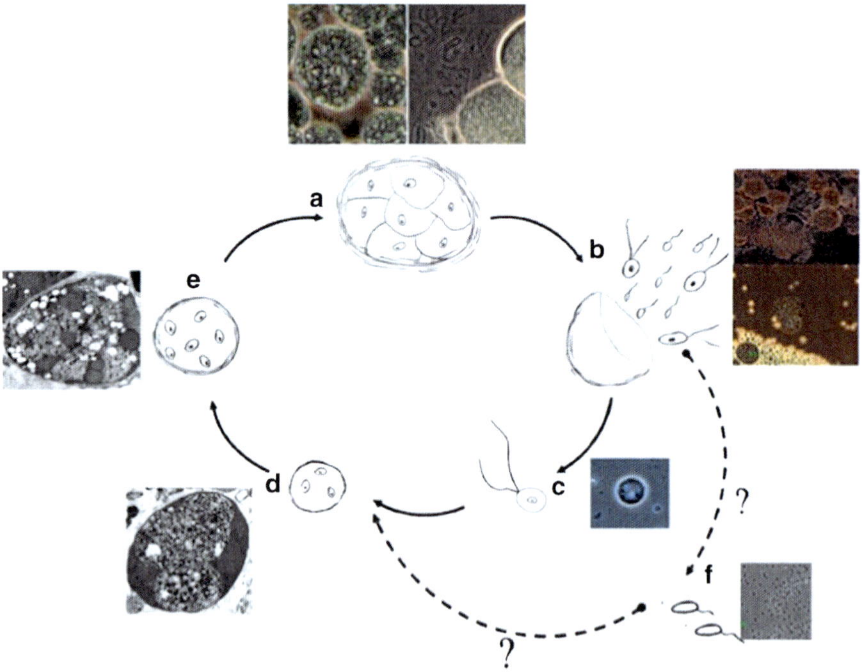

Fig. 10.2 Possible life cycle of *Fng1*: The mature parent cells (**a**), 35 ± 15 μm in diameter released (**b**) bi-flagellated daughter cells (5 ± 1 μm in diameter) (**c**) and motile zoospores (**f**) through an opening in the cell wall. The flagellated daughter cells grow, generate a lamellar cell wall, and give rise by asymmetric multiple fission to asymmetric daughter cells (**d** and **e**)

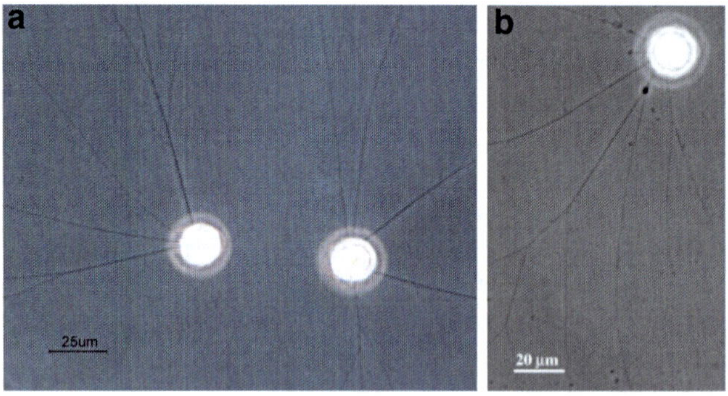

Fig. 10.3 Ectoplasmic network of strain *Fng1* grown on B1 liquid medium. Frames were captured using a light microscope at x1,000 magnification (a: picture by D. Orri; b: (Harel et al. 2008)

ability to develop ectoplasmatic net (Fig. 10.3) in certain conditions that may benefit it in the viscous environment of the coral mucus for material transfer and surface attachment. The rate of *Fng1* cells movement using the ectoplasmic net was measured as 0.25 μm/min.

10.3 Protists as a Nutrient Source

The corals may benefit from their association with these protists via direct predation or by utilization of some of their enzymatic and biochemical abilities, especially as some species are known to recycle organic substance and to produce a number of important nutrients (Lewis et al. 1999). Some Thraustochytrids, for example, are known to produce PUFA (polyunsaturated fatty acid) and a number of carotenoids (Carmona et al. 2003). Indeed, the morphology and electron density of some of their coccoid bodies suggest that they might be lipid bodies (Siboni et al. 2010). Moreover, the fatty acid analysis of the coral-associated thraustochytrid *Fng1* revealed above 75% unsaturated fatty acids, with ω3-PUFA as a significant component (Harel et al. 2008). It is therefore possible that thraustochytrids and other Stramenopiles found on the surface and in the mucus of corals may provide nutritional sources allowing the coral holobiont to survive stressful events such as bleaching (Harel et al. 2008). Indeed, one explanation as to why massive large polyped and solitary large polyped corals survive bleaching better than branched corals (Hoeksema 1991; Loya et al. 2001) may be the persistence of these microorganisms on the coral surface during bleaching. It is possible that these stramenopile protists may thus aid their coral hosts in acquiring enough nutrients from the environment or may themselves provide an additional food source during the bleaching events.

10.4 Negative Coral-Protist Associations

The association between marine invertebrates and plants and protist symbionts is not always a positive one. Indeed, a pathogenic species of the marine protist *Labyrinthula zosterae* has been identified as the etiological agent of several epidemics of wasting disease of eelgrass, *Zostera marina* (Muehlstein et al. 1991).

Coral-associated coccidian protozoan (named genotype N, phylum Apicomplexa) was studied using molecular methods in healthy and bleached Caribbean corals. 90% of healthy corals *Montastraea annularis* and *M. faveolata*, which were sampled, tested positive for protist genotype N (Upton and Peters 1996; Toller et al. 2002), which shared greatest similarity to *Toxoplasma*, *Neospora*, *Isospora*, and *Sarcocystis*, representative taxa from the apicomplexan class Coccidia (Levine 1982). The protist sequences vary (1.0–1.7%) from one coral colony to another (Toller et al. 2002). In the same study, a zooxanthellate coral species, *Balanophyllia elegans* and *Astrangia* sp., were also found to be associated with this protist genotype.

Protozoan infections have also been identified on corals held in aquaria, including the consumption of coral tissue by the ciliate *Helicostoma nonatum* (Borneman 2001). Coral-associated parasitic protists were also found on staghorn coral colonies, *Acropora muricata*, affected with brown band (BrB) disease in the

Great Barrier Reef and identified using microscopic and molecular approaches (Ulstrup et al. 2007; Bourne et al. 2008). 18S rRNA sequences of these ciliates retrieved from the brown band were found to be 95% similar to *Paraurónema longum*. The authors speculate that as the health of a coral deteriorates, necrotic tissue attracts the ciliate that feed on both bacteria and zooxanthellae associated with dead and dying coral tissue. At high densities, these ciliates may become the primary cause of tissue loss as they take up photocompetent zooxanthellae to alleviate potential oxygen limitations (Ulstrup et al. 2007; Bourne et al. 2008).

An additional general negative coral-protist association was observed on several Indo–Pacific reefs, infecting both branching and massive corals. In this case, a Heterotrichid folloculinid ciliate, named *Halofolliculina corallasia*, was found to cause a unique coral disease, termed Skeleton Eroding Band (SEB) (Antonius and Lipscomb 2000). This protist secretes a protective lorica and forms an advancing black line that denudes the coral of its tissues, leaving behind the skeleton and, if advancement is unchecked, finally killing the coral. SEB infects and damages a wide variety of branching and massive reef corals (24 species), including the species *Stylophora pistillata*, *Pocillopora damicornis*, *P. verrucosa*, *P. eydouxi*, *Montipora monasteriata*, *Acropora aspera*, *A. humilis*, *A. formosa*, *A. noblis*, *A. tenuis*, *A. valida*, *A. florida*, *A. hyacinthus*, *A. clathrata*, *A. downingi*, *Leptoseris explanata*, *Pachyseris rugosa*, *Hydnophora microconos*, *Favia stelligera*, *Favites abdita*, *Goniastrea retlformis*, *Leptastrea purpurea*, *Cyphastrea chalcidicum*, and *C. serailia* (Antonius and Lipscomb 2000). SEB was also found to infect Caribbean hard coral species of six families (Acroporidae, Agariciidae, Astrocoeniidae, Faviidae, Meandrinidae, and Poritidae; up to 25 scleractinian species) and even calcifying hydrozoan milleporids (Cróquer et al. 2006).

10.5 Unresolved Questions and Future Research

Additional research should be carried out in order to further understand the role of protist, in general, and Thraustochytrids, in particular, in the coral holobiont and the reef ecosystem, to identify whether it is a commensalistic/symbiotic/opportunistic/antagonistic/predation relationship. Better understanding of coral-protist classification, diversity, unique properties, and their adaptation to survive competition can assist in biotechnological applications. Since many of the PUFA, docosahexaenoic acid and carotenoids produced by these thraustochytrids are known to be important in human health and aquaculture, these Thraustochytrids could be of great biotechnological interest (reviewed in Raghukumar 2008).

Acknowledgements This work was supported by ISF Grant number 1169/07. The authors also thank the Inter-University Institute in Eilat for use of their facilities.

References

Anderson RS, Kraus BS, McGladdery SE, Reece KS, Stokes NA (2003) A thraustochytrid protist isolated from *Mercenaria mercenaria*: molecular characterization and host defense responses. Fish Shellfish Immunol 15:183–194

Antonius A, Lipscomb D (2000) First protozoan coral-killer identified in the Indo-Pacific. Atoll Res Bull 493:1–21

Austin B (1988) Marine microbiology. Cambridge University Press, Cambridge

Borneman EH (2001) Aquarium corals: selection husbandry and natural history. TFH Publications, Neptune City

Bourne DG, Boyett HV, Henderson ME, Muirhead A, Willis BL (2008) Identification of a ciliate (Oligohymenophorea: Scuticociliatia) associated with brown band disease on corals of the Great Barrier Reef. Appl Environ Microbiol 74(3):883–888

Bremer GB, Talbot G (1995) Cellulolytic enzyme activity in the marine protist *Schizochytrium Aggregatum*. Bot Mar 38:37–41

Carmona ML, Naganuma T, Yamaoka Y (2003) Identification by HPLC-MS of carotenoids of the Thraustochytrium CHN-1 strain isolated from the Seto Inland Sea. Biosci Biotechnol Biochem 67(4):884–888

Cavalier-Smith T (1997) Amoeboflagellates and mitochondrial cristae in eukaryote evolution: megasystematics of the new protozoan subkingdoms Eozoa and Neozoa. Arch Protistenkunde 147:237–258

Coleman NK, Vestal JR (1987) An epifluorescent microscopy study of enzymatic hydrolysis of fluorescein diacetate associated with the ectoplasmic net elements of the protist *Thraustochytrium striatum*. Can J Microbiol 33:841–843

Cróquer A, Bastidas C, Lipscomp D, Rodríguez-Martínez RE, Jordan-Dahlgren E, Guzman HM (2006) First report of folliculinid ciliates affecting Caribbean scleractinian corals. Coral Reefs 25(2):187–191

Darley WM, Porter D, Fuller MS (1973) Cell wall composition and synthesis via Golgi-directed scale formation in the marine eucaryote, *Schizochytrium aggregatum*, with a note on *Thraustochytrium* sp. Arch Mikrobiol 90:89–106

Dinsdale EA, Pantos O, Smriga S, Edwards RA, Angly F, Wegley L, Hatay M, Hall D, Brown E, Haynes M, Krause L, Sala E, Sandin SA, Thurber RV, Willis BL, Azam F, Knowlton N, Rohwer F (2008) Microbial ecology of four coral atolls in the Northern Line Islands. PLoS ONE 3(2):e1584

Ellis LL, Bishop SH (1989) Isolation of cell lines with limited growth potential from marine bivalves. In: Mitsuhashi J (ed) Invertebrate cell system application, vol 2. CRC Press, Boca Raton, pp 243–251

Erez J, Lazar B, Genin A, Dubinsky Z (1991) The biogeochemical interactions of coral reefs with their adjacent sea. In: Proceedings of the 12th Conference of the Interuniversity Institute, The H. Steinitz Marine Biology Laboratory, Eilat, pp 49–58

Fan KW, Vrijmoed LLP, Jones EBG (2002) Physiological studies of subtropical mangrove thraustochytrids. Bot Mar 45:50–57

Frank U, Rabinowitz C, Rinkevich B (1994) In vitro establishment of continuous cell cultures and cell lines from ten colonial cnidarians. Mar Biol 120:491–499

Goldberg MW (2002) Feeding behavior, epidermal structure and mucus cytochemistry of the scleractinian *Mycetophyllia reesi*, a coral without tentacles. Tissue Cell 34:232–245

Harel M, Ben-Dov E, Rasoulouniriana D, Siboni N, Kramarsky-Winter E, Loya Y, Barak Z, Wiesman Z, Kushmaro A (2008) A new Thraustochytrid, strain Fng1, isolated from the surface mucus of the hermatypic coral Fungia granulose. FEMS Microbiol Ecol 64:378–387

Hoeksema BW (1991) Control of bleaching in mushroom coral populations (Scleractinia, Fungiidae) in the Java Sea: stress tolerance and interference by life-history strategy. Mar Ecol Prog Ser 74(2–3):225–237

Honda D, Yokochi T, Nakahara T, Erata M, Higashihara T (1998) *Schizochytrium limacinum* sp. nov., a new thraustochytrid from a mangrove area in the west Pacific Ocean. Mycol Res 102:439–448

Ilan M, Contini H, Carmeli S, Rinkevich B (1996) Progress towards cell cultures from a marine sponge that produces bioactive compounds. J Mar Biotechnol 4:145–149

Johnston IS, Rohwer F (2007) Microbial landscapes on the outer tissue surfaces of the reef-building coral *Porites compressa*. Coral Reefs 26:375–383

Kimura H, Sato M, Sugiyama C, Naganuma T (2001) Coupling of thraustochytrids and POM, and of bacterio- and phytoplankton in a semi-enclosed coastal area: implication for different substrate preference by the planktonic decomposers. Aquat Microb Ecol 25:293–300

Knowlton N, Rohwer F (2003) Multispecies microbial mutualisms on coral reefs: the host as a habitat. Am Nat 162(4 Suppl):S51–S62

Kramarsky-Winter E, Harel M, Siboni N, Ben Dov E, Brickner I, Loya Y, Kushmaro A (2006) Identification of a protist-coral association and its possible ecological role. Mar Ecol Prog Ser 317:67–73

Kvingedal R, Owens L, Jerry DR (2006) A new parasite that infects eggs of the mud crab, Scylla serrata, in Australia. J Invertebr Pathol 93:54–59

Leander C, Porter D (2001) The *Labyrinthulomycota* is comprised of three distinct lineages. Mycologia 93:459–464

Leander CA, Porter D, Leander BS (2004) Comparative morphology and molecular phylogeny of aplanochytrids (*Labyrinthulomycota*). Eur J Protistol 40:317–328

Levine ND (1982) Apicomplexa. In: Parker SP (ed) Synopsis and classification of living organisms, vol 1. McGraw-Hill, New York, pp 750–787

Lewis ET, Peter DN, McMeekin AT (1999) The biotechnological potential of thraustochytrids. Mar Biotechnol 1:580–587

Loya Y, Sakai K, Yamazato K, Nakano Y, Sambali H, van Woesik R (2001) Coral bleaching: the winners and the losers. Ecol Lett 4(2):122–131

Mendoza L, Taylor JW, Ajello L (2002) The class mesomycetozoea: a heterogeneous group of microorganisms at the animal-fungal boundary. Annu Rev Microbiol 56:315–344

Moss ST (1986) Biology and phylogeny of the Labyrinthulales and Thraustochytriales. In: Moss ST (ed) The biology of marine fungi. Cambridge University Press, Cambridge, pp 105–129

Muehlstein LK, Porter D, Short FT (1991) *Labyrinthula zosterae* sp. nov., the causative agent of Wasting disease of eelgrass *Zostera marina*. Mycologia 83(2):180–191

Patterson DJ (1999) The diversity of Eukaryotes. Am Nat 65:96–124

Perkins FO (1973) Observation of thraustochytriaceous (Phycomycetes) and labyrinthulid (Rhizopodea) ectoplasmic nets on natural and artificial substrates – an electron microscopy study. Can J Bot 51:485–491

Porter D (1972) Cell division in the marine slime mold, *Labyrinthula* sp., and the role of the bothrosome in extracellular membrane production. Protoplasma 74(4):427–448

Porter D (1990) Phylum Labyrinthulomycota. In: Margulis L, Corliss JO, Melkonian M, Chapman DJ (eds) Handbook of Protoctista. Jones and Bartlett, Boston

Raghukumar S (1987) Occurrence of the Thraustochytrid, Corallochytrium limacisporum gen. et sp. nov. in the coral reef lagoons of the Lakshadweep Islands in the Arabian Sea. Bot Mar 30 (1):83–90

Raghukumar S (1988) Detection of the thraustochytrid protist *Ulkania visurgensis* in a hydroid, using immunofluorescence. Mar Biol 97(2):253–258

Raghukumar S (2002) Ecology of the marine protists, the Labyrinthulomycetes (Thraustochytrids and Labyrinthulids). Eur J Protistol 8(2):127–145

Raghukumar S (2008) Thraustochytrid marine protists: production of pufas and other emerging technologies. Mar Biotechnol 10:631–640

Raghukumar S, Balasubramanian R (1991) Occurrence of thraustochytrid fungi in corals and coral mucus, Indian Journal of Marine Sciences 20:176–181

Raghukumar S, Sharma S, Raghukumar C, Sathe-Pathak V, Chandramohan D (1994) Thraustochytrid and fungal component of marine detritus. IV. Laboratory studies on decomposition of leaves of the mangrove Rhizophora apiculata Blume. J Exp Mar Biol Ecol 183:113–131

Raghukumar S, Anil AC, Khandeparker L, Patil JS (2000) Thraustochytrid protists as a component of marine microbial films. Mar Biol 136:603–609

Rinkevich B (1999) Cell cultures from marine invertebrates: obstacles, new approaches and recent improvements. J Biotechnol 70:133–153

Rinkevich B, Ilan M, Blisko R (1998) Further steps in the initiation of cell cultures from embryos and adult sponge colonies, in Vitro. Cell Dev Biol Anim 34(10):753–756

Rohwer F, Kelly S (2004) Culture-independent analyses of coral-associated microbes. In: Rosenberg E, Loya Y (eds) Coral health and disease. Springer, Heidelberg, pp 265–278

Rohwer F, Breitbart M, Jara J, Azam A, Knowlton N (2001) Diversity of bacteria associated with the Caribbean coral *Montastraea franksi*. Coral Reefs 20:85–91

Rohwer F, Seguritan V, Azam A, Knowlton N (2002) Diversity and distribution of coral-associated bacteria. Mar Ecol Prog Ser 243:1–10

Schlichter D, Brendelberger H (1998) Plasticity of the scleractinian body plan: functional morphology and trophic specialization of *Mycedium elephantotus* (Pallas, 1766). Facies 39:227–241

Sharma S, Raghukumar C, Raghukumar S, Sathe-Pathak V, Chandramohan D (1994) Thraustochytrid and fungal component of marine detritus, II. Laboratory studies on decomposition of the brown alga *Sargassum cinereum* J Ag. J Exp Mar Biol Ecol 175:227–242

Sherr BF, Sherr EB, Caron DA, Vaulot D, Worden AZ (2007) Oceanic protists. Oceanography 20:130–134

Siboni N, Rasoulouniriana D, Ben-Dov E, Kramarsky-Winter E, Sivan A, Loya Y, Hoegh-Guldberg O, Kushmaro A (2010) Stramenopile microorganisms associated with the massive coral *Favia* sp. J Eukaryot Microbiol 57(3):236–244

Sumathi JC, Raghukumar S, Kasbekar DP, Raghukumar C (2006) Molecular evidence of fungal signatures in the marine protist *Corallochytrium limacisporum* and its implications in the evolution of animals and fungi. Protist 157(4):363–376

Thorsen MS (1999) Abundance and biomass of the gut-living microorganisms (bacteria, protozoa and fungi) in the irregular sea urchin *Echinocardium cordatum* (Spatangoida: Echinodermata). Mar Biol 133:353–360

Toller W, Rowan R, Knowlton N (2002) Genetic evidence for a protozoan (phylum Apicomplexa) associated with corals of the *Montastraea annularis* species complex. Coral Reefs 21 (2):143–146

Ulstrup KE, Kuhl K, Bourne DG (2007) Zooxanthellae harvested by ciliates associated with brown band syndrome of coral remain photosynthetically competent. Appl Environ Microbiol 73:1968–1975

Upton SJ, Peters EC (1986) A new and unusual species of coccidium (Apicomplexa: Agamococcidiorida) from Caribbean scleractinian corals, Journal of Invertebrate pathology 47 (2):184–193

Veron J (2000) Corals of the world. AIMS, Townsville, pp 85–269

Wagner-Merner DT, Duncan WR, Lawrence JM (1980) Preliminary comparison of Traustochytriaceae in the guts of a regular and irregular echinoid. Bot Mar 23:95–97

Zilber-Rosenberg I, Rosenberg E (2008) Role of microorganisms in the evolution of animals and plants: the hologenome theory of evolution. FEMS Microbiol Rev 32(5):723–735

Zuendorf A, Bunge J, Behnke A, Barger KJ, Stoeck T (2006) Diversity estimates of microeukaryotes below the chemocline of the anoxic Mariager Fjord, Denmark. FEMS Microbiol Ecol 58(3):476–491

Chapter 11
Effect of Ocean Acidification on the Coral Microbial Community

Dalit Meron, Lena Hazanov, Maoz Fine, and Ehud Banin

Abstract An emerging environmental stress that can impact coral reefs worldwide is ocean acidification, a result of increasing atmospheric CO_2 concentration. Surface seawater pH is 0.1 units lower than pre-industrial values and is predicted to decrease by up to 0.4 units by the end of the century. This change in pH may result in changes in the physiology of ocean organisms, in particular organisms that build their skeletons/shells from calcium carbonate, such as corals. This physiological change may also affect other members of the coral holobiont, which in turn may influence coral physiology and health. This chapter reviews the laboratory and field studies to date that have examined the effect of pH on the coral microbial community (i.e., bacteria, archaea, fungi, and endolithic communities). The impact of this change on the coral host is also discussed.

11.1 Introduction

Corals represent a fascinating example of a holobiont (Rohwer et al. 2002; Rosenberg et al. 2007), as they are associated with a dynamic and highly diverse consortium of microorganisms including *Bacteria*, *Archaea*, *Fungi*, and symbiotic algae (*zooxanthellae*). The association between the coral and its microbial community is often species-specific, but is also a dynamic symbiosis, which can shift following physiological or environmental (e.g., temperature, light intensity, pollution, and salinity) change (Rosenberg et al. 2007; Garren et al. 2009; Walsh et al. 2005).

D. Meron • E. Banin
The Mina and Everard Goodman Faculty of Life Sciences, Bar-Ilan University, Ramat Gan, Israel
e-mail: banine@mail.biu.ac.il

L. Hazanov • M. Fine
The Mina and Everard Goodman Faculty of Life Sciences, Bar-Ilan University, Ramat Gan, Israel

The Interuniversity Institute for Marine Science in Eilat, Eilat, Israel

E. Rosenberg and U. Gophna (eds.), *Beneficial Microorganisms in Multicellular Life Forms*, 163
DOI 10.1007/978-3-642-21680-0_11,

The coral–microbiota interaction may directly influence the coral physiology and health. The holobiont theory suggests that a shift in coral resident microbial community structure may contribute to a more rapid and versatile adaptation of the holobiont to changes in the environmental conditions (Reshef et al. 2006; Rosenberg et al. 2007, 2009).

11.2 Ocean Acidification

Ocean acidification is the result of increasing atmospheric CO_2 concentration and its absorption in ocean waters. Atmospheric CO_2 now exceeds 380 ppm, more than 80 ppm. above the maximum values of the past 740,000 years (Feely et al. 2004; Petit et al. 1999). This increase has already resulted in a change in seawater pH (decreased by 0.1 pH unit) since pre-industrial times and is predicted to further decrease in upcoming years (IPCC 2007). The ocean absorption of anthropogenic CO_2 also reduces the saturation state of seawater with respect to calcite and aragonite, the most common forms of calcium carbonate secreted by calcifying marine biota. Calcification is strongly dependent on the carbonate saturation state of seawater (Caldeira 2007). Thus, the drastic change in pH may dramatically influence the physiology of corals and other oceanic organisms, in particular organisms that deposit calcium carbonate to build their skeletons/shells (Gattuso et al. 1998; Orr et al. 2005). In fact, dissolution of coral skeletons (Fine and Tchernov 2007), increase of coral bleaching, and a decrease in calcification and net productivity (Anthony et al. 2008) were reported as a result of exposure to reduced pH in laboratory experiments. In addition to coral physiology, recent studies have also indicated changes in the coral microbial communities following exposure to reduced pH (Meron et al. 2011; Vega–Thurber et al. 2009; Tribollet et al. 2009). These changes are further discussed in this chapter.

11.3 The Effect of Decreasing pH on Changes in Microbial Communities Lessons from Laboratory Studies

Several laboratory studies to date have examined the affect of pH on the coral microbial community (Tribollet et al. 2009; Vega–Thurber et al. 2009; Meron et al. 2011). Despite differences in coral species and experimental setup these studies show that the change in pH seems to impact the entire microbial community associated with the coral. Meron et al. (2011) examined the effect of pH on the coral *Acropora eurystoma* maintained at ambient and pH 7.3 in laboratory conditions for 2 months (Meron et al. 2011). Cluster analyses of Denaturing gradient gel electrophoresis (DGGE) and 16S rRNA sequences obtained from clone libraries showed that *A. eurystoma* harbors a specific microbial community that is distinct from the surrounding water (Meron et al. 2011), as reported in other corals from various locations (Bourne and Munn 2005; Guppy and Bythell 2006; Ritchie 2006; Kooperman et al. 2007; Lampert et al. 2008; Arboleda and Reichardt 2009). Furthermore, the bacterial communities were clustered according to the pH treatment, more so than the coral fraction (i.e., skeleton, tissue, and mucus) (Fig. 11.1a, and Meron et al. 2011) highlighting the impact of pH on the community. While the *Cyanobacteria* and *Gammaproteobacteria* remained the two dominant groups in the two pH treatments, the *Alphaproteobacteria* increased in their overall presence (dominant clones belonged to the *Rhodobacteraceae*). Two other groups whose gene frequency was reduced following exposure to lower pH were the *Deltaproteobacteria* (dominant clones belonged to *Desulfobacter* species) and *Bacteroidetes*. It is interesting to note that although *Bacteroidetes* have been previously detected in healthy corals (Sekar et al. 2008), an increase in *Bacteroidetes* was associated with Black Band Disease (Cooney et al. 2002; Frias–Lopez et al. 2002, 2004). *Bacteroidetes* were also reported to increase in corals exposed to various stress conditions (Vega–Thurber et al. 2009).

Overall, the distribution of the bacterial groups shifted due to a decrease in pH and the Shannon–Wiener index, increased at the lower pH treatment compared to ambient conditions. These results may indicate an intermediate disturbance (Connell 1978), leading to an increase in biodiversity. It is important to emphasize that the corals in the study did not show any sign of stress or disease throughout the experiment. Yet, it is possible that changes in the host metabolism and hence changes in nutrients available to the microbiota have led to shifts in the microbial community rather than direct response to reduced pH. It is also possible that changes in keystone microbial species or dominant species lead to a cascade of

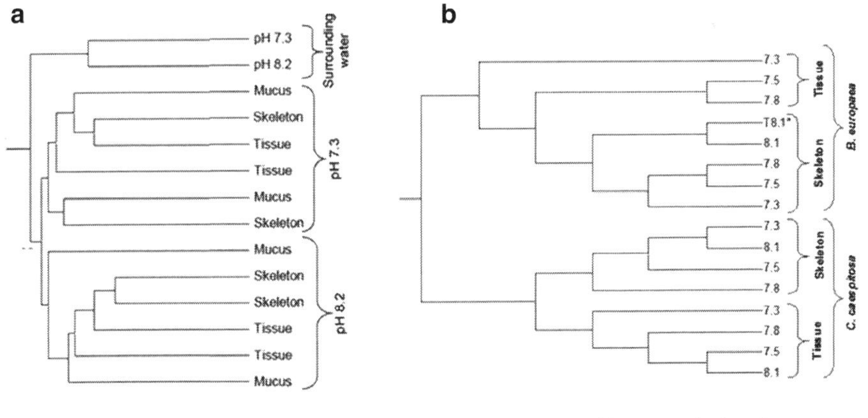

Fig. 11.1 Impact of pH on coral associated bacterial diversity. Cluster analysis of sequences obtained from microbial community 16 S rRNA clone libraries from (**a**) *A. eurystoma* (laboratory study) and (**b**) *B. europaea and C. caespitosa* (field study) following exposure to low pH
*Note that T8.1 (tissue from ambient pH in *B. europaea*) is clustered with all other skeleton fraction communities. Cluster analysis was performed using Jackknife Environment Clusters (Unifrac Metric analysis) (Lozupone and Knight 2005)

changes in the microbial community. One of the changes that was observed in Meron et al. (2011) laboratory study was an increase in bacterial species belonging to families of known coral pathogens (such as *Vibrionaceae* and *Alteromonadaceae*) (Ritchie 2006; Bourne et al. 2007; Arboleda and Reichardt 2009; Sunagawa et al. 2009) or families that were previously isolated from diseased, injured, or stressed marine invertebrates (*Rhodobacteraceae*) (Sekar et al. 2006; Sunagawa et al. 2009). Similar results were also reported by Vega–Thurber et al. who examined in laboratory conditions the impact of pH on the coral *Porites compressa* under extreme pH and high rate of pH change. Vega–Thurber et al. also carried out a metagenomic analysis which revealed changes in several pathways upon exposure to decreased pH conditions, including an increase in genes associated with stress response (e.g., antibiotics and toxins) and decrease in genes required for adhesion (Vega–Thurber et al. 2009). These results are supported by Meron et al. that reported an increase in bacteria that possess antimicrobial activity in corals maintained at the low pH. It is interesting to note that most of these antimicrobial active bacteria belonged to the *Vibrionaceae* and *Rhodobacteraceae* groups suggesting that this may be one of the mechanisms allowing these species to increase their distribution within the coral microbial community (Fig. 11.2) (Meron et al. 2011). Taken together, the results of the laboratory studies may suggest that although the corals do not show any physiological signs of stress upon exposure to low pH, there is a clear shift to bacterial groups that are known to be associated with stressed and diseased corals, and this may serve as a preliminary indicator for the stress of the holobiont.

Another important coral-associated microbial group having a role as microborers is the euendoliths (including algae, *Cyanobacteria,* and fungi). These microborers dissolve calcium carbonate to carbonate ions while residing in the coral skeleton (Golubic et al. 1981). Although the boring activity exhibited by euendoliths is much greater in dead substrates, it can also dissolve carbonates of

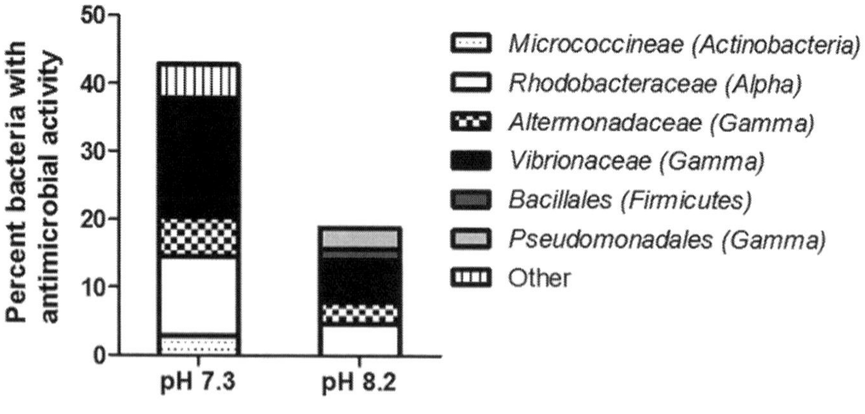

Fig. 11.2 Antimicrobial activity of bacteria isolated from coral mucus exposed to low pH. Percent bacteria with antimicrobial activity isolated from *A. eurystoma* mucous exposed to different pH treatments are presented

living organisms (Le Campion–Alsumard et al. 1995a; Tribollet and Payri 2001). Tribollet et al. (2009) examined the effects of elevated pCO_2 (750 ppmv predicted by 2100) on dissolution of dead coral carbonates (*Porites lobata*) by microbial euendoliths (Tribollet et al. 2009). Although elevated pCO_2 had no effect on net photosynthetic rates of the endolithic community (Tribollet et al. 2006), an enhanced growth of euendoliths under increased pCO_2 was observed. Depth of penetration of *Ostreobium quekettii* filaments, which dominated the euendolithic communities, grew faster under elevated pCO_2 (750 ppmv) than at ambient pCO_2. Tribollet et al. (2009) estimated that ubiquitous euendoliths dissolve about 3×10^{12} mol $CaCO_3$ a^{-1} in coral reefs today and estimate that biogenic dissolution by euendoliths should reach at least 4.44×10^{12} mol $CaCO_3$ a^{-1} by 2100, a change that may have a dramatic impact on coral reefs (Tribollet et al. 2009).

The impact of pH on the fungal population associated with corals was also examined under laboratory conditions. The study examined the fungal population associated with *Acropora loripes* and the results demonstrated an increase in both the number of coral fragments yielding fungal growth (Fig. 11.3a) and the total number of fungal isolates harvested from the corals maintained at the reduced pH treatment (Fig. 11.3b). In addition, an evident shift in the fungal community structure, from filamentous fungi to yeasts, was observed in the reduced pH treatments. This again might indicate a sign of stress induced on the host, reducing its fitness and subsequently compromising its defenses (Ross et al. 1996; Lenihan et al. 1999). It has been shown that corals are able to impair fungal penetration from skeleton to its tissue by localized calcification as a defense strategy (Le Campion–Alsumard et al. 1995b). However, under ocean acidification conditions and reduced calcification abilities, anti-fungal localized calcification may be too slow or not dense enough to prevent filament penetration. The apparent replacement of certain species by others might be derived from optimization of growth conditions for one species in terms of pH, while at the same time producing less

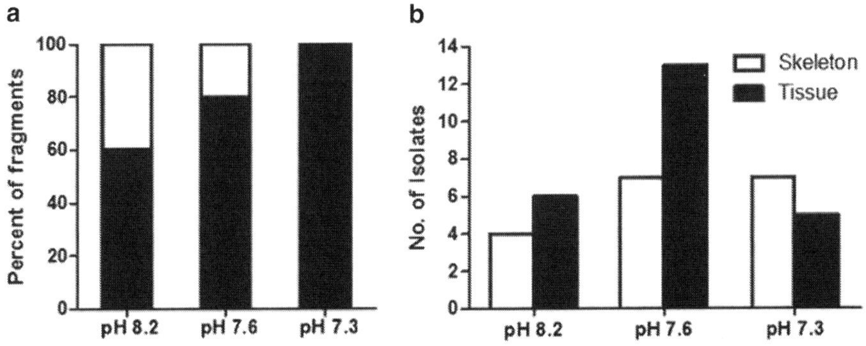

Fig. 11.3 Impact of pH on the coral fungal community. (**a**) Portion of corals yielded fungal isolates and (**b**) the number of isolated fungi from skeleton and tissue of corals subjected to various pH treatments. The absence or presence of fungi in all individuals is represented by empty or filled bars, respectively

favorable conditions for others. Interestingly, in ambient pH, more fungal isolates were obtained from the tissue fraction than from the skeleton, while the opposite trend was observed in the lower pH, where the number of isolates in the skeleton exceeded that of the tissue. It is believed that secondary metabolites produced by marine-fungi act as a defense mechanism and a way of adaptation to unique environments (Fenical and Jensen 1993; Jensen and Fenical 2002). These metabolites might alter the fungal microenvironment depending on the nature of the product.

11.4 The Effect of Decreasing pH on Changes in Microbial Communities Lessons from Field Study

To address the impact of pH on the coral microbial community under natural field conditions, a natural pH gradient in the Mediterranean Sea off the coast of Ischia (Gulf of Naples, 40°˜043.84′N; 013°˜ 57.08′E), Italy (Hall–Spencer et al. 2008), was chosen as a field site. The pH gradient in this site is formed by a volcanic CO_2 vent. Two species of corals (*Balanophyllia europaea and Cladocora caespitosa*) naturally living near the vent, were transplanted to various locations along a 300 m gradient according to pH (function of distance from the CO_2 vent) (i.e., ambient pH (out of the vent) and mean pH_T 7.8, 7.5 and 7.3). The corals were incubated for 7 months. Similar to the laboratory studies (Meron et al. 2011, see Sect. 11.3), each coral species harbored a distinct bacterial community and exposure to low pH caused a shift in the bacterial diversity (Fig. 11.1b) and increase in Shannon–Wiener index. However, unlike the laboratory studies, the bacterial diversity was divided mainly according to the fraction (i.e., tissue or skeleton) and not the pH treatment. This suggests a stronger association of the bacterial community with spatial location (i.e., fraction) rather than the surrounding seawater pH. It is important to note that a stronger pH influence was observed on *B. europaea*. Mainly, at the ambient pH, the tissue and skeleton fractions clustered together with the skeletons from all treatments, while the tissues from the lower pH sites formed a separate cluster. This might be explained by the fact that the tissue, i.e., the external fraction, is more exposed to the environment. A similar trend was also observed in the archaeal (Fig. 11.4) communities of *B. europaea* and *C. caespitosa*, which showed an inner division according to the change in pH while no such clustering was detected in the skeleton fractions. Furthermore, the archaeal community associated with the tissue was divided mainly by coral species than by the pH treatment. It is important to note that a separate cluster was observed in the normal pH, outside the vent area in both corals, although in *B. europaea*, the difference between the samples was minor compared to *C. caespitosa* (Fig. 11.4).

When examining the bacteriological markers of stress (i.e., increase in communities associated with diseased or stressed corals), no signs were observed in *C. caespitosa* as opposed to *B. europaea*, in which some signs such as increase in

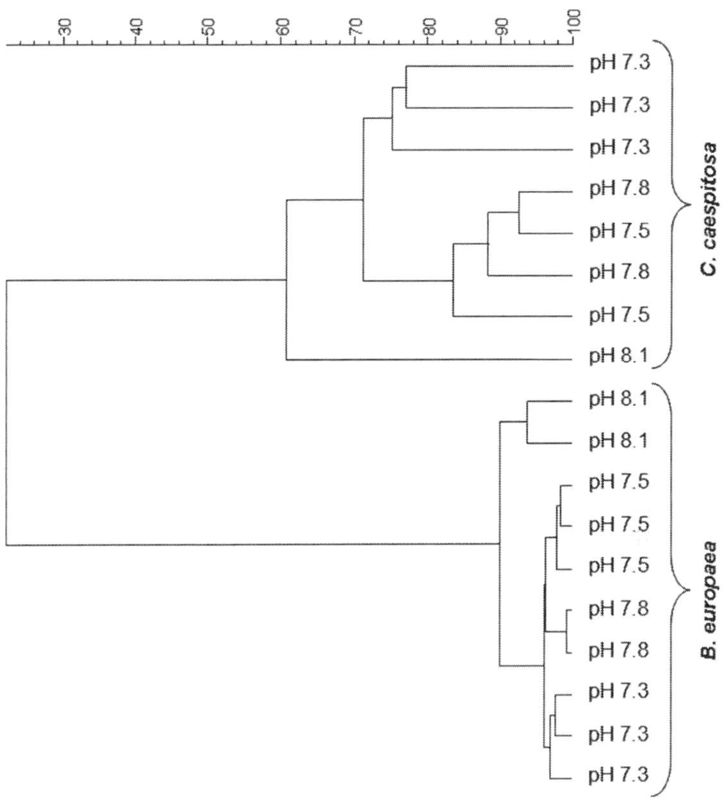

Fig. 11.4 Changes in tissue archaeal community following exposure to low pHs. Cluster analysis of DGGE patterns of archaeal communities in *B. europaea* and *C. caespitosa* exposed to ambient pH (out of the vent) and mean pH_T 7.8, 7.5 and 7.3. The cluster analysis was performed using the Fingerprint II software, Pearson correlation (0–100%)

Bacteroidetes and *Vibrionales* were detected in response to reduction in pH (similar to the changes observed in the laboratory studies). It is important to emphasize that both corals did not show any sign of disease or physiological stress following the incubation at the lower pH sites.

Overall, our results suggest no significant sign of stress upon exposure to reduced pH in the field. Yet, it is important to note that there are no corals or any other calcifying organisms growing naturally along the pH gradient. This may be a result of competition with sea grass and algae, which outcompete almost all other organisms under these high pCO_2/high light conditions. It should also be emphasized that the difference in results between the laboratory and field studies may be related to the fact that different corals were examined (*A. eurystoma* in laboratory vs. *C. caespitosa* and *B. europaea* in field studies); this is supported by the fact that different influences have been seen between the different coral species in the field study and this should be addressed in future work (see below). However, another important difference is the limited ability of the laboratory conditions to mimic the changing and dynamic parameters in the field, particularly when discussing microbial communities that are very dynamic and can rapidly change. There is no doubt that laboratory experiments are crucial to begin and understand the impact of reduced pH on marine organisms, but the results and conclusions may better be challenged in experiments performed in the natural environment.

11.5 Conclusions and Unresolved Questions

The current and predicted increase in atmospheric CO_2 concentration and ocean acidification highlight the need to elucidate the impact of these parameters on the marine environment. It seems that in addition to the influence on the coral host, changes in the microbial communities are inevitable. Controlled laboratory experiments demonstrated opposite results regarding the physiological state of the coral, while the microbial community showed a clear increase in bacteria associated with stress or diseased corals (Vega–Thurber et al. 2009; Meron et al. 2011). The field study further demonstrated changes in the microbial community, but these changes varied between the corals examined. One major drawback of these studies is that different corals were examined and that corals do not grow normally at the studied field site. Therefore, a comparison of the same coral species in natural and laboratory conditions should be done in order to understand if and how the laboratory results simulate the field process. Furthermore, if available, a coral naturally living in a pH gradient should be examined.

Another important question that has yet to be examined in the context of the microbial community is what impact will returning the coral to its original pH after exposure to a lower pH have on the coral microbial community. Similar experimental approach was carried out by Fine and Tchernov (2007) who examined the impact of the shift from low pH back to ambient pH on coral calcification and demonstrated that the corals returned to calcify and build their skeleton (Fine and Tchernov 2007). It will be interesting to see if and how the microbial community, stress signs, and coral physiology change upon the shift back. Will the community return to the previous state? Will it keep the new community or rather shift to yet another new steady state (Fig. 11.5)?

Finally, in order to understand the significance of the changes and the role of the symbiont microbial communities to the health, tolerance, and adaptation of the coral host, it is more important to know what the different bacteria are actually doing in the host rather than who they are. It may well be that although the taxonomic diversity changes, the functional diversity does not change and one species is replaced by a different species that carries the same functional activity. To address this question, more detailed metagenomic studies should be conducted in corals maintained in natural pH gradient field sites as was done by Vega–Thurber et al. (2009) in laboratory conditions. Such studies that also combine monitoring of the coral physiology can begin to address if there are any changes in function of the

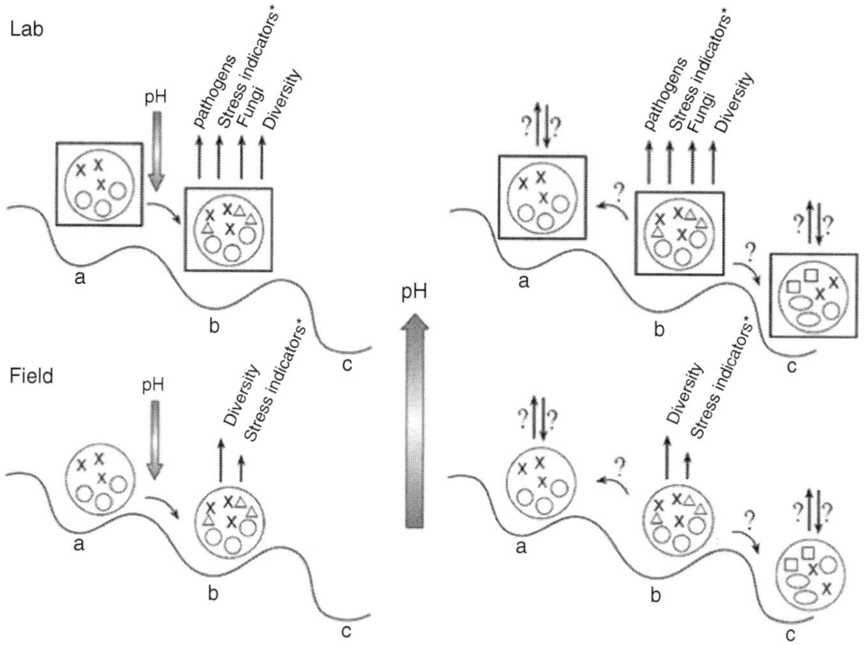

Fig. 11.5 Summary of the Influence of the decrease in pH on coral microbial community based on laboratory and field studies. (*Left*) The results observed upon shift to lower pH in laboratory and field studies. (*Right*) A prediction of the impact on the microbial community upon a shift back to high pH (**a**) change back to the community that was originally present at high pH; (**b**) no change or (**c**) a change to yet another new microbial community. *"Stress indicators" refers to changes in bacterial communities/processes that may induce stress

associated bacterial community, and how this contributes to physiological state and health of the coral host. This information will greatly enhance our understanding on how ocean acidification will impact coral reefs.

References

Anthony KR, Kline DI, Diaz-Pulido G, Dove S, Hoegh-Guldberg O (2008) Ocean acidification causes bleaching and productivity loss in coral reef builders. Proc Natl Acad Sci USA 105:17442–17446

Arboleda M, Reichardt W (2009) Epizoic communities of prokaryotes on healthy and diseased scleractinian corals in Lingayen Gulf, Philippines. Microb Ecol 57:117–128

Bourne DG, Munn CB (2005) Diversity of bacteria associated with the coral *Pocillopora damicornis* from the Great Barrier Reef. Environ Microbiol 7:1162–1174

Bourne D, Lida Y, Uthicke S, Smith-Keune C (2007) Changes in coral-associated microbial communities during a bleaching event. ISME J 2:350–363

Caldeira K (2007) What corals are dying to tell us about CO_2 and ocean acidification. Oceanography 20:188–195

Connell H (1978) Diversity in tropical rain forests and coral reefs. Science 199:1302–13100

Cooney RP, Pantos O, Le Tissier MD, Barer MR, O'Donnell AG, Bythell JC (2002) Characterization of the bacterial consortium associated with black band disease in coral using molecular microbiological techniques. Environ Microbiol 4:401–413

Feely RA, Sabine CL, Lee K, Berelson W, Kleypas J, Fabry VJ, Millero FJ (2004) Impact of anthropogenic CO_2 on the $CaCO_3$ system in the oceans. Science 305:362–366

Fenical W, Jensen PR (1993) Marine microorganisms: a new biomedical resource. In: Attaway DH, Zaborsky OR (eds) Marine biotechnology. Plenum Press, New York, pp 419–457

Fine M, Tchernov D (2007) Scleractinian coral species survive and recover from decalcification. Science 315:1811

Frias-Lopez J, Zerkle AL, Bonheyo GT, Fouke BW (2002) Partitioning of bacterial communities between seawater and healthy, black band diseased, and dead coral surfaces. Appl Environ Microbiol 68:2214–2228

Frias-Lopez J, Klaus JS, Bonheyo GT, Fouke BW (2004) Bacterial community associated with black band disease in corals. Appl Environ Microbiol 70:5955–5962

Garren M, Raymundo L, Guest J, Harvell CD, Azam F (2009) Resilience of coral-associated bacterial communities exposed to fish farm effluent. PLoS ONE 4:e7319

Gattuso JP, Frankignoulle M, Bourge I, Romaine S, Buddemeier RW (1998) Effect of calcium carbonate saturation of seawater on coral calcification. Glob Planet Change 18(18):37–46

Golubic S, Friedmann I, Schneider J (1981) The lithobiontic ecological niche, with special reference to microorganisms. J Sediment Res 51:475–478

Guppy R, Bythell JC (2006) Environmental effects on bacterial diversity in surface mucus layer of the reef coral *Montastraea faveolata*. Mar Ecol Prog Ser 328:133–142

Hall-Spencer JM, Rodolfo-Metalpa R, Martin S, Ransome E, Fine M, Turner SM, Rowley SJ, Tedesco D, Buia MC (2008) Volcanic carbon dioxide vents show ecosystem effects of ocean acidification. Nature 454:96–99

IPCC – The International Panel on Climate Change 2007 (Report). http://www.ipcc.ch

Jensen PR, Fenical W (2002) Secondary metabolites from marine fungi. In: Hyde KD (ed) Fungi in marine environments. Fungal Diversity Press, Hong Kong

Kooperman N, Ben-Dov E, Kramarsky-Winter E, Barak Z, Kushmaro A (2007) Coral mucus-associated bacterial communities from natural and aquarium environments. FEMS Microbiol Lett 276:106–113

Lampert Y, Kalman D, Nitzan Y, Dubinsky Z, Behar A, Hill RT (2008) Phylogenetic diversity of bacteria associated with the mucus of Red Sea corals. FEMS Microbiol Ecol 64:187–198

Le Campion-Alsumard T, Golubic S, Hutchings PA (1995a) Microbial endoliths in skeletons of live and dead corals: *Porites lobata* (Moorea, French Polynesia). Mar Ecol Prog Ser 117:149–157

Le Campion-Alsumard T, Golubic S, Priess K (1995b) Fungi in corals – symbiosis or disease – interaction between polyps and fungi causes pearl-like skeleton biomineralization. Mar Ecol Prog Ser 117:137–147

Lenihan HS, Micheli F, Shelton SW, Peterson CH (1999) The influence of multiple environmental stressors on susceptibility to parasites: an experimental determination with oysters. Limnol Oceanogr 44:910–924

Lozupone C, Knight R (2005) UniFrac: a new phylogenetic method for comparing microbial communities. Appl Environ Microbiol 71:8228–8235

Meron D, Atias E, Iasur-Kruh L, Elifantz H, Minz D, Fine M, Banin E (2011) The impact of reduced pH on the microbial community of the coral *Acropora* eurystoma. ISME J 5:51–60

Orr JC, Fabry VJ, Aumont O, Bopp L, Doney SC, Feely RA, Gnanadesikan A, Gruber N, Ishida A, Joos F, Key RM, Lindsay K, Mair-Reimer E, Matear R, Monfray P, Mouchet A, Najjar RG, Plattner GK, Rodgers KB, Sabine CL, Sarmiento JL, Schlitzer R, Slater RD, Totterdell IJ, Weirig MF, Yamanaka Y, Yool A (2005) Anthropogenic ocean acidification over the twenty-first century and its impact on calcifying organisms. Nature 437:681–686

Petit JR, Jouzel J, Raynaud D, Barkov NI, Barnola JM, Basile I, Bender M, Chappellaz J, Davis M, Delayque G, Delmotte M, Kotlyakov VM, Legrand M, Lipenkov VY, Lorius C, Pepin L,

Ritz C, Saltzman E, Stievenard M (1999) Climate and atmospheric history of the past 420,000 years from the Vostok ice core, Antarctica. Nature 399:429–436
Reshef L, Koren O, Loya Y, Zilber-Rosenberg I, Rosenberg E (2006) The coral probiotic hypothesis. Appl Environ Microbiol 8:2067–2073
Ritchie KB (2006) Regulation of microbial population by coral surface mucus and mucus-associated bacteria. Mar Ecol Prog Ser 322:1–14
Rohwer F, Seguritan V, Azam F, Knowlton N (2002) Diversity and distribution of coral-associated bacteria. Mar Ecol Prog Ser 243:1–10
Rosenberg E, Koren O, Reshef L, Efrony R, Zilber-Rosenberg I (2007) The role of microorganisms in coral health, disease and evolution. Nat Rev Microbiol 5:355–362
Rosenberg E, Kushmaro A, Kramarsky-Winter E, Banin E, Yossi L (2009) The role of microorganisms in coral bleaching. ISME J 3:139–146
Ross PS, De Swart RL, Van Loveren H, Osterhaus ADME, Vos JG (1996) The immunotoxicity of environmental contaminants to marine wildlife: a review. Annu Rev Fish Dis 6:151–165
Sekar R, Mills DK, Remily ER, Voss JD, Richardson LL (2006) Microbial communities in the surface mucopolysaccharide layer and the black band microbial mat of black band-diseased *Siderastrea siderea*. Appl Environ Microbiol 72:5963–5973
Sekar R, Kaczmarsky L, Richardson LL (2008) Microbial community composition of black band disease on the coral host *Siderastrea siderea* from three regions of the wider Caribbean. Mar Ecol Prog Ser 362:38–98
Sunagawa S, DeSantis TZ, Piceno YM, Brodie EL, DeSalvo MK, Voolstra CR, Weil E, Andersen GL, Medina M (2009) Bacterial diversity and White Plague Disease-associated community changes in the Caribbean coral *Montastraea faveolata*. ISME J 3:512–521
Tribollet A, Payri C (2001) Bioerosion of the crustose coralline alga *Hydrolithon onkodes* by microborers in the coral reefs of Moorea, French Polynesia. Oceanol Acta 24:329–342
Tribollet A, Atkinson M, Langdon C (2006) Effects of elevated pCO_2 on epilithic and endolithic metabolism of reef carbonates. Global Change Biol 12:2200–2208
Tribollet A, Godinot C, Atkinson M, Langdon C (2009) Effects of elevated pCO_2 on dissolution of coral carbonates by microbial euendoliths. Global Biogeochem Cycles 23:GB3008
Vega-Thurber R, Willner-Hall D, Rodriguez-Mueller B, Desnues C, Edwards RA, Angly F, Dinsadale E, Kelly L, Rohwer F (2009) Metagenomic analysis of stressed coral holobionts. Environ Microbiol 11:2148–2163
Walsh DA, Papke RT, Doolittle WF (2005) Archaeal diversity along a soil salinity gradient prone to disturbance. Environ Microbiol 10:1655–1666

Part IV
Microbes in Mammalian Health and Disease

Chapter 12
Toward the Educated Design of Bacterial Communities

Shiri Freilich and Eytan Ruppin

Abstract The composition of bacterial communities is a major factor in human health. Variations in the identity and abundance of species within a bacterial community affect its metabolic potential and hence have important medical implications. Such variations may also affect managed ecosystems such as bioreactors and agricultural fields. Thus, considering its medical and ecological implications, successful modeling of bacterial communities is likely to have broad consequences. Experimental and computational tools are now becoming available for the modeling of bacterial interactions, taking into account the identity of the interacting species and the available nutritional supplies. In this chapter, we review recent advances in the field of system biology and metabolic modeling, allowing a shift in focus from the species level to the community level, enabling a systematic exploration of species' ecology and lifestyle, inter-species interactions, and community structure. These new approaches will hopefully provide a platform for the educated design of bacterial consortia optimized toward bioremediation and bio-production applications.

12.1 Introduction

In natural environments, individual organisms do not live in isolation. Bacterial species occupy ecological niches, in most cases forming together dynamic communities. The composition and interactions within these communities change over time and in response to environmental stimuli. Two species co-inhabiting an ecological niche can compete for resources, cooperate to maximize resource

S. Freilich (✉) • E. Ruppin
Sackler Faculty of Medicine and The Blavatnik School of Computer Science, Tel Aviv University, Tel Aviv, Israel
e-mail: ruppin@post.tau.ac.il

E. Rosenberg and U. Gophna (eds.), *Beneficial Microorganisms in Multicellular Life Forms*, 177
DOI 10.1007/978-3-642-21680-0_12, © Springer-Verlag Berlin Heidelberg 2011

utilization, divide the available resources between them, or combine competition, cooperation, and division to various degrees. The relative fitness of a species in a certain ecological niche and, hence, the structure and composition of that particular community will be to a large extent determined by the ecological strategies taken by all organisms in the community and by their interactions.

The composition of bacterial communities is a major factor in human health. The microorganisms that live inside and on humans are estimated to outnumber human somatic and germ cells by a factor of ten. The genomes of these microbial symbionts provide functions that humans did not evolve on their own (Turnbaugh et al. 2007). The microbiome of the human gut, for example, is composed of different lineages with a capacity to communicate with one another and with the host; it consumes, stores, and redistributes energy; and it mediates physiologically important chemical transformations (Backhed et al. 2005). Variations in the identity and abundance of species within this community affect its metabolic potential and hence have important medical implications. Obesity, for example, is associated with changes in the relative abundance of bacterial phyla (Turnbaugh et al. 2006). Moreover, variations in the relative abundance of species in a community may shift the ecological balance of a given niche allowing, for example, the outburst of pathogens (Follows et al. 2007). Such variations may also affect managed ecosystems such as bioreactors and agricultural fields (Lenski et al. 1998; Sloan et al. 2007).

Experimental and computational tools are now becoming available for the modeling of bacterial interactions, taking into account the identity of the interacting species and the available nutritional supplies. Here, we provide a short review of these innovative methods, some developed by us, for processing high-throughput information into environmental and ecological information.

12.2 From Genomic to Environmental Representations

Progress in the reconstruction of genome-wide metabolic networks has led to the development of network-based computational approaches for linking an organism with its biochemical habitat. Briefly, the consecutive nature of the reactions in metabolic pathways means that they can be modeled in the form of a network of enzymes and chemical transformations, and graph theory can be used to represent and understand metabolism (Papin et al. 2003). The availability of many completely sequenced genomes together with several generic metabolic schemes (Kanehisa and Goto 2000) has led to the reconstruction of the metabolic networks of hundreds of species across the tree of life (Kreimer et al. 2008). A recently published "seed algorithm" (developed as part of our previous work (Borenstein et al. 2008)) allows to use metabolic networks for predicting the set of metabolites an organism consumes from its surroundings. This algorithm takes as an input the metabolic network of a given species and looks for source components (i.e., components with no incoming edges); those components cannot be synthesized by the species and are hence exogenously acquired. This set of compounds reflects the metabolic environment of a species (Lozada-Chavez et al. 2006). As described below, beyond predicting the biochemical habitat, such approaches are further exploited to study the interactions of microbes with other species thriving in similar habitats.

12.3 Using Environmental Information to Explore Genotype–Phenotype Relationship

The seed algorithm described above allows the computation of species-specific habitats. Aggregating the habitats computed for a set of species (for which a metabolic network is available), we obtained an ensemble of predicted natural habitats corresponding to hundreds of species across the tree of life. This ensemble represents the broadest ecological view provided by current data, allowing us to examine the viability of species across a wide-range of ecological niches. Traditionally, flux-balance approaches were used for estimating the ability of a species to grow on a given medium (Edwards and Palsson 2000), but the underlying stoichiometric metabolic models were until very recently available for only a few selected species. Topological-driven approaches (requiring only the network topological backbone and not a full-blown stoichiometric model) have been recently shown to be sufficient for estimating growth by studying the ability of an organism to successfully expand its metabolic network so it produces a set of target metabolites that are essential for growth (Ebenhoh et al. 2004; Wunderlich and Mirny 2006).

By studying species' viability across the collection of natural-like environments, we developed a new model for predicting robustness across species and across environments (Freilich et al. 2010a). For each species, we systematically explored its viability in each environment following simulating knock-outs in each of its genes, independently. We observe that variations among species in their level of network genetic robustness reflect adaptations to different ecological niches and lifestyles. The extent of conditional-dependent robustness – that is the fraction of genes that are non-essential under specific environments – is strongly associated with the environmental diversity of species (specialized or generalist). The extent of condition-independent robustness – that is the fraction of genes that are non-essential across all conditions examined – is associated with the corresponding metabolic capacities. That is, higher robustness is observed in fast growers or in organisms with an extensive production of secondary metabolites. The association between the level of robustness and the corresponding metabolic capacity of a species can be explained either by a selection to increase flux – similarly to the effect of gene dosage (Papp et al. 2004) – or by selection to optimize the metabolic efficiency under given conditions, for example by alternating between routes in accordance with the corresponding substrate concentrations (Helling 1994; Pfeiffer et al. 2001). Thus, the design of metabolic networks (as viewed by the presence of alternative pathways) represents a species-specific adaptation to both its needs and its environment.

12.4 From Environmental Information to Ecological Information

By describing the viability of all species over all environments, we have constructed the first large-scale ecological model describing environmental characteristics from both a species and a habitat point of view (Freilich et al. 2009). From a species perspective, the model describes the range of environments it can inhabit. In support of our observations, we observe that the species-specific level of environmental diversity is in strong agreement with acceptable, general measures of environmental diversity (Madan Babu et al. 2006; Parter et al. 2007). From an environmental perspective, it estimates for each environment the range of species it can contain. As an indication to the ecological plausibility of our model we observe that the level of population of different environments is compatible with ecological knowledge: soil bacteria inhabit the most densely populated environments, and obligatory symbionts inhabit sparsely populated environments (Parter et al. 2007; Mavromatis et al. 2007; Tringe et al. 2005) (Fig. 12.1). These reconstructions are then used to investigate the typical ecological strategies taken by organisms in terms of two basic species-specific measures. Characteristically, most bacterial organisms adopt one of the two main ecological strategies: (a) a specialized niche with little co-habitation, associated with a typical slow rate of growth versus and (b) ecological diversity with intense co-habitation, associated with a typical fast rate of growth. The pattern observed suggests a universal principle where metabolic flexibility is associated with a need to grow fast, possibly in the face of competition.

12.5 From Environmental Information to Community Information

This new ability to produce a quantitative description of the growth rate – metabolism – community relationship lays a computational foundation for the study of a variety of aspects of the communal metabolic life. As a first step, based on the topological properties of the metabolic networks, we systematically calculated a competition score to all pairwise combinations, describing their potential to share resources in an optimal media (Freilich et al. 2010b). Then we explored the competition – growth rate – respiration mode relationships in a collection of natural-like communities, constructed based on species co occurrence rate at the scientific literature. As a general trend we observe that most communities are either characterized by fast growth rate associated with low competition or slow growth potential associated with intense competition (Fig. 12.2). Fast-growing communities have a characteristic lower yield (Fig. 12.2). Notably, this division of clusters into low competition–fast growth–low yield versus high competition–slow growth–high yield category is in agreement with the r/K selection theory. The r/K

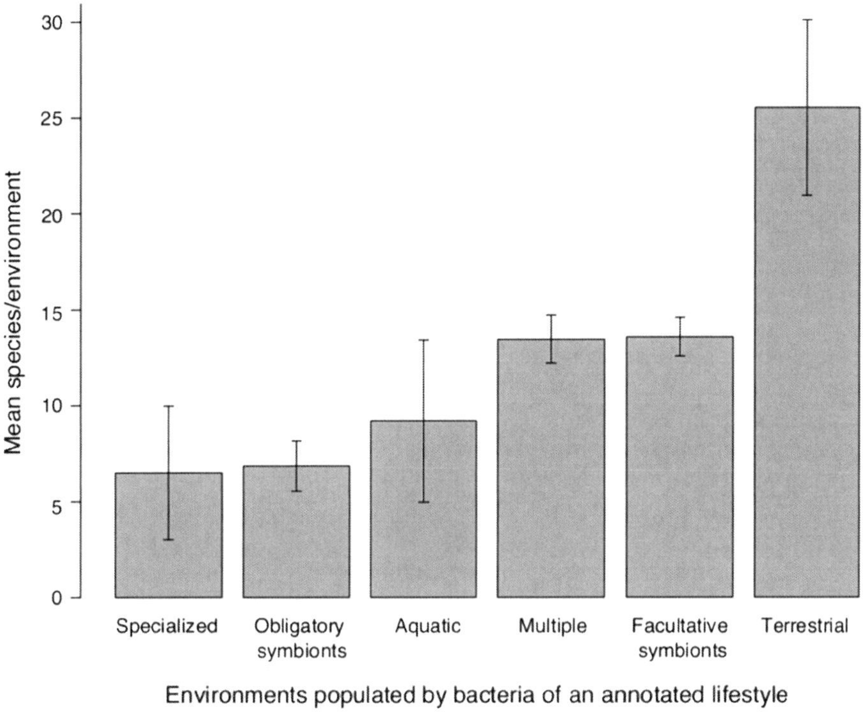

Fig. 12.1 Level of populations in predicted environments (Figure is taken from Freilich et al. (2009)). By applying a computational approach we constructed a set of natural-like environments. In each environment we systematically tested the viability of hundreds of bacterial species. Given a specific organism and an environment, an organism is considered viable in this environment and it can produce a defined group of essential metabolites required for growth (e.g., amino acids, nucleotides, lipids). This is examined by using a network expansion algorithm (Ebenhoh et al. 2004) that outputs an activated metabolic subnetwork, and verifying that the expanded subnetwork produces all essential metabolites. This process yields the *environmental viability matrix* whose rows denote the species and columns denote the metabolic environments, and binary entries denote the corresponding viability. From this matrix, we deduce for each environment the number of species it populates and for each species the number of environment it inhabits. Number of environments in each lifestyle (ordered as in the figure) are 6, 48, 5, 153, 121, 15 (environments can populate species of more than a single lifestyle). *Bars* show the standard error

selection theory, originally suggested for animals and plants (MacArthur and Wilson 1967), aims to explain the choice between slow and fast growth, given the environmental conditions and the level of competition encountered by a species. In general, r-strategies are adapted to maximize the rate of growth while K-strategies are adapted to compete and survive when resources are limited (Pianka 1970). When applied to bacterial species, r-selection is suggested to be typical of communities occupying rich metabolic environments where species can exhibit high growth rate and low yield; K-selection is suggested to be typical of species occupying less abundant environments, these species typically exhibit a lower

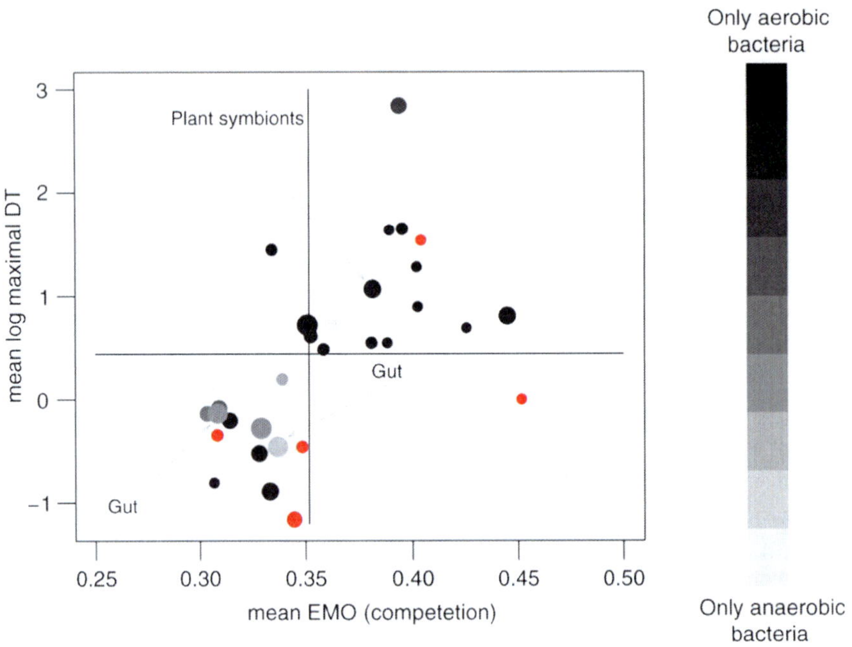

Fig. 12.2 Level of metabolic competition versus growth rate in 38 natural-like communities (Figure is taken from (Freilich et al. (2010b)). Sizes of dots correspond to the sizes of the clusters. Colors of dots correspond to the anaerobic/aerobic ratio. The anerobic to aerobic ratio in a cluster is used for estimating the typical growth yield – the efficiency of the conversion of substrate into biomass, whereas low yield is characteristic of anaerobic species and high yield is characteristic of aerobic species (Pfeiffer et al. 2001). *Red dots* correspond to cluster which do not contain any aerobic or anaerobic bacteria (alternative annotations include facultative, microaerophilic, and unknown species). The plot is divided according to median values of the axes (Abbreviations *EMO* effective metabolic overlap, *DT* duplication time)

growth potential but a higher capability to compete for substrates (Fierer and Jackson 2006; Torsvik and Ovreas 2002). Hence, the low competition–fast growth–low yield clusters correspond with the characteristics of r-selection whereas high competition–slow growth–high yield clusters correspond with K-selection. Clusters enriched with gut bacteria – a rich and highly populated metabolic environment – fall into the r-selection category (Fig. 12.2). Clusters which fall into the K-selection are typically composed of host-associated pathogens and symbionts; e.g., in cluster 13, the majority of species populate plant's tissues (9 out of 13 are plant's symbionts and pathogens; Fig. 12.2). The typical slow growth observed in these clusters possibly reflects the basic strategy of many host-associated bacteria that do not scavenge too much of the basic nutrients essential for the host metabolism since, otherwise, their host will soon starve to death and the bacteria will lose their protective niche (Joseph and Goebel 2007). Possibly, K-selection is also typical of free-living communities and the lack of such clusters in our data set is the outcome of biases in the collection of sequenced bacteria toward easily cultured, fast-grower species. For example, >99% of the soil bacteria (mostly, fastidious growers) have not been cultivated and characterized and hence the soil ecosystems are, to a large extent, uncharted (Torsvik and Ovreas 2002). The inclusion of uncultivated species in our data set can possibly lead to the identification of K-selection in communities dominated by free-living species. Overall, the computational-based predictions for the inter-relationship between growth rate, metabolic yield, and level of competition observed in these communities correspond with patterns expected according to the r/K-selection theory, thus providing the first large-scale account of this fundamental ecological theory and pointing at the importance of computational approaches for gaining a system-level understanding of the design principles of natural communities.

12.6 Using Metabolic Modeling for Simulating Growth in a Multi-species System

Metabolic modeling is one of the most advanced tools allowing processing genomic information into a system-level description of cellular activity (Palsson 2009). Briefly, metabolic modeling requires the processing of genomic driven information into a network description of metabolism. Given the metabolic network, modeling approaches such as Constrains Based Modeling (CBM) quantify the flow of material through metabolism pointing at the activity of the cell at a given time point. However, the scarcity of manually curated models has precluded such a community-scale study. Indeed previous computational studies of microbial ecology and metabolism, as described above, relied solely on network representations of enzymes and reactions (Papin et al. 2003; Freilich et al. 2009, 2010a; Ebenhoh and Handorf 2009). Very recently, an automatic high-throughput reconstruction pipeline has generated more than 100 genome-scale metabolic bacterial models spanning 13 bacterial divisions (Henry et al. 2010). Whereas most frequently, metabolic modeling approaches are used for predicting relative fitness at a cellular system by comparing its activity under optimal conditions to its activity following perturbations (modifications of the growth media or knockouts of reactions) (Edwards et al. 2002), recently, Stoichiometric-based metabolic models were shown to provide accurate predictions for the pattern of metabolic interactions in bacterial two-species systems (Wintermute and Silver 2010a; Rual et al. 2005; Stolyar et al. 2007).

In a recent work, making use of this innovative collection, we analyzed 118 constraint-based metabolic-models of bacteria to systematically simulate co-growth of their respective pairwise combinations (Freilich et al.). In this current study, we put forward a new conceptual framework for modeling bacterial metabolic interactions and study the competitive and cooperative interactions potentially occurring between all distinct 6,903 pairwise combinations of bacterial species, providing the first large-scale systematic description of species' interactions. Overall, we find that whereas the level of competition is linearly associated with the level of resource overlap (as expected), the dependency of cooperation on resource overlap demonstrates a quadratic-like association decaying at extreme levels of resource overlap. Interaction predictions derived from our framework are supported by *in vitro* experimental studies, designed to reveal induced shifts in the type of interactions following media modifications.

The ability to design novel interactions as well as to study existing ones suggests that microbial experiments can complement and extend classical plant and animal ecology, in which many of the principles of biological interactions were first described (Wintermute and Silver 2010b). The co-existence patterns between bacterial species are, thanks to metagenomic surveys, far more comprehensively described in comparison to the documentation of co-existence patterns in other kingdoms. Crossing bacterial co-existence patterns with computationally predicted interaction patterns, we conduct the first large-scale study probing the role of competition and cooperation in shaping bacterial community composition. We find high levels of competition between bacteria belonging to different, mutually-exclusive niches, in support of the role of competition in community assembly. Most cooperative interactions are directional (give-take) and seemingly yield a clear benefit only to the receiver side, raising the question of their advantage for giver species. Notably, addressing this quandary, we find that, within natural bacterial communities, bacterial species tend to form close cooperative loops resulting in an indirect benefit to all species involved. This may also suggest an explanation to the observed rise in the total number of bacteria when species' diversity increases. Finally, we demonstrate that the location of a species in the network of give-take interactions in its community affects its survival prospects in consortia.

From a computational perspective, this study presents the first application of the emerging field of genome scale metabolic modeling for the large-scale study of metabolic co-growth and interactions across bacterial species.

12.7 Unresolved Questions and Future Challenges

Up to date it was difficult to predict which bacteria can stably co-exist, let alone collaborate, making the design of beneficial microbial consortia extremely difficult. Our approach for using metabolic modeling for predicting pairwise interactions is obviously not without limitations; it is aimed solely at the metabolic dimension and lacks information on the true metabolic composition of the environments considered. Yet, it succeeds in uncovering fundamental ecological principles in a systematic fashion. The successful utilization of metabolic modeling for revealing the design principles of bacterial communities provides a computational basis for many exciting future applications. These include the design of "expert" communities for bioremediation, whereas currently the selection of member species is done by intelligent guesswork, as well as the rational design of probiotic administration, and the identification of species that may metabolically outcompete pathogenic species. Following are two of the many possible directions for future research and applications:

1. *Toward the educated design of select communities.* Data and approaches described in the previous sections will be applied for a more extensive study of selected communities. Here, we shortly review challenges and opportunities in the modeling of the gut microbiome. However, similar approaches can be taken toward the modeling of other selected communities.

In gut communities, DNA sequencing-based approaches have already shed some light on the composition and diversity of the microbiome (Turnbaugh et al. 2009; Kuczynski et al. 2010; Ley et al. 2008; Turnbaugh and Gordon 2009). A complementary approach for studying the microbiome at the phenotypic level is metabonomics. Recent metabonomic studies have shown that ^{1}H NMR-based metabolic analysis of fecal extracts can provide insights into interspecies metabolic differences (Saric et al. 2008), metabolic response to nutritional intervention (Jacobs et al. 2008), and carry diagnostic information for diseases, including Crohn's disease and ulcerative colitis (Marchesi et al. 2007). By integrating metagenomic and meta-proteomic information on species' distribution together with metabolic modeling, we aim to explore the function-composition relationship across bacterial communities.

Analysis will be based on the following information (for each sample/individual):

(a) Composition of community including information on the diversity of species and their relative abundance.
(b) Information on metabolic environment.
(c) Clinical status (including details on the medical status of the source, its age, and its gender, and details on the types of medications the source was taking).

Considering this biotic and a-biotic description of the community, metabolic-modeling will be used for exploring the following:

(a) *Susceptibility to pathogens of different communities.* Based on antagonistic and positive interactions it is possible to predict favorable and unfavorable community structure for the invasion of pathogens.
(b) *Relationship between antibiotic treatment and community structure.* By simulating the effect of antibiotics on community structure and the subsequent exclusion of key players it is possible to predict the composition of the emerging community.
(c) *Relationships between nutrition and community structure.* The efficiency of food processing of several diets by different communities can be predicted as a first step toward designing optimized gut communities.

For the gut community, the ultimate goal is to establish a theoretical framework that will allow the future manipulations of gut microbiota in order to promote human health. Similarly, the same approaches can be used for the educated design of bacterial communities in agricultural land or of communities involved in the degradation of pollutants.

2. *Exploring function differentiation within the holobiont*. The hologenome theory of evolution considers the holobiont (the animal or plant with all of its associated microorganisms) as a unit of selection in evolution (Rosenberg et al. 2007; Zilber-Rosenberg and Rosenberg 2008). The use of metabolic modeling for studying the dynamics in the interactions between a host and its community of microorganisms can shed light on the role of function differentiation with the hologenome community and its effect on the stability and resistance to pathogens. A possible case study is the coral holobiont. Whereas until recently, the genomic sequence resources available for corals were limited (van Oppen and Gates 2006), fragmented reads from pilot shotgun sequencing projects are now available for three corals (*Acropora millepora*, *Acropora palmata*, and *Porites lobata*). In addition, a large-scale, comprehensive, transcriptomic resource for *Aiptasia pallida* and its dinoflagellate symbiont was recently published (Sunagawa et al. 2009). In addition, the advent of next-generation sequencing methods allows the description of the holobiont-individual communities of microorganisms (Sunagawa et al. 2009). Integrating the genomic information into a framework of metabolic modeling allows exploring the functional role of different species in the holobiont as well as functional implications of variations in community structure.

References

Backhed F, Ley RE, Sonnenburg JL, Peterson DA, Gordon JI (2005) Host-bacterial mutualism in the human intestine. Science 307:1915–1920 (New York, NY)

Borenstein E, Kupiec M, Feldman MW, Ruppin E (2008) Large-scale reconstruction and phylogenetic analysis of metabolic environments. Proc Natl Acad. Sci. USA 105:14482–14487

Ebenhoh O, Handorf T (2009) Functional classification of genome-scale metabolic networks. EURASIP J Bioinform Syst Biol 2009:570456.

Ebenhoh O, Handorf T, Heinrich R (2004) Structural analysis of expanding metabolic networks. Genome Inf 15:35–45

Edwards JS, Palsson BO (2000) The Escherichia coli MG1655 in silico metabolic genotype: its definition, characteristics, and capabilities. Proc Natl Acad Sci USA 97:5528–5533

Edwards JS, Ramakrishna R, Palsson BO (2002) Characterizing the metabolic phenotype: a phenotype phase plane analysis. Biotechnol Bioeng 77:27–36

Fierer N, Jackson RB (2006) The diversity and biogeography of soil bacterial communities. Proc Natl Acad Sci USA 103:626–631

Follows MJ, Dutkiewicz S, Grant S, Chisholm SW (2007) Emergent biogeography of microbial communities in a model ocean. Science 315:1843–1846 (New York, NY)

Freilich S, Kreimer A, Borenstein E, Yosef N, Sharan R, Gophna U, Ruppin E (2009) Metabolic-network-driven analysis of bacterial ecological strategies. Genome Biol 10:R61

Freilich S, Kreimer A, Borenstein E, Gophna U, Sharan R, Ruppin E (2010a) Decoupling environment-dependent and independent genetic robustness across bacterial species. PLoS Comput Biol 6:e1000690

Freilich S, Kreimer A, Meilijson I, Gophna U, Sharan R, Ruppin E (2010b) The large-scale organization of the bacterial network of ecological co-occurrence interactions. Nucleic Acids Res 38:3857–3868

Freilich S, Zarecki R, Eilam O, Shtifman Segal O, Henry CS, Kupiec M, Gophna U, Sharan. R, Ruppin E Competitive and cooperative metabolic interactions in bacterial communities. Under Revision

Helling RB (1994) Why does Escherichia coli have two primary pathways for synthesis of glutamate? J Bacteriol 176:4664–4668

Henry CS, DeJongh M, Best AA, Frybarger PM, Linsay B, Stevens RL (2010) High-throughput generation, optimization and analysis of genome-scale metabolic models. Nat Biotechnol 28:977–982

Jacobs DM, Deltimple N, van Velzen E, van Dorsten FA, Bingham M, Vaughan EE, van Duynhoven J (2008) (1)H NMR metabolite profiling of feces as a tool to assess the impact of nutrition on the human microbiome. NMR Biomed 21:615–626

Joseph B, Goebel W (2007) Life of Listeria monocytogenes in the host cells' cytosol. Microbes Infect/Institut Pasteur 9:1188–1195

Kanehisa M, Goto S (2000) KEGG: kyoto encyclopedia of genes and genomes. Nucleic Acids Res 28:27–30

Kreimer A, Borenstein E, Gophna U, Ruppin E (2008) The evolution of modularity in bacterial metabolic networks. Proc Natl Acad Sci USA 105:6976–6981

Kuczynski J, Costello EK, Nemergut DR, Zaneveld J, Lauber CL, Knights D, Koren O, Fierer N, Kelley ST, Ley RE, Gordon JI, Knight R (2010) Direct sequencing of the human microbiome readily reveals community differences. Genome biol 11:210

Lenski RE, Mongold JA, Sniegowski PD, Travisano M, Vasi F, Gerrish PJ, Schmidt TM (1998) Evolution of competitive fitness in experimental populations of E. coli: what makes one genotype a better competitor than another? Antonie van Leeuwenhoek 73:35–47

Ley RE, Lozupone CA, Hamady M, Knight R, Gordon JI (2008) Worlds within worlds: evolution of the vertebrate gut microbiota. Nat Rev Microbiol 6:776–788

Lozada-Chavez I, Janga SC, Collado-Vides J (2006) Bacterial regulatory networks are extremely flexible in evolution. Nucleic Acids Res 34:3434–3445

MacArthur RH, Wilson EO (1967) The theory of island biogeography. Princeton University Press, Princeton

Madan Babu M, Teichmann SA, Aravind L (2006) Evolutionary dynamics of prokaryotic transcriptional regulatory networks. J Mol Biol 358:614–633

Marchesi JR, Holmes E, Khan F, Kochhar S, Scanlan P, Shanahan F, Wilson ID, Wang Y (2007) Rapid and noninvasive metabonomic characterization of inflammatory bowel disease. J Proteome Res 6:546–551

Mavromatis K, Ivanova N, Barry K, Shapiro H, Goltsman E, McHardy AC, Rigoutsos I, Salamov A, Korzeniewski F, Land M, Lapidus A, Grigoriev I, Richardson P, Hugenholtz P, Kyrpides NC (2007) Use of simulated data sets to evaluate the fidelity of metagenomic processing methods. Nat Method 4:495–500

Palsson B (2009) Metabolic systems biology. FEBS Lett 583:3900–3904

Papin JA, Price ND, Wiback SJ, Fell DA, Palsson BO (2003) Metabolic pathways in the post-genome era. Trends Biochem Sci 28:250–258

Papp B, Pal C, Hurst LD (2004) Metabolic network analysis of the causes and evolution of enzyme dispensability in yeast. Nature 429:661–664

Parter M, Kashtan N, Alon U (2007) Environmental variability and modularity of bacterial metabolic networks. BMC Evol Biol 7:169

Pfeiffer T, Schuster S, Bonhoeffer S (2001) Cooperation and competition in the evolution of ATP-producing pathways. Science 292:504–507

Pianka E (1970) On r- and K-selection. Am Nat 104:592–597

Rosenberg E, Koren O, Reshef L, Efrony R, Zilber-Rosenberg I (2007) The role of microorganisms in coral health, disease and evolution. Nat Rev 5:355–362

Rual JF, Venkatesan K, Hao T, Hirozane-Kishikawa T, Dricot A, Li N, Berriz GF, Gibbons FD, Dreze M, Ayivi-Guedehoussou N, Klitgord N, Simon C, Boxem M, Milstein S, Rosenberg J, Goldberg DS, Zhang LV, Wong SL, Franklin G, Li S et al (2005) Towards a proteome-scale map of the human protein-protein interaction network. Nature 437(7062):1173–1178

Saric J, Wang Y, Li J, Coen M, Utzinger J, Marchesi JR, Keiser J, Veselkov K, Lindon JC, Nicholson JK, Holmes E (2008) Species variation in the fecal metabolome gives insight into differential gastrointestinal function. J Proteome Res 7:352–360

Sloan WT, Woodcock S, Lunn M, Head IM, Curtis TP (2007) Modeling taxa-abundance distributions in microbial communities using environmental sequence data. Microb Ecol 53:443–455

Stolyar S, Van Dien S, Hillesland KL, Pinel N, Lie TJ, Leigh JA, Stahl DA (2007) Metabolic modeling of a mutualistic microbial community. Mol Syst Biol 3:92

Sunagawa S, Wilson EC, Thaler M, Smith ML, Caruso C, Pringle JR, Weis VM, Medina M, Schwarz JA (2009) Generation and analysis of transcriptomic resources for a model system on the rise: the sea anemone Aiptasia pallida and its dinoflagellate endosymbiont. BMC Genomics 10:258

Torsvik V, Ovreas L (2002) Microbial diversity and function in soil: from genes to ecosystems. Curr Opin Microbiol 5:240–245

Tringe SG, von Mering C, Kobayashi A, Salamov AA, Chen K, Chang HW, Podar M, Short JM, Mathur EJ, Detter JC, Bork P, Hugenholtz P, Rubin EM (2005) Comparative metagenomics of microbial communities. Science 308:554–557 (New York, NY)

Turnbaugh PJ, Gordon JI (2009) The core gut microbiome, energy balance and obesity. J Physiol 587:4153–4158

Turnbaugh PJ, Ley RE, Mahowald MA, Magrini V, Mardis ER, Gordon JI (2006) An obesity-associated gut microbiome with increased capacity for energy harvest. Nature 444:1027–1031

Turnbaugh PJ, Ley RE, Hamady M, Fraser-Liggett CM, Knight R, Gordon JI (2007) The human microbiome project. Nature 449:804–810

Turnbaugh PJ, Hamady M, Yatsunenko T, Cantarel BL, Duncan A, Ley RE, Sogin ML, Jones WJ, Roe BA, Affourtit JP, Egholm M, Henrissat B, Heath AC, Knight R, Gordon JI (2009) A core gut microbiome in obese and lean twins. Nature 457:480–484

van Oppen MJ, Gates RD (2006) Conservation genetics and the resilience of reef-building corals. Mol Ecol 15:3863–3883

Wintermute EH, Silver PA (2010a) Emergent cooperation in microbial metabolism. Mol Syst Biol 6:407

Wintermute EH, Silver PA (2010b) Dynamics in the mixed microbial concourse. Genes Dev 24:2603–2614

Wunderlich Z, Mirny LA (2006) Using the topology of metabolic networks to predict viability of mutant strains. Biophys J 91:2304–2311

Zilber-Rosenberg I, Rosenberg E (2008) Role of microorganisms in the evolution of animals and plants: the hologenome theory of evolution. FEMS Microbiol Rev 32:723–735

Chapter 13
Oral Microbes in Health and Disease

Gilad Bachrach, Marina Faerman, Ofir Ginesin, Amir Eini, Asaf Sol, and Shunit Coppenhagen-Glazer

Abstract Oral bacteria were the first bacteria reported over 300 years ago, yet oral microbiology is still a developing field. Recent advances in molecular biology greatly improve our understanding of the oral microbiota. Oral bacteria form multispecies biofilms on oral surfaces in order to resist salivary wash and mechanical cleaning. These multispecies environments dictate oral health or disease. More than 700 oral bacterial species have been identified so far. Current technology enables for the first time the definition of the microbial composition associated with oral health and disease. Understanding the changes occurring in the microbial composition during transition from oral health to disease should lead to novel approaches to prevent and to treat oral disease.

13.1 Oral Bacteria

The study of oral microbes was first reported by Antonie van Leeuwenhoek who in 1683 described oral bacteria ("very little living animalcules") from plaque sampled between his own teeth. Due to recent developments in nucleic acid amplification and analysis, more knowledge of oral microflora composition has been acquired in the past decade and a half than in the preceding 300 years. Consequently, it is now possible to detect and identify previously unaccounted rare species.

The warm, moist human oral environment comprising both soft and hard surfaces, presents a unique niche for microorganisms. The estimated number of oral phylotypes (organisms that share a common 16S rDNA sequence) ranges up to 19,000 (Keijser et al. 2008) but is accepted to be approximately 700 (Paster et al. 2006;

G. Bachrach (✉) • M. Faerman • O. Ginesin • A. Eini • A. Sol • S. Coppenhagen-Glazer
Institute of Dental Sciences, Hebrew University-Hadassah School of Dental Medicine, P.O.B 12272, Jerusalem 91120, Israel
e-mail: giladba@ekmd.huji.ac.il; marinaf@cc.huji.ac.il; ofirgi@gmail.com; amir_eini@hotmail.com; asaf.sol2@gmail.com; tinushy@yahoo.com

E. Rosenberg and U. Gophna (eds.), *Beneficial Microorganisms in Multicellular Life Forms*, DOI 10.1007/978-3-642-21680-0_13,

Zaura et al. 2009). To date, 756 phylotypes of bacteria have been detected in the human mouth representing at least 600 different species (Dewhirst et al. 2010) (see also http://www.homd.org/), many of which have not been cultivated yet. Estimates suggest that any healthy individual harbors approximately 100–250 phylotypes (Paster et al. 2006; Zaura et al. 2009). Relatively small numbers of fungi and viruses can also be found in the oral cavity, but they will not be dealt with in this review.

13.2 Oral Biofilms

The oral cavity is constantly washed by the salivary flow. Therefore, bacteria which are not attached to hard (tooth) or soft oral surfaces will be washed away. The surface-attached bacteria form communities called biofilms. All tooth surfaces in the oral cavity are coated with a layer of proteins and glycoproteins known as the acquired salivary pellicle. Accumulation of pellicle on clean enamel starts within 2–3 min and reaches a thickness of around 10 μm after approximately 30 min (Skjorland et al. 1995). The acquired pellicle serves as a dock for bacteria floating in saliva. Once sessile, these attach to and anchor other planktonic bacteria in a process called coadherence. An ordered attachment of new species to the developing biofilm leads to the development of multispecies communities of progressively more complex composition. Under certain conditions, some of these highly complex, older communities might become pathogenic and should be removed. Within minutes after oral hygiene treatments, the communities will reform and the process of biofilm generation will recommence and proceed until its next removal (Kolenbrander and London 1993; Kolenbrander et al. 2010).

13.3 Coadherence

Infectious processes initiate by the attachment of a pathogenic microorganism to a host surface and by subsequent colonization of the pathogen within the host. The earliest colonizers attach to the salivary pellicle coating the tooth surface during the first 4 h after professional teeth cleaning. These early colonizers (or pioneers) consist primarily of Gram-positive bacteria (Nyvad and Kilian 1987), mainly *streptococci* equipped for pellicle binding (Heddle et al. 2003; Jenkinson 1994).

If left undisturbed, an increasingly complex population of bacteria develops, shifting from a primarily Gram-positive to a primarily Gram-negative flora. The species' ability to attach to the constantly changing dental plaque (Kolenbrander and London 1993) as well as its metabolic requirements and communication abilities (Kolenbrander et al. 2005) are likely to dictate this colonization succession. Coadhesion (specific cell-to-cell recognition and binding among genetically distinct cell types) is hypothesized to be a significant attachment mechanism of non-pioneer colonizers to the developing plaque (Bos et al. 1996; Kolenbrander and London 1993; Lamont et al. 2002). Coadherence also creates spatial proximity that facilitates microbial communication and metabolic synergism i.e., cooperation of multiple species to metabolize complex nutrients (Egland et al. 2004). Some bacterial species can form biofilms in vitro only in the presence of one or more coaggregation (coadhesion in suspension) partners (Periasamy et al. 2009; Periasamy and Kolenbrander 2009a, b; Vesey and Kuramitsu 2004).

In the past, the "specific plaque hypothesis" guided dental research in a search for specific pathogen or pathogens involved in dental diseases. Currently, most investigators accept the "ecological plaque hypothesis" that disease is due to a change in local environmental conditions, which disrupts the natural balance between the plaque and the host, leading to the enrichment of organisms that can cause disease (Marsh 1994). The goal of identifying and characterizing the human oral microflora associated with dental health or disease now appears to be within reach.

13.4 Dental Caries

The two most prevalent dental diseases are dental caries and periodontal disease. Both ailments are bacteria induced. Dental caries is tooth decay caused by acids from bacterial metabolism of dietary sugars and is the most common cause for toothache and tooth loss. Tooth surfaces that are hard to clean by personal hygiene are at a higher risk to develop a caries lesion. For example, in the upper first molar, caries prevalence is only 2% on lingual *smooth* surface compared to 36% on occlusal *fissured* (grooved) surface (Li et al. 1993). The organisms involved in dental caries such as *mutans streptococci* and *Lactobacillus* spp. are typically efficient acid producers (acidogenic) and can tolerate low pH (aciduric). Continuous sugar consumption supplies these organisms with an energy source and enables them to reduce the surrounding's pH. This in turn will inhibit growth of competitor species with a lower aciduric potential and lead to the dominance of the cariogenic species. When localized acid production exceeds the pH buffering capacity of saliva, calcium and phosphate ions are released from the tooth to the surrounding environment adjacent to the enamel (Fig. 13.1) creating white spot lesions which can deteriorate to irreversible caries (Arends and Christoffersen 1986; Kolenbrander et al. 2009). Topical use of fluoride or calcium phosphate increases enamel tolerance to acidic conditions (Moberg Skold et al. 2005; Reynolds 2009). The total extent of exposure to acidic conditions is the primary determinant for whether demineralization of the tooth enamel will occur. Elevated proportions of *S. mutans* were found in early caries lesions. *S. mutans* were also capable of initiating caries in animal models confirming the cariogenic capacity of this species (Tanzer 1979).

Neighboring plaque bacteria can dampen and even counteract cariogenicity of dental pathogens. *Veillonella parvula*, a Gram-negative oral anaerobe, reduces the cariogenic potential of *S. mutans* by metabolizing the lactate produced by *S. mutans*

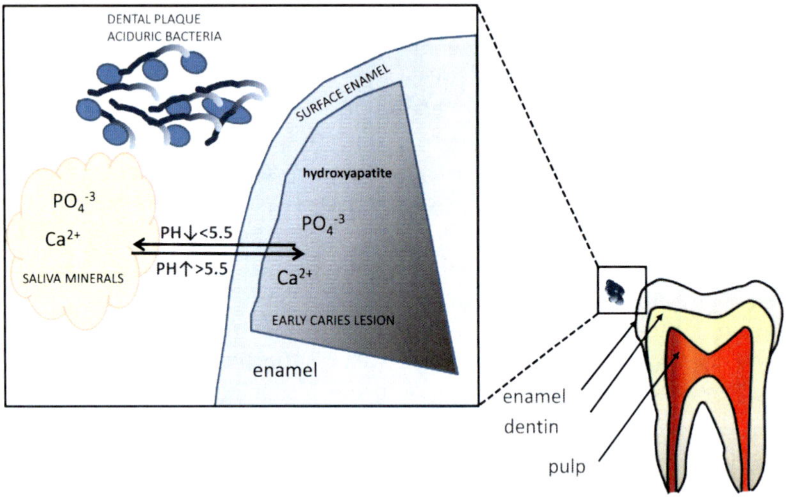

Fig. 13.1 White spot lesion formation. Local pH is lowered below ~5.5 by aciduric bacteria leading to the degradation of the hydroxyapatite composing the enamel crystal. This initial reaction is reversible at pH above ~5.5

to weaker acids, particularly propionic acid. Degradation of arginine into basic end products, such as urea or ammonium can potentially counter acid production by *mutans streptococci* and *lactobacilli*. High levels of arginolytic (arginine-degrading) bacteria give rise to increased pH and these bacteria are therefore considered beneficial (Kolenbrander et al. 2009).

The past "specific plaque hypothesis" guided caries research in a search for a specific pathogen or pathogens involved in caries. The current "ecological plaque hypothesis" associates disease with a shift in the balance of the resident microflora due to changes in the local environment. In the case of caries, repeated cycles of low pH following frequent sugar uptake favors cariogenic (acidogenic and aciduric) bacteria (Fig. 13.2.). Though high hopes were held for the elimination of a single species such as *S. mutans* by vaccination, it may not be effective in preventing dental caries in humans as this may give rise to one or more of the many remaining acidogens. Indeed, carious lesions which did not contain *S. mutans* were found (Aas et al. 2008).

13.5 Periodontal Disease

Another major class of oral pathology is inflammatory gum disease. Periodontitis (perio-around, dent-tooth) is a general term to describe a collection of bacteria induced inflammatory conditions ranging from gingivitis, a transient gingival (gum) inflammation (Fig. 13.3), to extensive destruction of the ligaments and the

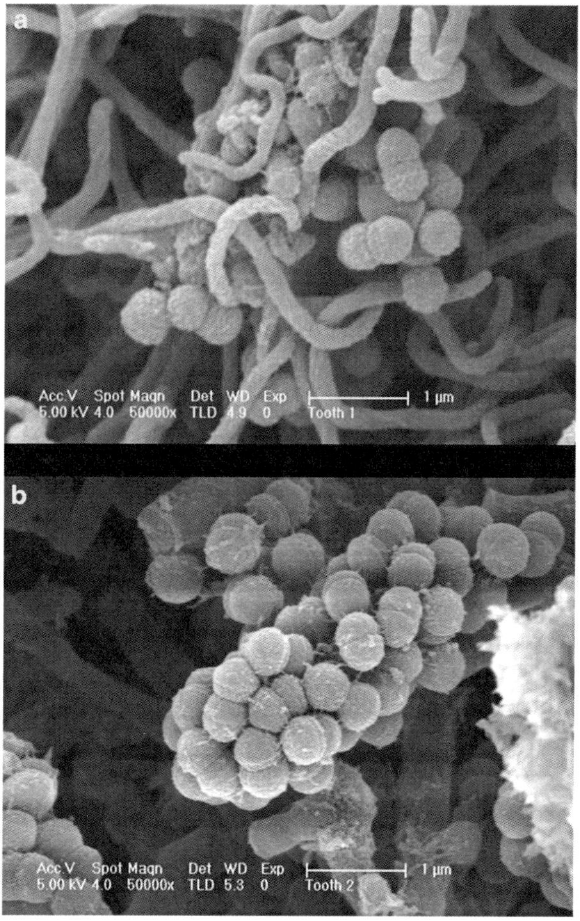

Fig. 13.2 Scanning electron microscopy of samples extracted from a carious tooth demonstrating coadherence among multiple oral species (**a**) including sites dominated by oral streptococci (**b**)

alveolar bone supporting the teeth (Fig. 13.4). Periodontitis is one of the most common chronic inflammatory diseases in humans (Pihlstrom et al. 2005). Accumulating evidence suggest a link between periodontitis and systemic disorders such as diabetes (Grossi and Genco 1998), atherosclerosis (Gibson and Genco 2007), and with preterm births (Han et al. 2004, 2009, 2010; Offenbacher et al. 1996; Ikegami et al. 2009).

When the dental plaque is allowed to expand along the tooth surface and to accumulate below the gumline, an increasingly complex population of bacteria develops. A community shift from primarily Gram-positive, oxygen-tolerant to a primarily Gram-negative anaerobic flora is observed. The species that expand in the subgingival (below the gumline) plaque include *Fusobacterium nucleatum, Porphyromonas gingivalis, Aggregatibacter actinomycetemcomitans, Tannerella forsythia* and tryponemes such as *T. denticola* (Socransky et al. 1998). This shift in plaque composition is often accompanied or perhaps preceded (Cobb et al. 2009) by the appearance of gingivitis that can progress to irreversible periodontitis

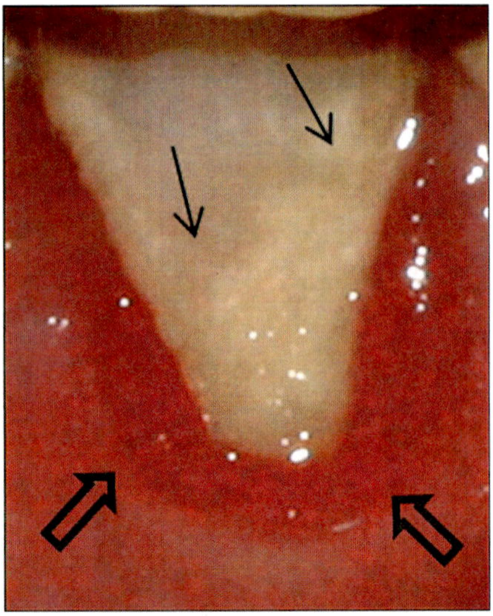

Fig. 13.3 Dental plaque (*solid arrows*) resulting in initial gingivitis (*hollow arrows*) around the gingiva of the inferior lateral incisor

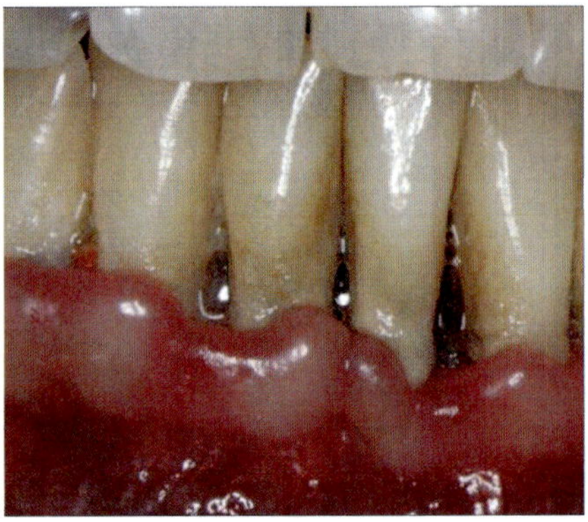

Fig. 13.4 Severe periodontitis: Poor dental hygiene and massive calculus accumulation, results in chronic inflammation characterized by inflamed gingiva and alveolar bone loss ("tooth elongation")

(Socransky et al. 1998). Changes occur in the local environmental conditions when transitioning from the supragingival (above gumline) to the subgingival region such as increased gingival crevicular fluid (GCF) during inflammation, low oxygen tension, and selection by the immune mediators. According to the ecological plaque hypothesis, these changes favor growth of proteolytic and obligatory anaerobic (mostly Gram-negative) species at the expense of those seen in healthy sites. This results in a shift in the overall balance of the subgingival microflora, and predisposes a site to disease (Marsh 1994).

Host susceptibility appears to be a major requirement for the development of periodontal disease. A small group of individuals does not develop periodontal disease even when refraining from dental hygiene (Loe et al. 1986). Hypo-responsiveness of the host's immunity associated with systemic conditions (diabetes or HIV infection)(Armitage 1999) or genetic disorders (such as Down syndrome or impaired neutrophil function) (Izumi et al. 1989; Kantarci et al. 2003) will facilitate bacterial invasion and is frequently associated with aggressive periodontitis. However, hyper-responsiveness of certain immunological pathways resulting in a bombardment of inflammatory mediators is considered the main contributor for the tissue destruction associated with chronic periodontal disease (Baker 2000; Dixon et al. 2004; Gronert et al. 2004; Kantarci et al. 2003). Inflammation can lead to the formation of gaps (pockets) of up to 8–10 mm depth, between the gum and the teeth. These gaps can increase and cause bone resorption and eventually tooth loss (Fig. 13.4). Thus, common chronic periodontitis results largely from an inappropriate immune response to anaerobic, proteolytic bacteria in subgingival stubborn biofilm.

Novel treatment strategies for periodontitis focus on the resolution of the gingival inflammation. In animal models, inflammation reduction was found to decrease the numbers of periopathogens in the infected site and to restore the damaged tissues (including the alveolar bone) (Hasturk et al. 2007). However, regular efficient removal of bacterial biofilms from the teeth remains the effective way to prevent and to treat periodontitis (Axelsson et al. 2004; Friedewald et al. 2009).

13.6 Biofilm Models

Understanding the oral ecological system is prerequisite to understanding dental diseases. Biofilms can be grown in static biofilm models (usually wells of a 96-well plate) or under flow conditions such as those created in the flow cell system (described in detail by Palmer (1999)) or in similar commercial options. The flow cell system involves inoculating bacteria into a track mounted on a coverslip. Medium such as diluted saliva is continuously pumped through the flowcell and the system is maintained at 37°C. If required, the flowcell can be run under anaerobic conditions. Biofilms in the flowcell can be stained non-disruptively and examined using confocal laser scanning microscopy.

An in vivo approach for studying dental plaque is the retrievable enamel chip model system (Palmer et al. 2001). In this model, a volunteer wears an acrylic appliance on both sides of the mandible, each containing three sterilized enamel chips prepared from extracted human third molars. Once the appliance is removed, and the enamel chips are retrieved, they are incubated with fluorescently conjugated specific antibodies or oligonucleotide probes and processed for microscopy without disturbance to spatial relationships within the native biofilm. The retrievable chip model produced definitive evidence as to the involvement of coadhesion mediators (adhesins and receptors) in the development of biofilms in vivo (Palmer et al. 2003). This model has also confirmed that the oral biofilm is generated by interdigitated mixed species rather than mono-species arranged side-by-side. These findings reinforce earlier cultivation-based observations according to which the initial colonization process follows a sequential rather than a random pattern (Palmer et al. 2003).

13.7 Methods of Studying the Composition of Oral Biofilms

Most knowledge on oral biofims is based on analysis of samples from healthy or diseased sites. Sampling is obtained from extracted teeth or from healthy or infected sites using dental instruments or paper points. Plaque samples were originally analyzed microscopically (as first performed by van Leeuwenhoek), and later by culturing techniques. For decades, cultivation has been the only method for identifying and quantifying microorganisms in a particular sample. The recent introduction of culture independent molecular methods has enabled rapid and sensitive identification of both cultivable and uncultivable bacteria. The first DNA based high-throughput method for plaque composition analysis was checkerboard DNA hybridization described in 1994 by Socransky and colleagues (Socransky et al. 1994). Using this procedure, up to 1,350 independent hybridizations (identification of 45 known species in 30 samples) can be simultaneously performed on a single membrane (Socransky et al. 1998). PCR amplification and sequencing of the 16S rRNA gene is commonly used to identify sampled bacteria. In this procedure, PCR primers are designed to react with the highly conserved sites flanking all eubacterial 16S rRNA genes. These "universal primers" are then used for PCR amplification of DNA extracted from samples. The PCR amplification products are cloned in *Escherichia coli* and sequenced (or sequenced directly using deep sequencing – see below). The microbial community species identity or closest relative is determined by comparison with known sequences in the ribosomal data base (http://rdp.cme.msu.edu/). By combining broad-range bacterial 16S rRNA gene PCR amplification with checkerboard DNA–DNA hybridization, Bruce Paster and colleagues introduced the reverse-capture, checkerboard DNA–DNA hybridization technique. Using this method, the presence of previously identified cultivable or uncultivable bacteria in numerous samples can be determined and semi-quantified simultaneously (Paster et al. 2002). A microarray targeting 16S rRNA species-specific regions for the rapid determination of bacterial profiles of clinical samples taken from the human oral cavity, esophagus, and lung has been created and validated (Human Oral Microbe Identification Microarray, HOMIM, http://mim.forsyth.org/homim.html).

Next generation sequencing is revolutionizing our ability to analyze the composition of the dental plaque and even to decipher metabolic interactions among its inhabitants through whole transcriptome analysis. The huge sequencing power (currently, up to hundreds of millions of bases in a single sequencing run) can be used for sequencing the entire genetic material recovered from environmental samples without need of cloning. Over 500 species-level phylotypes (16S rDNA sequences that clustered at 3% genetic difference) sampled from several intraoral niches of three healthy individuals, were identified using 454 pyro-sequencing (Zaura et al. 2009). On average, 266 species-level phylotypes per individual were found. Nine sequences were highly abundant in all three individuals: two were assigned to *Streptococcus*, two to the *Veillonellaceae* family, and one each to

Granulicatella, *Corynebacterium*, *Rothia*, *Porphyromonas*, and *Fusobacterium* (Kolenbrander et al. 2010; Zaura et al. 2009).

13.8 Co-evolution of Humans and the Human Cariogenic Microbial Flora

The composition of the microbiota associated with periodontitis and particularly with caries is bound to be a dynamic process, constantly adapting to changes in human nutritional habits. Currently, *S. mutans* is the microbe most associated with human caries. This organism has specialized in utilizing sucrose for colonization, biofilm formation, and robust acid production (via the glucosyltransferase enzymes). Caries was prevalent well before sucrose was introduced into the diet of man. Changes in human diet, including food source, food processing techniques, and most importantly introduction of sugar cane sucrose have had to result in alterations in the oral microbial composition, for example, dominance of *S. mutans* in caries lesions. It is also reasonable to speculate that changes in human nutritional habits affected cariogenic bacteria in the species level and that *S. mutans* has adapted to sucrose consumption by acquiring the glucosyltransferase sucrose-utilizing enzymes.

Indeed, examination of teeth recovered from archaeological sites reveals that dental caries and periodontitis have always inflicted mankind (Fig. 13.5). The bacterial species associated with caries and periodontitis in ancient times are yet unknown. However, current next generation sequencing techniques facilitate not only the study of microorganisms involved in contemporary diseases but also those that have inflicted diseases centuries and even millennia ago. 16S rDNA can be amplified from the trace amounts of DNA preserved in ancient plaque and sequenced to identify its bacterial composition. However, ancient DNA is always severely degraded and is present in low amounts. Endogenous and exogenous nucleases and hydrolysis and oxidation reactions fragment the DNA backbone. The average size of an ancient DNA segment is 100–200 bp. This DNA fragmentation excludes the possibility of whole (~1,500 bp) 16S rDNA amplification.

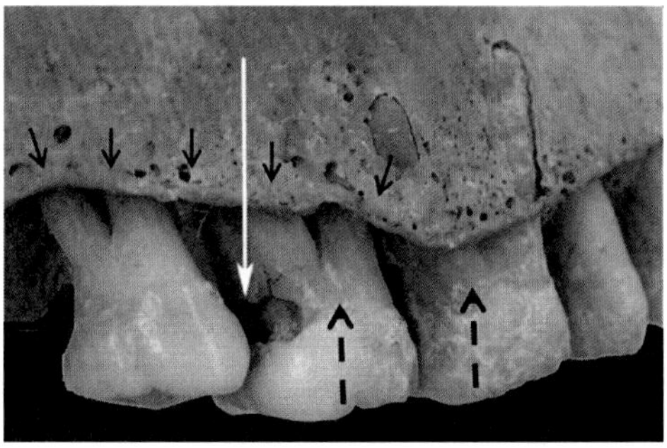

Fig. 13.5 Carious lesion (*white arrow*) and severe alveolar bone resorption (*periodontitis, black arrows*) in archaeological skeletal remains. *Broken arrow* depicts original level of alveolar bone prior to periodontitis

Short segments (100–200bp) in the V5 and the V3–4 regions of the 16S rDNA however, can be amplified and their sequence can be used to identify bacteria in the species level.

Sequencing amplified ancient DNA is also impeded by post-mortem cytosine to uracil deamination that can result in erroneous base incorporation (Brotherton et al. 2007). The Phusion polymerase (Finnzymes) which does not replicate uracil could potentially reduce this problem (Rasmussen et al. 2010). Use of this polymerase coupled with a comparison of the V3–4 and V5 amplified sequences to eliminate background sequences should enable reliable conclusions.

The wealth of sequence-data of ancient pathogens that can be generated using next generation sequencing should enable sufficient resolution not only to determine the species involved in caries formation in ancient times but also to reveal clues regarding the biochemical pathways involved in virulence. Recovered data will shed light on the co-evolution of oral pathogens with human dietary changes and will provide essential information on these preventable diseases.

13.9 Unresolved Questions and Future Research

The human oral microbiome is the most intensively studied and perhaps the best characterized ecosystem in humans (Kolenbrander et al. 2010). The composition of the oral microflora is being resolved and the specific compositions of bacterial communities involved in health and in disease are being defined. The interactions between community members and the stages leading from a commensal community to a pathogenic one need further resolution. The high-throughput capacities of the available second, third, and fourth generation DNA sequencing technologies offer tools for deciphering the commune-ecological phases required for induction of dental diseases. Further knowledge and understanding of the events leading to dental diseases should lead to improved approaches for their prevention and treatment.

References

Aas JA, Griffen AL, Dardis SR, Lee AM, Olsen I, Dewhirst FE, Leys EJ, Paster BJ (2008) Bacteria of dental caries in primary and permanent teeth in children and young adults. J Clin Microbiol 46:1407–1417

Arends J, Christoffersen J (1986) The nature of early caries lesions in enamel. J Dent Res 65:2–11

Armitage GC (1999) Development of a classification system for periodontal diseases and conditions. Ann Periodontol 4:1–6

Axelsson P, Nystrom B, Lindhe J (2004) The long-term effect of a plaque control program on tooth mortality, caries and periodontal disease in adults. Results after 30 years of maintenance. J Clin Periodontol 31:749–757

Baker PJ (2000) The role of immune responses in bone loss during periodontal disease. Microbes Infect 2:1181–1192

Bos R, van der Mei HC, Busscher HJ (1996) Co-adhesion of oral microbial pairs under flow in the presence of saliva and lactose. J Dent Res 75:809–815

Brotherton P, Endicott P, Sanchez JJ, Beaumont M, Barnett R, Austin J, Cooper A (2007) Novel high-resolution characterization of ancient DNA reveals C > U-type base modification events as the sole cause of post mortem miscoding lesions. Nucleic Acids Res 35:5717–5728

Cobb CM, Costerton JW, Van Dyke TE (2009) Have we become so focused on inflammation and host response that we have neglected the role of bacteria in initiating periodontal disease? Compend Contin Educ Dent 30(46):48

Dewhirst FE, Chen T, Izard J, Paster BJ, Tanner AC, Yu WH, Lakshmanan A, Wade WG (2010) The human oral microbiome. J Bacteriol 192:5002–5017

Dixon DR, Bainbridge BW, Darveau RP (2004) Modulation of the innate immune response within the periodontium. Periodontol 2000 35:53–74

Egland PG, Palmer RJ Jr, Kolenbrander PE (2004) Interspecies communication in *Streptococcus gordonii-Veillonella atypica* biofilms: signaling in flow conditions requires juxtaposition. Proc Natl Acad Sci USA 101:16917–16922

Friedewald VE, Kornman KS, Beck JD, Genco R, Goldfine A, Libby P, Offenbacher S, Ridker PM, Van Dyke TE, Roberts WC (2009) The American Journal of Cardiology and Journal of Periodontology editors' consensus: periodontitis and atherosclerotic cardiovascular disease. J Periodontol 80:1021–1032

Gibson FC 3rd, Genco CA (2007) *Porphyromonas gingivalis* mediated periodontal disease and atherosclerosis: disparate diseases with commonalities in pathogenesis through TLRs. Curr Pharm Des 13:3665–3675

Gronert K, Kantarci A, Levy BD, Clish CB, Odparlik S, Hasturk H, Badwey JA, Colgan SP, Van Dyke TE, Serhan CN (2004) A molecular defect in intracellular lipid signaling in human neutrophils in localized aggressive periodontal tissue damage. J Immunol 172:1856–1861

Grossi SG, Genco RJ (1998) Periodontal disease and diabetes mellitus: a two-way relationship. Ann Periodontol 3:51–61

Han YW, Redline RW, Li M, Yin L, Hill GB, McCormick TS (2004) *Fusobacterium nucleatum* induces premature and term stillbirths in pregnant mice: implication of oral bacteria in preterm birth. Infect Immun 72:2272–2279

Han YW, Shen T, Chung P, Buhimschi IA, Buhimschi CS (2009) Uncultivated bacteria as etiologic agents of intra-amniotic inflammation leading to preterm birth. J Clin Microbiol 47:38–47

Han YW, Fardini Y, Chen C, Iacampo KG, Peraino VA, Shamonki JM, Redline RW (2010) Term stillbirth caused by oral *Fusobacterium nucleatum*. Obstet Gynecol 115:442–445

Hasturk H, Kantarci A, Goguet-Surmenian E, Blackwood A, Andry C, Serhan CN, Van Dyke TE (2007) Resolvin E1 regulates inflammation at the cellular and tissue level and restores tissue homeostasis in vivo. J Immunol 179:7021–7029

Heddle C, Nobbs AH, Jakubovics NS, Gal M, Mansell JP, Dymock D, Jenkinson HF (2003) Host collagen signal induces antigen I/II adhesin and invasin gene expression in oral *Streptococcus gordonii*. Mol Microbiol 50:597–607

Ikegami A, Chung P, Han YW (2009) Complementation of the fadA mutation in Fusobacterium nucleatum demonstrates that the surface-exposed adhesin promotes cellular invasion and placental colonization. Infect Immun 77:3075–3079

Izumi Y, Sugiyama S, Shinozuka O, Yamazaki T, Ohyama T, Ishikawa I (1989) Defective neutrophil chemotaxis in Down's syndrome patients and its relationship to periodontal destruction. J Periodontol 60:238–242

Jenkinson HF (1994) Adherence and accumulation of oral *streptococci*. Trends Microbiol 2:209–212

Kantarci A, Oyaizu K, Van Dyke TE (2003) Neutrophil-mediated tissue injury in periodontal disease pathogenesis: findings from localized aggressive periodontitis. J Periodontol 74:66–75

Keijser BJ, Zaura E, Huse SM, van der Vossen JM, Schuren FH, Montijn RC, ten Cate JM, Crielaard W (2008) Pyrosequencing analysis of the oral microflora of healthy adults. J Dent Res 87:1016–1020

Kolenbrander PE, London J (1993) Adhere today, here tomorrow: oral bacterial adherence. J Bacteriol 175:3247–3252

Kolenbrander PE, Egland PG, Diaz PI, Palmer RJ Jr (2005) Genome-genome interactions: bacterial communities in initial dental plaque. Trends Microbiol 13:11–15

Kolenbrander PE, Jakubovics NS, Bachrach G (2009) Oral microbiology. In: Schaechter M (ed) Encyclopedia of microbiology. Elsevier, Oxford, pp 566–588

Kolenbrander PE, Palmer RJ Jr, Periasamy S, Jakubovics NS (2010) Oral multispecies biofilm development and the key role of cell-cell distance. Nat Rev Microbiol 8:471–480

Lamont RJ, El-Sabaeny A, Park Y, Cook GS, Costerton JW, Demuth DR (2002) Role of the *Streptococcus gordonii* SspB protein in the development of *Porphyromonas gingivalis* biofilms on streptococcal substrates. Microbiology 148:1627–1636

Li SH, Kingman A, Forthofer R, Swango P (1993) Comparison of tooth surface-specific dental caries attack patterns in US schoolchildren from two national surveys. J Dent Res 72:1398–1405

Loe H, Anerud A, Boysen H, Morrison E (1986) Natural history of periodontal disease in man. Rapid, moderate and no loss of attachment in Sri Lankan laborers 14 to 46 years of age. J Clin Periodontol 13:431–445

Marsh PD (1994) Microbial ecology of dental plaque and its significance in health and disease. Adv Dent Res 8:263–271

Moberg Skold U, Petersson LG, Lith A, Birkhed D (2005) Effect of school-based fluoride varnish programmes on approximal caries in adolescents from different caries risk areas. Caries Res 39:273–279

Nyvad B, Kilian M (1987) Microbiology of the early colonization of human enamel and root surfaces In vivo. Scand J Dent Res 95:369–380

Offenbacher S, Katz V, Fertik G, Collins J, Boyd D, Maynor G, McKaig R, Beck J (1996) Periodontal infection as a possible risk factor for preterm low birth weight. J Periodontol 67:1103–1113

Palmer RJ Jr (1999) Microscopy flowcells: perfusion chambers for real-time study of biofilms. Method Enzymol 310:160–166

Palmer RJ Jr, Wu R, Gordon S, Bloomquist CG, Liljemark WF, Kilian M, Kolenbrander PE (2001) Retrieval of biofilms from the oral cavity. Method Enzymol 337:393–403

Palmer RJ Jr, Gordon SM, Cisar JO, Kolenbrander PE (2003) Coaggregation-mediated interactions of *streptococci* and *actinomyces* detected in initial human dental plaque. J Bacteriol 185:3400–3409

Paster BJ, Russell MK, Alpagot T, Lee AM, Boches SK, Galvin JL, Dewhirst FE (2002) Bacterial diversity in necrotizing ulcerative periodontitis in HIV-positive subjects. Ann Periodontol 7:8–16

Paster BJ, Olsen I, Aas JA, Dewhirst FE (2006) The breadth of bacterial diversity in the human periodontal pocket and other oral sites. Periodontol 2000 42:80–87

Periasamy S, Kolenbrander PE (2009a) *Aggregatibacter actinomycetemcomitans* builds mutualistic biofilm communities with *Fusobacterium nucleatum* and *Veillonella species* in saliva. Infect Immun 77:3542–3551

Periasamy S, Kolenbrander PE (2009b) Mutualistic biofilm communities develop with *Porphyromonas gingivalis* and initial, early and late colonizers of enamel. J Bacteriol 191 (22):6804–6811

Periasamy S, Chalmers NI, Du-Thumm L, Kolenbrander PE (2009) *Fusobacterium nucleatum* ATCC 10953 requires *Actinomyces naeslundii* ATCC 43146 for growth on saliva in a three-species community that includes *Streptococcus oralis* 34. Appl Environ Microbiol 75:3250–3257

Pihlstrom BL, Michalowicz BS, Johnson NW (2005) Periodontal diseases. Lancet 366:1809–1820

Rasmussen M, Li Y, Lindgreen S, Pedersen JS, Albrechtsen A, Moltke I, Metspalu M, Metspalu E, Kivisild T, Gupta R (2010) Ancient human genome sequence of an extinct Palaeo-Eskimo. Nature 463:757–762

Reynolds EC (2009) Casein phosphopeptide – amorphous calcium phosphate: the scientific evidence. Adv Dent Res 21:25–29

Skjorland KK, Rykke M, Sonju T (1995) Rate of pellicle formation in vivo. Acta Odontol Scand 53:358–362

Socransky SS, Smith C, Martin L, Paster BJ, Dewhirst FE, Levin AE (1994) "Checkerboard" DNA-DNA hybridization. Biotechniques 17:788–792

Socransky SS, Haffajee AD, Cugini MA, Smith C, Kent RL Jr (1998) Microbial complexes in subgingival plaque. J Clin Periodontol 25:134–144

Tanzer JM (1979) Essential dependence of smooth surface caries on, and augmentation of fissure caries by, sucrose and *Streptococcus mutans* infection. Infect Immun 25:526–531

Vesey PM, Kuramitsu HK (2004) Genetic analysis of *Treponema denticola* ATCC 35405 biofilm formation. Microbiology 150:2401–2407

Zaura E, Keijser BJ, Huse SM, Crielaard W (2009) Defining the healthy "core microbiome" of oral microbial communities. BMC Microbiol 9:259

Chapter 14
The Role of the Rumen Microbiota in Determining the Feed Efficiency of Dairy Cows

Itzhak Mizrahi

Abstract Dairy cattle hold enormous significance for man as they convert the energy stored in indigestible plant mass into milk and meat, which are digestible and consumed worldwide. The efficiency with which an individual cow converts the energy stored in its feed into food products, termed feed efficiency, is thought to be stable; however, for reasons poorly understood, it varies considerably between cows. One factor that could markedly affect dairy cow feed efficiency is its residing microbiota which, in ruminants, is responsible for most of the food's digestion and absorption. The microbes are mainly situated within the first compartment of the digestive tract – the reticulorumen. Recent studies connecting methane emission, clustering, and diversity of microbial populations with the cow's feed efficiency strongly imply that the reticulorumen microbiota is correlated with cows' energy utilization. Hence, understanding the relationship between the reticulorumen microbiota and cows' feed efficiency may favor more energy-efficient microbiomes, and therefore increase fiber degradation, elevate reticulorumen microbial protein concentration, reduce methane emission, and consequentially improve cow productivity.

14.1 Ruminants

Ruminants hold enormous significance for man as they can convert energy stored in plant mass, which is mainly indigestible for humans, to the digestible food products. A significant proportion of domesticated animal species worldwide – the source for most meat and dairy products – are ruminants. Chief among these are dairy cattle. Ruminants are herbivores, and their digestive system allows them to

---

I. Mizrahi (✉)
Department of Ruminant Science, Institute of Animal Sciences, Agricultural Research Organization, Volcani Center, PO Box 6, Bet Dagan 50250, Israel
e-mail: itzhakm@volcani.agri.gov.il

E. Rosenberg and U. Gophna (eds.), *Beneficial Microorganisms in Multicellular Life Forms*,
DOI 10.1007/978-3-642-21680-0_14,

absorb and digest large amounts of plant material. Upon consumption, the feed is packed into the first two stomach chambers, collectively termed the reticulorumen; ruminants then regurgitate the partially digested feed (cud) and chew it again (Hoover and Miller 1991). This process decreases the size and increases the surface area of the plant fiber particles interacting with reticulorumen-residing microorganisms, which are mainly responsible for breaking down the plant fibers (see further on) (Stewart 1986; Dehority 1991; McAllister et al. 1994; Miron et al. 2001). The regurgitation process also mixes together the microorganisms, digesta, and saliva, further increasing fiber break-down (McAllister et al. 1994; Merchen et al. 1997). The saliva is composed of a bicarbonate phosphate buffer at pH ~ 8 which assists in keeping the reticulorumen pH stable (Kay 1966). After the feed plant fibers have been degraded, the digesta passes into the next stomach chamber, the omasum, where water and many of the inorganic mineral elements are absorbed into the bloodstream. Then, the digesta moves to the true stomach, the abomasum. The abomasum is the direct equivalent of the human monogastric stomach and the digesta is digested there in much the same way as in non-ruminant mammals (Merchen et al. 1997).

14.2 Ruminants' Efficiency of Energy Utilization

The ability of cattle to utilize and divert the energy stored in its feed to milk and meat production is governed by its energetic efficiency. Energetic efficiency is defined by the ratio between the energetic value of the animal and its products and the energetic value of the feed or diet; it thus includes the energy invested in maintenance as well as the energy invested in production of dairy and meat (Johnson et al. 2003). In general, the energy is preferentially utilized for maintenance over production (Tolkamp 2010). The energy from the consumed feeds that is not retained in the body or milk is lost in the form of feces, urine, heat, and combustion of gases such as methane. Different methods are used to evaluate an animal's energetic efficiency, among them the residual feed intake (RFI) method (also known as net feed efficiency method) (Koch et al. 1963). This method evaluates energetic efficiency according to the difference between the animal's actual feed intake and its estimated feed intake over a specified period of time (Koch et al. 1963; Archer et al. 1999). Animals that have low RFI values are considered to be more energetically efficient than those with high values. The fact that this method is independent of growth and body size makes it suitable for comparisons between animals (Archer et al. 1999; Moore et al. 2009).

The energetic efficiency varies considerably between breeds, as well as between different individuals from the same breed (Ferrell and Jenkins. 1985; Thiessen and Taylor 1985; Taylor et al. 1986; Solis et al. 1988; Aharoni et al. 2006). The nature of this variation is not entirely clear: various factors are thought to affect the animal's RFI and hence its energy utilization (Johnson et al. 2003). The presence of a moderate genetic component affecting energy utilization was demonstrated by elevating feed efficiency via successful selection of animals according to their RFI as well as specific genomic markers (Hotovy et al. 1991; Archer et al. 1999; Herd and Archer 2003; Richardson et al. 2004; Crews 2005). Differences between high and low RFI animals have also been reported in terms of metabolic activity, digestibility, and methane production (Richardson et al. 1996; Arthur et al. 2001; Basarab et al. 2003; Nkrumah et al. 2004, 2006). These differences could be related to various factors, one of which might be the reticulorumen microbiota, upon which the animal's digestion and absorption of feed are largely dependent. Hence, the reticulorumen microbiota may play an important role in energy harvesting from the feed, affecting the animal's energy utilization.

14.3 Rumen Microbiology and Metabolism

As mentioned, the reticulorumen is a two-chamber stomach composed of the reticulum and the rumen. The reticulorumen functions as a pre-gastric anaerobic fermentation chamber. It is inhabited by a high density of resident microbiota, consisting of bacteria, protozoa, archaea, and fungi, which degrade the consumed plant materials (Flint 1997). In general, plant materials are composed of organic matter such as proteins, lipids, and nucleic acids and also a large proportion of carbohydrate polymers such as celluloses, hemicelluloses, pectins, fructosans, starches, and other polysaccharides which are mostly indigestible by the animal (Flint et al. 2008). The reticulorumen microorganisms utilize specialized enzymes and enzyme complexes to degrade these polysaccharides into monomeric or dimeric sugars and subsequently ferment them (Flint et al. 2008). The degradation and fermentation take place in a coordinated and complex manner in which the substrate for one microorganism is the product of another (Dehority 1991; Mackie 2002). This enables the degradation of all plant materials except lignin, which is a phenolic compound present in the cell wall. The biodegradation and metabolism of plant feed by the reticulorumen microorganisms make these otherwise indigestible polymers accessible for absorption and utilization.

These indigestible polymers are converted by the reticulorumen microorganisms into two main digestible forms. The first is the microbial proteins synthesized by these microorganisms to build their cells – these proteins are digested in the abomasum (true stomach) as the microbes travel along the animal's digestive tract (Kay 1969). The microbial protein so digested makes up 50–80% of the total protein absorbed by the animal (Storm and Ørskov 1983). The second is volatile fatty acids (VFAs), which are final fermentation products along with methane and carbon dioxide. VFAs are short-chain fatty acids that are absorbed by the animal in the reticulorumen and serve in important biochemical processes, such as gluconeogenesis in the case of propionate and fatty acid synthesis in the case of acctate and butyrate (Russell and Wilson 1996). Hence, the VFAs serve as an energy source for the animal. In a recent study, the authors estimated that VFAs retain approximately 70% of the energy content of the plant material's polymers (Weimer et al. 2009).

Some of the energy lost during the conversion can be attributed to methane production which, as mentioned above, is one of the end products of reticulorumen fermentation. Methane is a gas, produced in the rumen by specialized methanogenic archaea which utilize the hydrogen from microbial fiber degradation and fermentation to reduce carbon dioxide to methane (Thauer et al. 2008). The methane is eructated into the atmosphere along with its retained energy, which is lost from the cow's reticulorumen. This process results in a loss of 5–19% of the energy content of the feed (Johnson and Johnson 1995) and has wide environmental implications: methane is a very potent greenhouse gas (23 times more potent than CO_2) and in some countries, ruminants are responsible for up to 60% of its emission (Wuebbles and Hayhoe 2002). A recent study in which animals with different RFI values (energetic efficiencies) were compared reported that animals with low energetic efficiency exhibit significantly more methane emission (Nkrumah et al. 2006). The authors speculated that the differences in methane emission are the outcome of differences in the methanogenic populations.

Hence, the relative proportion of digestion and energy harvested from the feed and absorbed by the animal may be largely affected by its reticulorumen-residing microbial activity and composition. In order to address these issues, traditional microbiological methods, such as pure culture methods, are insufficient as only a small fraction of the microorganisms in an ecological niche can be cultivated (Amann et al. 1990). In the past few years, culture-free methods have dramatically improved our ability to gain a broad overview of the microbiota in different environments, including the reticulorumen (Deng et al. 2008). Powerful next-generation sequencing techniques have boosted these methods, allowing the sequencing of thousands of sequences in parallel. These possibilities enable studying the microbiota as a whole, in terms of composition, genes, and metabolic activities, termed microbiome and metagenome studies. In a recent metagenome study of the reticulorumen fiber-adherent microbiome, a comparative metagenomic approach revealed forage-specific glycoside hydrolases, and provided insights into the fiber-colonization steps by the reticulorumen microorganisms (Brulc et al. 2009). In the near future, these techniques and studies are likely to help elucidate the cow's reticulorumen microbiota relationship in regard to feed efficiency.

14.4 Microbiome Capacity for Energy Harvesting

The notion of the residing microbiota affecting their hosts' ability to harvest energy from its food has been the topic of intense research in the last few years (Ley et al. 2006; Turnbaugh et al. 2006; Turnbaugh and Gordon 2009). In these studies, the authors managed to establish a strong correlation between bacterial gut divisions and the ability to harvest energy from food, and consequently increase body weight and fat. In one such study, the authors taxonomically analyzed the microbiomes of genetically obese mice and their lean littermates, as well as those of obese and lean humans, and found that obesity is associated with changes in the relative abundance of two dominant bacterial taxonomic divisions. Using comparative metagenomic analysis, the authors predicted increased capacity for dietary energy harvesting in the obese microbiome as it was enriched with sequences encoding many enzymes involved in the initial steps of breaking down otherwise indigestible dietary polysaccharides. This prediction was subsequently validated by biochemical assays in which fermentation end products were measured and found to be more concentrated in the obese microbiome. This research demonstrated that changes in the microbiota can affect the metabolic potential to harvest energy from the diet and subsequently affect the host's fat production.

The relationship between the reticulorumen microbiota and the host's feed efficiency was examined in a few recent studies in beef cattle (Guan et al. 2008; Zhou et al. 2009). In one study, the authors used a PCR-denaturing gradient gel electrophoresis (DGGE) method to compare bacterial reticulorumen population composition in 18 steers from different breeds which had different RFI values. The authors reported the clustering of specific bacterial populations according to the animals' energy efficiency, accompanied by a correlation with VFA concentration. However, the identity, activity, and coding capacity of these populations – a significant part of the factors determining the microbial-feed efficiency relationship – were not studied (Guan et al. 2008). In a second study, the methanogenic populations of animals with different RFIs were compared using 16S clone libraries. Efficient animals showed less species diversity and there was a correlation between the composition of methanogens and the host's efficiency (Zhou et al. 2009). In two following studies, the ruminal methanogenic and bacterial populations of 56 beef cattle which differed in feed efficiency, as well as diet were analyzed using PCR-DGGE profiles (Hernandez-Sanabria et al. 2010; Zhou et al. 2010). The results indicated that the methanogenic PCR-DGGE pattern was associated with the feed efficiency of the host. Diet-associated bands and feed efficiency-related bands were identified. However, the size of the methanogenic population did not correlate with differences in feed efficiency, diet, or metabolic measurements (Zhou et al. 2010). Bacterial PCR-DGGE bands were only related to feed efficiency associated traits and metabolites (Hernandez-Sanabria et al. 2010). Taken together these suggest a link between microbiota and the animals' feed efficiency.

14.5 Unresolved Questions and Future Research

Indeed, as described by the recent findings, a link between the rumen microbiota and the feed efficiency of the animal seems to exist. Nevertheless, a few unresolved questions rise from these findings. Following are three cardinal unresolved questions:

1. Is the link between the rumen microbiota and the feed efficiency causative? And if so, which is the cause and which is the effect?
2. What are the microbial aspects governing this link and what is the impact of each one of them on the animal's feed efficiency? More specifically, can this impact be attributed to specific taxa, combination of taxa, or the total microbial gene pool level?
3. What is the relationship between the animal's genetics and the microbiota in regards to its feed efficiency?

In order to answer these questions and to thoroughly understand the relationship between the animal's reticulorumen microbiota and energy utilization, the former has to be studied more thoroughly in terms of its relation to the animal's energy-utilization efficiency. Such a study should address several aspects of microbial communities in respect to the feed efficiency of the animal and its genetics: (1) The composition of microbial populations has to be better identified. (2) The complexity and diversity of the populations have to be defined. (3) The genes encoded by the reticulorumen microorganisms have to be further explored, as it is likely that gene families and coding functions are more strongly correlated with energy utilization than species composition (Turnbaugh and Gordon 2009). These different aspects of the microbial communities are more likely to be elucidated using next generation sequencing techniques that allow a higher scale of resolution and a broader view of the microbes and their encoded genes. These examinations have to be followed by a study of metabolic parameters such as digestibility, VFA content, methane emission, etc.

Together, these efforts may gain an understanding of the relationship between feed efficiency in dairy cows and their microbiota composition, coding capacity, and activity. This information holds tremendous importance for agriculture as it will enable determining the most energy-efficient microbiome in terms of microbial divisions and coding capacity. This knowledge may allow the improvement of handling and feeding procedures to favor this efficient microbiome, thereby improving animal cost effectiveness and productivity. Another benefit, of a more environmental nature, will be reduction of methane emission. Methane is a very potent green house gas that large portion of its emission worldwide is attributed to ruminants. It is established that by increasing the animals' feed efficiency, methane emission is expected to decrease.

References

Aharoni Y, Brosh A et al (2006) The efficiency of utilization of metabolizable energy for milk production: a comparison of Holstein with F1 Montbeliarde 3 Holstein cows. Anim Sci 82:101–109

Amann RI, Binder BJ et al (1990) Combination of 16S rRNA-targeted oligonucleotide probes with flow cytometry for analyzing mixed microbial populations. Appl Environ Microbiol 56(6):1919–1925

Archer JA, Richardson EC et al (1999) Potential for selection to improve efficiency of feed use in beef cattle. Aust J Agric Res 50(2):147–162
Arthur PF, Archer JA et al (2001) Genetic and phenotypic variance and covariance components for feed intake, feed efficiency, and other postweaning traits in Angus cattle. J Anim Sci 79(11):2805–2811
Basarab JA, Price MA et al (2003) Residual feed intake and body composition in young growing cattle. Can J Anim Sci 83:189–204
Brulc JM, Antonopoulos DA et al (2009) Gene-centric metagenomics of the fiber-adherent bovine rumen microbiome reveals forage specific glycoside hydrolases. Proc Natl Acad Sci USA 106(6):1948–1953
Crews DH Jr (2005) Genetics of efficient feed utilization and national cattle evaluation: a review. Genet Mol Res 4(2):152–165
Dehority BA (1991) Effects of microbial synergism on fibre digestion in the rumen. Proc Nutr Soc 50(2):149–159
Deng W, Xi D et al (2008) The use of molecular techniques based on ribosomal RNA and DNA for rumen microbial ecosystem studies: a review. Mol Biol Rep 35(2):265–274
Ferrell CL, Jenkins TG (1985) "Energy utilization by Hereford and Simmental males and females". Anim Prod 41:53–61
Flint HJ (1997) The rumen microbial ecosystem – some recent developments. Trends Microbiol 5(12):483–488
Flint HJ, Bayer EA et al (2008) Polysaccharide utilization by gut bacteria: potential for new insights from genomic analysis. Nat Rev Microbiol 6(2):121–131
Guan LL, Nkrumah JD et al (2008) Linkage of microbial ecology to phenotype: correlation of rumen microbial ecology to cattle's feed efficiency. FEMS Microbiol Lett 288(1):85–91
Herd RM, Archer JA (2003) "Reducing the cost of beef production through genetic improvement in residual feed intake: opportunity and challenges to application". J Anim Sci 81:9–17
Hernandez-Sanabria E, Guan LL et al (2010) Correlation of particular bacterial PCR-denaturing gradient gel electrophoresis patterns with bovine ruminal fermentation parameters and feed efficiency traits. Appl Environ Microbiol 76(19):6338–6650
Hoover WH, Miller TK (1991) Rumen digestive physiology and microbial ecology. Vet Clin North Am Food Anim Pract 7(2):311–325
Hotovy SK, Johnson KA et al (1991) Variation among twin beef cattle in maintenance energy requirements. J Anim Sci 69(3):940–946
Johnson KA, Johnson DE (1995) Methane emissions from cattle. J Anim Sci 73(8):2483–2492
Johnson DE, Ferrell CL et al (2003) The history of energetic efficiency research: where have we been and where are we going? J Anim Sci 81:27–38
Kay RN (1966) The influence of saliva on digestion in ruminants. World Rev Nutr Diet 6:292–325
Kay RN (1969) Digestion of protein in the intestines of adult ruminants. Proc Nutr Soc 28 (1):140–151
Koch RM, Swiger LA et al (1963) Efficiency of feed use in beef cattle. J Anim Sci 22:486–494
Ley RE, Turnbaugh PJ et al (2006) Microbial ecology: human gut microbes associated with obesity. Nature 444(7122):1022–1023
Mackie RI (2002) Mutualistic fermentative digestion in the gastrointestinal tract: diversity and evolution. Integr Comp Biol 2:42319–42323
McAllister TA, Bae HD et al (1994) Microbial attachment and feed digestion in the rumen. J Anim Sci 72(11):3004–3018
Merchen NR, Elizalde JC et al (1997) Current perspective on assessing site of digestion in ruminants. J Anim Sci 75(8):2223–2234
Miron J, Ben-Ghedalia D et al (2001) Invited review: adhesion mechanisms of rumen cellulolytic bacteria. J Dairy Sci 84(6):1294–1309
Moore SS, Mujibi FD et al (2009) Molecular basis for residual feed intake in beef cattle. J Anim Sci 87(14 Suppl):E41–E47

Nkrumah JD, Basarab JA et al (2004) Different measures of energetic efficiency and their phenotypic relationships with growth, feed intake, and ultrasound and carcass merit in hybrid cattle. J Anim Sci 82(8):2451–2459
Nkrumah JD, Okine EK et al (2006) Relationships of feedlot feed efficiency, performance, and feeding behavior with metabolic rate, methane production, and energy partitioning in beef cattle. J Anim Sci 84(1):145–153
Richardson EC, Herd RM et al (1996) Possible physiological indicators of net feed conversion efficiency. Proc Aust Soc Anim Prod 21:901–908
Richardson EC, Herd RM et al (2004) Metabolic differences in Angus steers divergently selected for residual feed intake. Aust J Exp Agric 44:441–452
Russell JB, Wilson DB (1996) Why are ruminal cellulolytic bacteria unable to digest cellulose at low pH? J Dairy Sci 79(8):1503–1509
Solis JC, Byers FM et al (1988) Maintenance requirements and energetic efficiency of cows of different breed types. J Anim Sci 66(3):764–773
Stewart CS (1986) Rumen function with special reference to fibre digestion. Soc Appl Bacteriol Symp Ser 13:263–286
Storm E, Ørskov ER (1983) The nutritive value of rumen microorganisms in ruminant. 1. Large-scale isolation and chemical composition of rumen microorganisms. Br J Nutr 50:463–470
Taylor SCS, Thiessen RB et al (1986) "Inter-breed relationship of maintenance efficiency to milk yield in cattle". Anim Prod 43:37–61
Thauer RK, Kaster AK et al (2008) Methanogenic archaea: ecologically relevant differences in energy conservation. Nat Rev Microbiol 6(8):579–591
Thiessen RB, Taylor St CS (1985) "Multibreed comparisons of British cattle. Variation in relative growth rate, relative food intake, and food conversion efficiency". Anim Prod 41:193–199
Tolkamp BJ (2010) Efficiency of energy utilisation and voluntary feed intake in ruminants. Animal 4(7):1084–1092
Turnbaugh PJ, Gordon JI (2009) The core gut microbiome, energy balance and obesity. J Physiol 587(Pt 17):4153–4158
Turnbaugh PJ, Ley RE et al (2006) An obesity-associated gut microbiome with increased capacity for energy harvest. Nature 444(7122):1027–1031
Weimer PJ, Russell JB et al (2009) Lessons from the cow: what the ruminant animal can teach us about consolidated bioprocessing of cellulosic biomass. Bioresour Technol 100 (21):5323–5331
Wuebbles DJ, Hayhoe K (2002) "Atmospheric methane and global change". Earth-Sci Rev 57:117–210
Zhou M, Hernandez-Sanabria E et al (2009) Assessment of the microbial ecology of ruminal methanogens in cattle with different feed efficiencies. Appl Environ Microbiol 75(20):6524–6533
Zhou M, Hernandez-Sanabria E et al (2010) Characterization of variation in rumen methanogenic communities under different dietary and host feed efficiency conditions, as determined by PCR-denaturing gradient gel electrophoresis analysis. Appl Environ Microbiol 76(12):3776–3786

Chapter 15
The Intestinal Microbiota and Intestinal Disease: Irritable Bowel Syndrome

Nirit Keren and Uri Gophna

Abstract Irritable bowel syndrome (IBS) is a functional bowel disorder characterized by chronic abdominal pain and discomfort associated with alterations in bowel habits. Some IBS symptoms can be attributed to microbial activity, and some IBS patients have small intestinal bacterial overgrowth, leading many to believe that modified microbial communities may contribute to the etiology of IBS. Here, we review the recent advances in linking bacteria and archaea to IBS symptoms and the attempts to treat IBS symptoms with probiotics.

15.1 Introduction

Irritable bowel syndrome (IBS) is a functional bowel disorder characterized by chronic abdominal pain and discomfort associated with alterations in bowel habits in the absence of any detectable organic disease (Thompson et al. 1999). Throughout the world about 0.8–20% of adults and adolescents have symptoms consistent with IBS, and most studies find a female predominance(Saito et al. 2002; Gwee 2005; Longstreth 2005, Chang et al. 2010; Park et al. 2010; Lix et al. 2010). IBS symptoms come and go over time, and often overlap with other functional disorders (Whitehead et al. 2002), impair the quality of life (Wilson et al. 2004), and result in high healthcare costs (Longstreth et al. 2003). IBS contributes both direct and indirect costs to the total healthcare bill. Total direct cost estimates per patient per year ranges from $US348 to $US8750 (calculated for the year 2002). The average number of days off work per year due IBS symptoms was between 8.5 and 21.6, and indirect costs ranged from $US355 to $US3344 (Maxion-Bergemann et al. 2006).

N. Keren • U. Gophna (✉)
Department of Molecular Microbiology and Biotechnology, Tel Aviv University, Ramat Aviv, Tel Aviv 69978, Israel
e-mail: nirit.keren@clalit.org.il; urigo@tauex.tau.ac.il

E. Rosenberg and U. Gophna (eds.), *Beneficial Microorganisms in Multicellular Life Forms*, 211
DOI 10.1007/978-3-642-21680-0_15, © Springer-Verlag Berlin Heidelberg 2011

Table 15.1 Diagnostic criteria[a] for irritable bowel syndrome

Recurrent abdominal pain or discomfort[b] at least 3 days per month in the last 3 months associated with *2 or more* of the following:
1. Improvement with defecation
2. Onset associated with a change in frequency of stool
3. Onset associated with a change in form (appearance) of stool

[a]Criteria fulfilled for the last 3 months with symptom onset at least 6 months prior to diagnosis
[b]Discomfort means an uncomfortable sensation not described as pain. In pathophysiology research and clinical trials, a pain/discomfort frequency of at least 2 days a week during screening evaluation for subject eligibility

Table 15.2 Subtyping irritable bowel syndrome (IBS) by the predominant stool pattern (Rome III criteria)

1. IBS with constipation (IBS-C) – hard or lumpy stools[a] ≥25% and loose (mushy) or watery stools [b] <25% of bowel movements[c].
2. IBS with diarrhea (IBS-D) – loose (mushy) or watery stools[b] ≥ 25% and hard or lumpy stool[a] <25% of bowel movements[c].
3. Mixed IBS (IBS-M) – hard or lumpy stools[a] ≥ 25% and loose (mushy) or watery stools [b] ≥25% of bowel movements[c].
4. Unsubtyped IBS (IBS-U) – insufficient abnormality of stool consistency to meet criteria IBS-C, D or M[c].

[a]Bristol Stool Form Scale 1–2 [separate hard lumps like nuts (difficult to pass) or sausage shaped but lumpy]
[b]Bristol Stool Form Scale 6–7 (fluffy pieces with ragged edges, a mushy stool or watery, no solid pieces, entirely liquid)
[c]In the absence of use of antidiarrheals or laxatives

The Rome III diagnostic criteria for IBS are described in Table 15.1 (Longstreth et al. 2006). Supportive symptoms that are not part of the diagnostic criteria include abnormal stool frequency ([a] ≤3 bowel movements per week or [b] >3 bowel movements per day), abnormal stool form ([c] lumpy/hard stool or [d] loose/watery stool), [e] defecation straining, [f] urgency, or also a feeling of incomplete bowel movement, passing mucus, and bloating.

The bowel habits in IBS patients show considerable inter- and intra-individual variability (Mearin et al. 2004). Hence, IBS patients are divided into different subgroups based on their predominant bowel pattern - Rome III criteria, subtype IBS-patients exclusively on their stool form (Longstreth et al. 2006; Table 15.2).

15.2 Microbial Alterations in IBS

Since IBS is a chronic intestinal disorder, it is logical to assume that the microbiota may be altered in IBS patients and may even have a causative role. Given that IBS is so common, the idea that one could treat potentially IBS with probiotics, attracted much attention, both in academia and industry, motivating many studies on the intestinal microbiota in this disorder.

Several lines of indirect evidence support microbial irregularities in IBS patients:

1. IBS patients have a higher prevalence of small intestinal bacterial overgrowth (SIBO), at least in its milder form (Lin 2004; Posserud et al. 2007, Shah et al. 2010). These increased microbial concentrations have been associated with abnormalities in small intestinal motor function (Pimentel et al. 2002), and eradication of SIBO has been associated with symptomatic relief (Pimentel et al. 2003; Attar et al. 1999).
2. Methanogenic archaeal activity, indentified by detecting methane in the breath or the feces, is moderately but consistently correlated with constipation (Attaluri et al. 2010; Hwang et al. 2010; Kunkel et al. 2011; Oxley et al. 2010; Fiedorek et al. 1990; Oufir et al. 2000; Pimentel et al. 2003, 2006; Soares et al. 2005; Majewski and McCallum 2007; Morken et al. 2007; Bratten et al. 2008; Parodi et al. 2009). The predominant human methanogen is *Methanobrevibacter smithii* (Oxley et al. 2010). It is still unclear whether or not methane producers affect bowel movements or the relatively slow transit time enables colonization of these archaea. An *in-vivo* study performed by Pimental and coworkers demonstrated a strong effect of methane infusion on intestinal transit in dogs and guinea pigs (Pimentel et al. 2006). Thus, methanogens could be responsible for at least some IBS symptoms. The fact that intestinal fermentation that produces hydrogen is higher in IBS patients compared to controls (King et al. 1998) also implies that a higher concentration of this substrate is available to methanogens in these patients.
3. Consumption of antibiotics in IBS patients is correlated with alleviation of some of the disorder's symptoms (Pimentel et al. 2003; Sharara et al. 2006), even in patients who did not have SIBO at baseline (Pimentel et al. 2011; Dear et al. 2005). This improvement was correlated with a decrease in breath hydrogen excretion.
4. The prevalence of IBS is higher in urban than in rural areas (Usai et al. 2010; Pan et al. 2000; Sperber et al. 2005). Although microbial exposure is by no means the only difference between rural and urban styles of life, the hygiene hypothesis is often brought up to explain the increase in IBS, along with other intestinal diseases, such as Crohn's disease and ulcerative colitis (Ponsonby et al. 2009).

Most cultivation-based studies of the intestinal microbiota did not find substantial differences between IBS patients and controls, but did report some increase in facultative aerobes in fecal samples of IBS patients. An early study reported less coliforms, lactobacilli, and bifidobacteria in fecal samples of IBS patients compared to controls (Balsari et al. 1982), while later study detected less bifidobacteria in IBS patients, but found no difference in lactobacilli, *Bacteroides* species *Clostridum perfringens* and enterococci (Si et al. 2004). This study, however, showed an increase in the Enterobacteriaceae, somewhat contradicting the 1982 report (Si et al. 2004). A study combining cultivation and PCR-DGGE found no significant difference in the abundance of *Bacteroides* species, bifidobacteria, spore-forming bacteria, lactobacilli, enterococci, or yeasts, but an increase in coliforms and aerobe:anaerobe ratio in the IBS patients group (Matto et al. 2005). Similarly, Carroll and coworkers showed using cultivation-based methods, and Quantitative real-time PCR (qPCR) assays of selected bacterial groups, that their levels were mostly unchanged in IBS, but more aerobic bacteria and *Lactobacillus* species were present in fecal samples of IBS patients suffering from diarrhea (Carroll et al. 2010). Nevertheless, a Japanese study did find higher viable counts of *Lactobacillus* and *Veillonella* (by qPCR) and fewer *Clostridium* species in IBS patients than controls (Tana et al. 2010), and a FISH-based study showed a reduction of bifidobacteria in IBS patients (Kerckhoffs et al. 2009), while many other groups of bacteria were unchanged (*Bacteroides-Prevotella* group, *Clostridium coccoides-Eubacterium rectale* group, *Clostridium histolyticum* group, *Clostridium lituseburense* group, *Clostridium difficile, Faecalibacterium prausnitzii*, and the *Lactobacillus-Enterococcus* group).

16S rRNA gene fingerprinting by PCR DGGE of duodenal and fecal samples found no clear difference between IBS patients and healthy controls (Codling et al. 2010; Malinen et al. 2005) but did show significantly increased variation in the gut microbiota of healthy volunteers than that of IBS (Codling et al. 2010). Importantly, although PCR-DGGE and RT-PCR-DGGE analysis with universal bacterial primers also did not reveal any IBS-specific amplicons, IBS patients had lower temporal stability of their microbiota composition than healthy controls (Matto et al. 2005; Maukonen et al. 2006).

16S rRNA gene sequencing can often reveal alterations in the microbiota with improved sensitivity. Thus, extensive high-throughput 16S rDNA cloning and sequencing of 3753 clones derived from patients with IBS and controls showed that the patients differed significantly in fecal microbial composition (Kassinen et al. 2007). The genera found to be enriched in IBS, *Coprococcus*, *Collinsella*, and *Coprobacillus* were further quantified by qPCR. In a follow-up study, only patients suffering from diarrhea were compared to controls and 16S rDNA clone libraries revealed microbial communities enriched in members of the phyla Proteobacteria and Firmicutes (especially sequences belonging to the family Lachnospiraceae), but reduced in the number of Actinobacteria and Bacteroidetes compared to controls) Krogius-Kurikka et al. 2009. These studies were continued by quantifying a set of 14 16S rRNA gene phylotypes by qPCR and comparing the sequence data of 16S rRNA gene clone libraries of healthy controls and symptomatically sub-grouped IBS patients (Lyra et al. 2009). This analysis showed little differences between the groups: A phylotype with 85% similarity to *Clostridium thermosuccinogenes* and a phylotype with 94% similarity to *Ruminococcus torques* were significantly higher in IBS patients with diarrhea than controls. A phylotype with 93% similarity to *R. torques* was enriched in control samples when compared with IBS with mixed symptoms. Additionally, a *Ruminococcus bromii*-like phylotype was associated with IBS patients suffering from constipation in comparison to control subjects. Comparing the clostridial fecal microbiota of IBS subjects and healthy controls by denaturing gradient gel electrophoresis (DGGE) and a multiplexed and quantitative hybridization-based technique, transcript analysis with the aid of affinity capture (TRAC), showed that *C. coccoides-E. rectale* group, which corresponds to clostridial phylogenetic clusters XIVa and XIVb, was significantly lower in the constipation-type IBS subjects than in the control subjects (Maukonen et al. 2006).

15.3 IBS and Infection

Approximately 25% of patients with IBS have a history of infection (Barbara et al. 2004; Parry and Forgacs 2005; Spiller 2003). This condition, termed post-infectious IBS (PI-IBS), manifested in altered gut physiology, has been described, which persists even after the infection and associated subtle inflammation in the colonic epithelium subside (Vallance et al. 2001; Akiho et al. 2005; Barbara et al. 1997). This inflammatory process has been associated with increases in enterochromaffin cells and intraepithelial lymphocytes (Dunlop et al. 2003a; Spiller et al. 2000). Persistent chronic diarrhea is a common problem associated with IBS, and is more frequently associated with PI-IBS (Dunlop et al. 2003b).

A rat model for PI-IBS showed that 3 month following a *Campylobacter jejuni* infection, more than half of the animals experienced changes in stool consistency, significantly higher than controls, and many of them had SIBO (Pimentel et al. 2008). A study that experimented on intestinal mouse tissue after infection with the parasitic nematode *Trichinella spiralis*, as a model for PI-IBS, showed an enhanced ileal ion secretion in response to a primary and secondary bile acids, compared to normal tissue (Kalia et al. 2008). A decrease in active bile re-absorption in the ileum was also observed. Bile acid malabsorption, leading to higher bile levels in the colon, can therefore be a cause of chronic diarrhea in PI-IBS. Taken together, these studies support a causative role for infectious disease in inducing IBS symptoms.

15.4 IBS and Probiotics

Probiotics are defined as mono- or mixed cultures of live micro-organisms which, when applied to animal or man, beneficially affect the host by improving the properties of the indigenous microbiota (Havenaar et al. 1992). Certain probiotics possess potent antibacterial and antiviral properties. Probiotic antibacterial activity may derive from the direct secretion of substances that harm pathogenic species directly, such as bacteriocins (Cotter et al. 2005), the elaboration of proteases that cleave secreted bacterial toxins (Castagliuolo et al. 1999) or through their ability to adhere to epithelial cells and thus exclude pathogens, known as colonization resistance (Zareie et al. 2006). The antiviral properties of some probiotic organisms, including the stimulation of interferon production, together with the well-documented efficacy of certain probiotics in the therapy of rotavirus-mediated diarrhea (Allen et al. 2010) suggested a potential for a role for these agents in PI-IBS. The efficacy of some probiotics in preferentially relieving "gas-related" symptoms may be related to qualitative changes in the colonic microbiota, as described earlier, or through the suppression of SIBO (Nobaek et al. 2000; Kim et al. 2003, 2005; Kajander et al. 2005; Niv et al. 2005; Tsuchiya et al. 2004; Sen et al. 2002; Bausserman and Michail 2005), for which some encouraging reports exist (Quigley and Quera 2006).

Probiotics could also have a role in terms of anti-inflammatory activity (McCarthy et al. 2003; Hart et al. 2004; O'Mahony et al. 2005; Rachmilewitz et al. 2004). Some bifidobacterial species have been shown to promote secretion of the anti-inflammatory cytokine, IL-10, and suppress the secretion of the pro-inflammatory cytokine, IL-12, in dendritic and mononuclear cells derived from the human peripheral blood (Hart et al. 2004; O'Mahony et al. 2005). In experimental animal models of inflammatory bowel disease, various probiotic strains and probiotic cocktails have demonstrated potent anti-inflammatory properties, suppressing mucosal inflammation and restoring cytokine balance towards an anti-inflammatory state (McCarthy et al. 2003; Rachmilewitz et al. 2004).

Probiotics have the potential to influence both motility and visceral sensation. In a murine model of PI-IBS, the probiotic *Lactobacillus paracasei* prevented and reversed muscle hypercontractility upon an intestinal infestation (Verdu et al. 2004). Effect of probiotic *Lactobacillus* species (*L. paracasei*, *Lactobacillus farciminis*, and *Lactobacillus reuteri*) on visceral hypersensitivity was also reported in murine models (Verdu et al. 2006; Ait-Belgnaoui et al. 2006; Kamiya et al. 2006) and linked to anti-inflammatory (Verdu et al. 2006), barrier-enhancing (Ait-Belgnaoui et al. 2006) and neuromodulatory (Verdu et al. 2006; Ait-Belgnaoui et al. 2006) activities. A resolution of abdominal pain was detected in patients treated with *Lactobacillus plantarum* 299V (Niedzielin et al. 2001).

Lactobacillus casei GG and *L. plantarum* have been shown to reduce bloating (O'Sullivan and O'Morain 2000; Nobaek et al. 2000). *B. infantis* 35624 reduced the cardinal symptoms of the IBS (abdominal pain/discomfort, distension/ bloating and difficult defecation) (Whorwell et al. 2006). The effect of the probiotic strain is not necessarily dependent on active bacterial metabolism, since heat-killed *Lactobacillus acidophilus* reduced an IBS symptom index (Halpern et al. 1996).

It is perhaps unsurprising that there are great differences in the immunomodulatory effects of probiotic species belonging to the groups lactobacilli and bifidobacteria, as well as substantial differences within these groups (McCarthy et al. 2003; Hart et al. 2004; O'Mahony et al. 2005; Madsen et al. 2001) or antibacterial activity (Madsen et al. 2001; O'Mahony et al. 2006; Jijon et al. 2004). Currently, the prevailing view, as stated by the American College of Gastroenterology Task Force on the management of IBS, is that *Lactobacillus* species-based therapies have not shown efficacy in IBS management, while *Bifidobacterium*-based approaches have demonstrated some efficacy (Aragon et al. 2010). Nevertheless, one should keep in mind that even the most promising studies generally show improvement in one or two clinical categories, out of several examined.

15.5 Concluding Remarks

The study of microbial involvement in IBS remains in its infancy, yet already the accumulating body of evidence hints that specific bacteria species may contribute to IBS symptoms. Future studies should first rule out SIBO as well as additional physiological problems that may be identified by colonoscopy, and then focus on relatively homogenous symptom groups, which should improve the consistency and magnitude of the results obtained. However, one should consider that even when significant differences between cases and controls are observed, these changes are not necessarily causative. To advance from evidence for association to causation is likely to require work in animal models, which are currently restricted to PI-IBS. Thus, alterations in the microbiota that are observed in IBS can often be attributed to the IBS symptoms, such as diarrhea, and the magnitude of their contribution is difficult to estimate.

Probiotics may have a future therapeutic role in treating IBS. Preliminary studies of *Bifidobacterium*-based probiotics are encouraging, but larger, well-controlled studies are required to establish these findings, and to help develop more efficacious treatment protocols, recommended dosage and potentially different requirements based on IBS sub-type (Aragon et al. 2010).

15.6 Unresolved Questions and Future Research

Despite findings that support the idea that there are alterations in gut microbiota in IBS patients compared to controls (e.g. -higher temporal instability and less intra-individual variation), the studies published so far contradict one another in terms of what the changes in the microbiota are. Hence, there is a need for studies including larger groups of patients, division of the three IBS subgroups, and elimination of other possible causes of IBS symptoms prior to diagnosis (including SIBO, lactose intolerance, etc.). Due to the relatively high prevalence of IBS in the population, the high healthcare costs and the knowledge of risk factors for IBS (women who live in urban areas), prospective studies on a large cohort should be made in this field. Thus the field should not rely too heavily on animal models of IBS (especially since animal models are restricted to PI-IBS). Although prospective studies are challenging and require a large investment of time and resources, one should take into account the relatively high incidence of IBS (around 1.5%), which should make this strategy a viable approach.

References

Ait-Belgnaoui A, Han W, Lamine F, Eutamene H, Fioramonti J, Bueno L, Theodorou V (2006) Lactobacillus farciminis treatment suppresses stress induced visceral hypersensitivity: a possible action through interaction with epithelial cell cytoskeleton contraction. Gut 55:1090–1094

Akiho H, Deng Y, Blennerhassett P, Kanbayashi H, Collins SM (2005) Mechanisms underlying the maintenance of muscle hypercontractility in a model of postinfective gut dysfunction. Gastroenterology 129:131–141

Allen SJ, Okoko B, Martinez E, Gregorio G, Dans LF (2010) Probiotics for treating infectious diarrhoea. Cochrane Database Syst Rev 11:CD003048

Aragon G, Graham DB, Borum M, Doman DB (2010) Probiotic therapy for irritable bowel syndrome. Gastroenterol Hepatol (NY) 6:39–44

Attaluri A, Jackson M, Valestin J, Rao SS (2010) Methanogenic flora is associated with altered colonic transit but not stool characteristics in constipation without IBS. Am J Gastroenterol 105:1407–1411

Attar A, Flourie B, Rambaud JC, Franchisseur C, Ruszniewski P, Bouhnik Y (1999) Antibiotic efficacy in small intestinal bacterial overgrowth-related chronic diarrhea: a crossover, randomized trial. Gastroenterology 117:794–797

Balsari A, Ceccarelli A, Dubini F, Fesce E, Poli G (1982) The fecal microbial population in the irritable bowel syndrome. Microbiologica 5:185–194

Barbara G, Vallance BA, Collins SM (1997) Persistent intestinal neuromuscular dysfunction after acute nematode infection in mice. Gastroenterology 113:1224–1232

Barbara G, De Giorgio R, Stanghellini V, Cremon C, Salvioli B, Corinaldesi R (2004) New pathophysiological mechanisms in irritable bowel syndrome. Aliment Pharmacol Ther 20 (Suppl 2):1–9

Bausserman M, Michail S (2005) The use of Lactobacillus GG in irritable bowel syndrome in children: a double-blind randomized control trial. J Pediatr 147:197–201

Bratten JR, Spanier J, Jones MP (2008) Lactulose breath testing does not discriminate patients with irritable bowel syndrome from healthy controls. Am J Gastroenterol 103:958–963

Carroll IM, Chang YH, Park J, Sartor RB, Ringel Y (2010) Luminal and mucosal-associated intestinal microbiota in patients with diarrhea-predominant irritable bowel syndrome. Gut Pathog 2:19

Castagliuolo I, Riegler MF, Valenick L, LaMont JT, Pothoulakis C (1999) Saccharomyces boulardii protease inhibits the effects of Clostridium difficile toxins A and B in human colonic mucosa. Infect Immun 67:302–307

Chang FY, Lu CL, Chen TS (2010) The current prevalence of irritable bowel syndrome in Asia. J Neurogastroenterol Motil 16:389–400

Codling C, O'Mahony L, Shanahan F, Quigley EM, Marchesi JR (2010) A molecular analysis of fecal and mucosal bacterial communities in irritable bowel syndrome. Dig Dis Sci 55:392–397

Cotter PD, Hill C, Ross RP (2005) Bacteriocins: developing innate immunity for food. Nat Rev Microbiol 3:777–788

Dear KL, Elia M, Hunter JO (2005) Do interventions which reduce colonic bacterial fermentation improve symptoms of irritable bowel syndrome? Dig Dis Sci 50:758–766

Dunlop SP, Jenkins D, Neal KR, Spiller RC (2003a) Relative importance of enterochromaffin cell hyperplasia, anxiety, and depression in postinfectious IBS. Gastroenterology 125:1651–1659

Dunlop SP, Jenkins D, Spiller RC (2003b) Distinctive clinical, psychological, and histological features of postinfective irritable bowel syndrome. Am J Gastroenterol 98:1578–1583

Fiedorek SC, Pumphrey CL, Casteel HB (1990) Breath methane production in children with constipation and encopresis. J Pediatr Gastroenterol Nutr 10:473–477

Gwee KA (2005) Irritable bowel syndrome in developing countries – a disorder of civilization or colonization? Neurogastroenterol Motil 17:317–324

Halpern GM, Prindiville T, Blankenburg M, Hsia T, Gershwin ME (1996) Treatment of irritable bowel syndrome with Lacteol Fort: a randomized, double-blind, cross-over trial. Am J Gastroenterol 91:1579–1585

Hart AL, Lammers K, Brigidi P, Vitali B, Rizzello F, Gionchetti P, Campieri M, Kamm MA, Knight SC, Stagg AJ (2004) Modulation of human dendritic cell phenotype and function by probiotic bacteria. Gut 53:1602–1609

Havenaar R, Ten Brink B, Huis in't Veld J (1992) Selection of strains for probiotic use. In: Fuller R (ed) Probiotics: the scientific basis. Chapman and Hall, London, pp 209–224

Hwang L, Low K, Khoshini R, Melmed G, Sahakian A, Makhani M, Pokkunuri V, Pimentel M (2010) Evaluating breath methane as a diagnostic test for constipation-predominant IBS. Dig Dis Sci 55:398–403

Jijon H, Backer J, Diaz H, Yeung H, Thiel D, McKaigney C, De Simone C, Madsen K (2004) DNA from probiotic bacteria modulates murine and human epithelial and immune function. Gastroenterology 126:1358–1373

Kajander K, Hatakka K, Poussa T, Farkkila M, Korpela R (2005) A probiotic mixture alleviates symptoms in irritable bowel syndrome patients: a controlled 6-month intervention. Aliment Pharmacol Ther 22:387–394

Kajander K, Myllyluoma E, Rajilic-Stojanovic M, Kyronpalo S, Rasmussen M, Jarvenpaa S, Zoetendal EG, de Vos WM, Vapaatalo H, Korpela R (2008) Clinical trial: multispecies probiotic supplementation alleviates the symptoms of irritable bowel syndrome and stabilizes intestinal microbiota. Aliment Pharmacol Ther 27:48–57

Kalia N, Hardcastle J, Keating C, Grasa L, Pelegrin P, Bardhan KD, Grundy D (2008) Intestinal secretory and absorptive function in Trichinella spiralis mouse model of postinfective gut dysfunction: role of bile acids. Gut 57:41–49

Kamiya T, Wang L, Forsythe P, Goettsche G, Mao Y, Wang Y, Tougas G, Bienenstock J (2006) Inhibitory effects of Lactobacillus reuteri on visceral pain induced by colorectal distension in Sprague-Dawley rats. Gut 55:191–196

Kassinen A, Krogius-Kurikka L, Makivuokko H, Rinttila T, Paulin L, Corander J, Malinen E, Apajalahti J, Palva A (2007) The fecal microbiota of irritable bowel syndrome patients differs significantly from that of healthy subjects. Gastroenterology 133:24–33

Kerckhoffs AP, Samsom M, van der Rest ME, de Vogel J, Knol J, Ben-Amor K, Akkermans LM (2009) Lower Bifidobacteria counts in both duodenal mucosa-associated and fecal microbiota in irritable bowel syndrome patients. World J Gastroenterol 15:2887–2892

Kim HJ, Camilleri M, McKinzie S, Lempke MB, Burton DD, Thomforde GM, Zinsmeister AR (2003) A randomized controlled trial of a probiotic, VSL#3, on gut transit and symptoms in diarrhoea-predominant irritable bowel syndrome. Aliment Pharmacol Ther 17:895–904

Kim HJ, Vazquez Roque MI, Camilleri M, Stephens D, Burton DD, Baxter K, Thomforde G, Zinsmeister AR (2005) A randomized controlled trial of a probiotic combination VSL# 3 and placebo in irritable bowel syndrome with bloating. Neurogastroenterol Motil 17:687–696

King TS, Elia M, Hunter JO (1998) Abnormal colonic fermentation in irritable bowel syndrome. Lancet 352:1187–1189

Krogius-Kurikka L, Lyra A, Malinen E, Aarnikunnas J, Tuimala J, Paulin L, Makivuokko H, Kajander K, Palva A (2009) Microbial community analysis reveals high level phylogenetic alterations in the overall gastrointestinal microbiota of diarrhoea-predominant irritable bowel syndrome sufferers. BMC Gastroenterol 9:95

Kunkel D, Basseri RJ, Makhani MD, Chong K, Chang C, Pimentel M (2011) Methane on breath testing is associated with constipation: a systematic review and meta-analysis. Dig Dis Sci 56 (6):1612–1618

Lin HC (2004) Small intestinal bacterial overgrowth: a framework for understanding irritable bowel syndrome. JAMA 292:852–858

Lix LM, Yogendran MS, Shaw SY, Targownick LE, Jones J, Bataineh O (2010) Comparing administrative and survey data for ascertaining cases of irritable bowel syndrome: a population-based investigation. BMC Health Serv Res 10:31

Longstreth GF (2005) Definition and classification of irritable bowel syndrome: current consensus and controversies. Gastroenterol Clin North Am 34:173–187

Longstreth GF, Wilson A, Knight K, Wong J, Chiou CF, Barghout V, Frech F, Ofman JJ (2003) Irritable bowel syndrome, health care use, and costs: a U.S. managed care perspective. Am J Gastroenterol 98:600–607

Longstreth GF, Thompson WG, Chey WD, Houghton LA, Mearin F, Spiller RC (2006) Functional bowel disorders. Gastroenterology 130:1480–1491
Lyra A, Rinttila T, Nikkila J, Krogius-Kurikka L, Kajander K, Malinen E, Matto J, Makela L, Palva A (2009) Diarrhoea-predominant irritable bowel syndrome distinguishable by 16S rRNA gene phylotype quantification. World J Gastroenterol 15:5936–5945
Madsen K, Cornish A, Soper P, McKaigney C, Jijon H, Yachimec C, Doyle J, Jewell L, De Simone C (2001) Probiotic bacteria enhance murine and human intestinal epithelial barrier function. Gastroenterology 121:580–591
Majewski M, McCallum RW (2007) Results of small intestinal bacterial overgrowth testing in irritable bowel syndrome patients: clinical profiles and effects of antibiotic trial. Adv Med Sci 52:139–142
Malinen E, Rinttila T, Kajander K, Matto J, Kassinen A, Krogius L, Saarela M, Korpela R, Palva A (2005) Analysis of the fecal microbiota of irritable bowel syndrome patients and healthy controls with real-time PCR. Am J Gastroenterol 100:373–382
Matto J, Maunuksela L, Kajander K, Palva A, Korpela R, Kassinen A, Saarela M (2005) Composition and temporal stability of gastrointestinal microbiota in irritable bowel syndrome – a longitudinal study in IBS and control subjects. FEMS Immunol Med Microbiol 43:213–222
Maukonen J, Satokari R, Matto J, Soderlund H, Mattila-Sandholm T, Saarela M (2006) Prevalence and temporal stability of selected clostridial groups in irritable bowel syndrome in relation to predominant faecal bacteria. J Med Microbiol 55:625–633
Maxion-Bergemann S, Thielecke F, Abel F, Bergemann R (2006) Costs of irritable bowel syndrome in the UK and US. Pharmacoeconomics 24:21–37
McCarthy J, O'Mahony L, O'Callaghan L, Sheil B, Vaughan EE, Fitzsimons N, Fitzgibbon J, O'Sullivan GC, Kiely B, Collins JK, Shanahan F (2003) Double blind, placebo controlled trial of two probiotic strains in interleukin 10 knockout mice and mechanistic link with cytokine balance. Gut 52:975–980
Mearin F, Baro E, Roset M, Badia X, Zarate N, Perez I (2004) Clinical patterns over time in irritable bowel syndrome: symptom instability and severity variability. Am J Gastroenterol 99:113–121
Morken MH, Berstad AE, Nysaeter G, Berstad A (2007) Intestinal gas in plain abdominal radiographs does not correlate with symptoms after lactulose challenge. Eur J Gastroenterol Hepatol 19:589–593
Niedzielin K, Kordecki H, Birkenfeld B (2001) A controlled, double-blind, randomized study on the efficacy of Lactobacillus plantarum 299V in patients with irritable bowel syndrome. Eur J Gastroenterol Hepatol 13:1143–1147
Niv E, Naftali T, Hallak R, Vaisman N (2005) The efficacy of Lactobacillus reuteri ATCC 55730 in the treatment of patients with irritable bowel syndrome–a double blind, placebo-controlled, randomized study. Clin Nutr 24:925–931
Nobaek S, Johansson ML, Molin G, Ahrne S, Jeppsson B (2000) Alteration of intestinal microflora is associated with reduction in abdominal bloating and pain in patients with irritable bowel syndrome. Am J Gastroenterol 95:1231–1238
O'Mahony L, McCarthy J, Kelly P, Hurley G, Luo F, Chen K, O'Sullivan GC, Kiely B, Collins JK, Shanahan F, Quigley EM (2005) Lactobacillus and bifidobacterium in irritable bowel syndrome: symptom responses and relationship to cytokine profiles. Gastroenterology 128:541–551
O'Mahony L, O'Callaghan L, McCarthy J, Shilling D, Scully P, Sibartie S, Kavanagh E, Kirwan WO, Redmond HP, Collins JK, Shanahan F (2006) Differential cytokine response from dendritic cells to commensal and pathogenic bacteria in different lymphoid compartments in humans. Am J Physiol Gastrointest Liver Physiol 290:G839–G845
O'Sullivan MA, O'Morain CA (2000) Bacterial supplementation in the irritable bowel syndrome. A randomised double-blind placebo-controlled crossover study. Dig Liver Dis 32:294–301
Oufir LE, Barry JL, Flourie B, Cherbut C, Cloarec D, Bornet F, Galmiche JP (2000) Relationships between transit time in man and in vitro fermentation of dietary fiber by fecal bacteria. Eur J Clin Nutr 54:603–609

Oxley AP, Lanfranconi MP, Wurdemann D, Ott S, Schreiber S, McGenity TJ, Timmis KN, Nogales B (2010) Halophilic archaea in the human intestinal mucosa. Environ Microbiol 12:2398–2410

Pan G, Lu S, Ke M, Han S, Guo H, Fang X (2000) Epidemiologic study of the irritable bowel syndrome in Beijing: stratified randomized study by cluster sampling. Chin Med J (Engl) 113:35–39

Park DW, Lee OY, Shim SG, Jun DW, Lee KN, Kim HY, Lee HL, Yoon BC, Choi HS (2010) The differences in prevalence and sociodemographic characteristics of irritable bowel syndrome according to Rome II and Rome III. J Neurogastroenterol Motil 16:186–193

Parodi A, Dulbecco P, Savarino E, Giannini EG, Bodini G, Corbo M, Isola L, De Conca S, Marabotto E, Savarino V (2009) Positive glucose breath testing is more prevalent in patients with IBS-like symptoms compared with controls of similar age and gender distribution. J Clin Gastroenterol 43:962–966

Parry S, Forgacs I (2005) Intestinal infection and irritable bowel syndrome. Eur J Gastroenterol Hepatol 17:5–9

Pimentel M, Soffer EE, Chow EJ, Kong Y, Lin HC (2002) Lower frequency of MMC is found in IBS subjects with abnormal lactulose breath test, suggesting bacterial overgrowth. Dig Dis Sci 47:2639–2643

Pimentel M, Chow EJ, Lin HC (2003) Normalization of lactulose breath testing correlates with symptom improvement in irritable bowel syndrome: a double-blind, randomized, placebo-controlled study. Am J Gastroenterol 98:412–419

Pimentel M, Lin HC, Enayati P, van den Burg B, Lee HR, Chen JH, Park S, Kong Y, Conklin J (2006) Methane, a gas produced by enteric bacteria, slows intestinal transit and augments small intestinal contractile activity. Am J Physiol Gastrointest Liver Physiol 290:G1089–G1095

Pimentel M, Chatterjee S, Chang C, Low K, Song Y, Liu C, Morales W, Ali L, Lezcano S, Conklin J, Finegold S (2008) A new rat model links two contemporary theories in irritable bowel syndrome. Dig Dis Sci 53:982–989

Pimentel M, Lembo A, Chey WD, Zakko S, Ringel Y, Yu J, Mareya SM, Shaw AL, Bortey E, Forbes WP (2011) Rifaximin therapy for patients with irritable bowel syndrome without constipation. N Engl J Med 364:22–32

Ponsonby AL, Catto-Smith AG, Pezic A, Dupuis S, Halliday J, Cameron D, Morley R, Carlin J, Dwyer T (2009) Association between early-life factors and risk of child-onset Crohn's disease among Victorian children born 1983–1998: a birth cohort study. Inflamm Bowel Dis 15:858–866

Posserud I, Stotzer PO, Bjornsson ES, Abrahamsson H, Simren M (2007) Small intestinal bacterial overgrowth in patients with irritable bowel syndrome. Gut 56:802–808

Quigley EM, Quera R (2006) Small intestinal bacterial overgrowth: roles of antibiotics, prebiotics, and probiotics. Gastroenterology 130:S78–S90

Rachmilewitz D, Katakura K, Karmeli F, Hayashi T, Reinus C, Rudensky B, Akira S, Takeda K, Lee J, Takabayashi K, Raz E (2004) Toll-like receptor 9 signaling mediates the anti-inflammatory effects of probiotics in murine experimental colitis. Gastroenterology 126:520–528

Saito YA, Schoenfeld P, Locke GR 3rd (2002) The epidemiology of irritable bowel syndrome in North America: a systematic review. Am J Gastroenterol 97:1910–1915

Sen S, Mullan MM, Parker TJ, Woolner JT, Tarry SA, Hunter JO (2002) Effect of Lactobacillus plantarum 299v on colonic fermentation and symptoms of irritable bowel syndrome. Dig Dis Sci 47:2615–2620

Shah ED, Basseri RJ, Chong K, Pimentel M (2010) Abnormal breath testing in IBS: a meta-analysis. Dig Dis Sci 55:2441–2449

Sharara AI, Aoun E, Abdul-Baki H, Mounzer R, Sidani S, Elhajj I (2006) A randomized double-blind placebo-controlled trial of rifaximin in patients with abdominal bloating and flatulence. Am J Gastroenterol 101:326–333

Si JM, Yu YC, Fan YJ, Chen SJ (2004) Intestinal microecology and quality of life in irritable bowel syndrome patients. World J Gastroenterol 10:1802–1805

Soares AC, Lederman HM, Fagundes-Neto U, de Morais MB (2005) Breath methane associated with slow colonic transit time in children with chronic constipation. J Clin Gastroenterol 39:512–515

Sperber AD, Friger M, Shvartzman P, Abu-Rabia M, Abu-Rabia R, Abu-Rashid M, Albedour K, Alkranawi O, Eisenberg A, Kazanoviz A, Mazingar L, Fich A (2005) Rates of functional bowel disorders among Israeli Bedouins in rural areas compared with those who moved to permanent towns. Clin Gastroenterol Hepatol 3:342–348

Spiller RC (2003) Postinfectious irritable bowel syndrome. Gastroenterology 124:1662–1671

Spiller RC, Jenkins D, Thornley JP, Hebden JM, Wright T, Skinner M, Neal KR (2000) Increased rectal mucosal enteroendocrine cells, T lymphocytes, and increased gut permeability following acute Campylobacter enteritis and in post-dysenteric irritable bowel syndrome. Gut 47:804–811

Tana C, Umesaki Y, Imaoka A, Handa T, Kanazawa M, Fukudo S (2010) Altered profiles of intestinal microbiota and organic acids may be the origin of symptoms in irritable bowel syndrome. Neurogastroenterol Motil 22(512–519):e114–e515

Thompson WG, Longstreth GF, Drossman DA, Heaton KW, Irvine EJ, Muller-Lissner SA (1999) Functional bowel disorders and functional abdominal pain. Gut 45(Suppl 2):II43–II47

Trauner M, Boyer JL (2003) Bile salt transporters: molecular characterization, function, and regulation. Physiol Rev 83:633–671

Tsuchiya J, Barreto R, Okura R, Kawakita S, Fesce E, Marotta F (2004) Single-blind follow-up study on the effectiveness of a symbiotic preparation in irritable bowel syndrome. Chin J Dig Dis 5:169–174

Usai P, Manca R, Lai MA, Russo L, Boi MF, Ibba I, Giolitto G, Cuomo R (2010) Prevalence of irritable bowel syndrome in Italian rural and urban areas. Eur J Intern Med 21:324–326

Vallance BA, Blennerhassett PA, Huizinga JD, Collins SM (2001) Mast cell-independent impairment of host defense and muscle contraction in T. spiralis-infected W/W(V) mice. Am J Physiol Gastrointest Liver Physiol 280:G640–G648

Verdu EF, Bercik P, Bergonzelli GE, Huang XX, Blennerhasset P, Rochat F, Fiaux M, Mansourian R, Corthesy-Theulaz I, Collins SM (2004) Lactobacillus paracasei normalizes muscle hypercontractility in a murine model of postinfective gut dysfunction. Gastroenterology 127:826–837

Verdu EF, Bercik P, Verma-Gandhu M, Huang XX, Blennerhassett P, Jackson W, Mao Y, Wang L, Rochat F, Collins SM (2006) Specific probiotic therapy attenuates antibiotic induced visceral hypersensitivity in mice. Gut 55:182–190

Whitehead WE, Palsson O, Jones KR (2002) Systematic review of the comorbidity of irritable bowel syndrome with other disorders: what are the causes and implications? Gastroenterology 122:1140–1156

Whorwell PJ, Altringer L, Morel J, Bond Y, Charbonneau D, O'Mahony L, Kiely B, Shanahan F, Quigley EM (2006) Efficacy of an encapsulated probiotic Bifidobacterium infantis 35624 in women with irritable bowel syndrome. Am J Gastroenterol 101:1581–1590

Wilson A, Longstreth GF, Knight K, Wong J, Wade S, Chiou CF, Barghout V, Frech F, Ofman JJ (2004) Quality of life in managed care patients with irritable bowel syndrome. Manag Care Interface 17(24–28):34

Zareie M, Johnson-Henry K, Jury J, Yang PC, Ngan BY, McKay DM, Soderholm JD, Perdue MH, Sherman PM (2006) Probiotics prevent bacterial translocation and improve intestinal barrier function in rats following chronic psychological stress. Gut 55:1553–1560

Chapter 16
Intestinal Microbiota and Intestinal Disease: Inflammatory Bowel Diseases

Amir Kovacs and Uri Gophna

Abstract The two major inflammatory bowel diseases (IBD), Crohn's disease (CD), and ulcerative colitis (UC) are idiopathic relapsing disorders characterized by chronic inflammation of the gastrointestinal tract. It is increasingly evident that the commensal intestinal microbiota plays a role in the pathogenesis of IBD, as multiple lines of evidence, both from rodent and human studies, support microbial involvement in the etiology of these diseases. In general, it is thought that IBD are driven by an irregular immune response to the commensal microbiota in genetically susceptible individuals. A leading hypothesis, concerning the nature of the role that bacteria play in the pathogenesis of IBD, suggests that the disease state is promoted by *dysbiosis*, a shift in the balance of healthy microbiota in favor of pro-inflammatory microbial species. Numerous studies have described a reduction in the biodiversity of the Firmicutes phylum in CD patients, particularly clostridial species. This phylogenetic group contains many bacteria that produce butyrate, a short chain fatty acid considered to have anti-inflammatory properties. Moreover, recent data suggest that clostridial species are involved in multiple regulatory processes of the innate immune system. Further research, elucidating the interactions between the gut microbiota and the immune system could potentially provide the key for understanding IBD.

16.1 Inflammatory Bowel Diseases

Inflammatory bowel diseases (IBD), which comprise Crohn's disease (CD) and ulcerative colitis (UC), are characterized by chronic inflammation of the gastrointestinal tract. These diseases share similar symptoms, such as persistent diarrhea,

A. Kovacs • U. Gophna (✉)
Department of Molecular Microbiology & Biotechnology, Tel Aviv University, Ramat Aviv, Tel Aviv 69978, Israel
e-mail: amir.kovacs@gmail.com; UriGo@tauex.tau.ac.il

E. Rosenberg and U. Gophna (eds.), *Beneficial Microorganisms in Multicellular Life Forms*, 223
DOI 10.1007/978-3-642-21680-0_16,

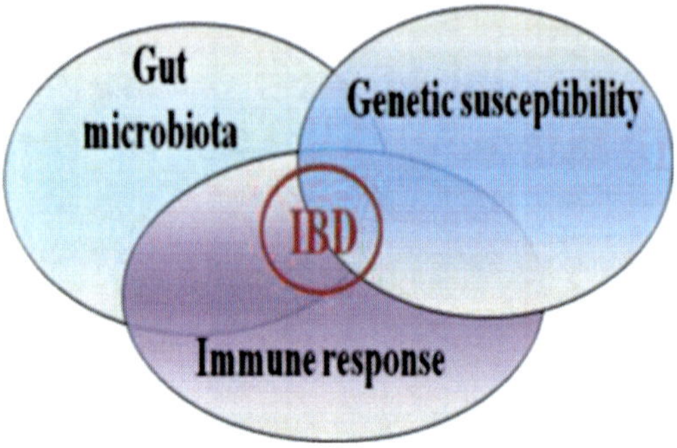

Fig. 16.1 IBD are multifactorial diseases resulting from the interaction between microbes, host genetics, and immunity

blood in the stool, abdominal pain, and bloating (Hendrickson et al. 2002). However, while CD can affect any part of the gastrointestinal tract, from the esophagus to the rectum, and is characterized by transmural inflammation, UC is a diffuse mucosal disease, which affects the colon exclusively.

The number of IBD patients is estimated to be 1.4 million persons in the United States and 2.2 million persons in Europe. Although these areas have been historically associated with IBD and exhibit the highest prevalence and incidence rates, regions such as Africa, Asia, and South America have shown increased incidence and prevalence during the past decade. This underscores the fact that the occurrence of IBD is a dynamic process, affected by environmental changes (Loftus 2004).

As the etiology of IBD is unknown, they are incurable, and current therapies are aimed at relieving the symptoms and include different agents, such as antibiotics, steroids, 5-aminosalicylic acid, thiopurines, and anti TNFα agents (Leenen and Dieleman 2007). Better understanding of the etiology and pathogenesis of these diseases is essential for the development of new and improved therapeutic approaches. It is clear that both genetics and environmental factors (Hviid et al. 2011) contribute to the formation of IBD. Extensive research in the field of IBD has established the general notion that IBD are driven by an irregular immune response to the commensal bacteria in genetically susceptible individuals (Xavier and Podolsky 2007; Fig. 16.1).

This chapter will address the role of the commensal bacteria in the gut – the gut microbiota in the etiology and pathogenesis of IBD.

16.2 The Gut Microbiota Is Essential to Induce Experimental Colitis

The strongest evidence implicating bacteria in the pathogenesis of IBD comes from research in mouse models. Studies in germ-free, specific pathogen-free, and gnotobiotic mice that have been re-colonized with a defined microbiota have proved the importance of bacteria in the initiation and development of chronic intestinal disease (Uhlig and Powrie 2009). For example, experiments with germ-free IL-10 knock-out mice showed that these mice do not develop colitis, thus stressing the fact that bacteria are essential for inflammation, even in the presence of an immune predisposition (Sellon et al. 1998). Additional experiments applied mono-associations as well as dual-associations with specific commensal bacteria, and showed different levels of inflammation for different bacterial species (Kim et al. 2005, 2007).

16.3 The Concept of Dysbiosis

While there is little doubt that the gut microbiota plays a central role in the pathogenesis of IBD, as had been well established by the mouse model studies, the exact nature of this involvement is still in question. In general, there are three hypothetical types of roles which bacteria can play in the etiology of IBD.

1. Specific bacterial pathogen is causing the disease in an infectious manner - the most persistent hypothesis implicates *Mycobacterium avium subsp. paratuberculosis* as a direct pathogen in CD (Quirke 2001; Greenstein 2003; Sartor 2005).
2. The gut microbiota as a whole causes inflammation in genetically susceptible individuals, regardless of its composition. Genetic studies in humans showed that the products of genes associated with disease susceptibility are part of the innate immune system and are involved in recognition of general bacterial motifs. These include NOD2/CARD15 and Toll-like receptors (Girardin et al. 2003; Inohara et al. 2003; Franchimont et al. 2004). According to this hypothesis, the bacterial antigens that meet the poorly regulated immune response and the permeable epithelium are the cause for inflammation (Macdonald and Monteleone 2005; Tannock 2010).
3. A leading hypothesis in the field of IBD research suggests that the disease state is created by dysbiosis, a shift in the balance of healthy microbiota in favor of pro-inflammatory microbial species, which can lead to intestinal inflammation (Sartor 2001; Tamboli et al. 2004).

16.4 Changes in the Bacterial Community Composition in IBD

Recent years have seen a surge in studies supporting the dysbiosis hypothesis indirectly by showing that the microbiotas of IBD patients are indeed significantly modified when compared to controls (Macfarlane et al. 2004; Ott et al. 2004; Gophna et al. 2006; Sokol et al. 2006, 2008a; Frank et al. 2007). One observation that was repeatedly found in several independent studies is a reduction in the diversity and abundance of members of the phylum Firmicutes (Manichanh et al. 2006; Baumgart et al. 2007). This observation is especially interesting, since it holds the potential to directly link a quantitative deficiency in the normal microbiota to the pathogenesis of IBD. Many members of the class Clostridia in the Firmicutes are producers of the short chain fatty acid butyrate. This short chain fatty acid is not only an important energy source for colonocytes (Roediger 1982), but also possesses anti-inflammatory properties, including the ability to reduce LPS-induced cytokine response (Segain et al. 2000). Butyrate has anti-inflammatory effects that result from inhibition of activation of the transcription factor NFκB and IκBa degradation, and consequent reduced formation of pro-inflammatory cytokines (Luhrs et al. 2002a, b). Several studies demonstrated that butyrate exerts immunomodulatory effects such as downregulation of T-cell responses, induction of Th1 cell anergy, and modulation of antigen-presentation-associated molecules (Gilbert and Weigle 1993; Jackson et al. 2002). Diakos et al. (2002) demonstrated that butyrate exerts a strong inhibitory effect on T-cells. Butyrate abrogates the production of IL-2 by a novel mechanism in which it completely inhibits the nuclear binding of the transcription factor, nuclear factor of activated T-cells (NF-AT). NF-AT plays a key role in the activation of many early-immune-response genes. Given the multitude of immunomodulatory roles of butyrate, a deficiency in bacteria that produce it could predispose some individuals to subsequent intestinal inflammation.

16.5 Differences Exist in the Bacterial Etiology Between Crohn's Disease and Ulcerative Colitis

CD and UC display distinctly different pathologies and immunologies, and therefore, it is reasonable to assume a distinctly different bacterial etiology, if one such exists, for each of these diseases. Several studies have shown that differences also exist in the microbiota composition between CD and UC. One such study stands out, since it involves the analysis of biopsies from intestinal sites of newly diagnosed and untreated CD and UC patients (Bibiloni et al. 2006). 16S rRNA gene clone libraries analysis revealed an increased prevalence of unclassified *Bacteroidetes* in samples collected from CD patients in comparison to those from healthy subjects or UC patients. In addition, unclassified members of the phylum *Verrucomicrobia* were only detected in biopsies from CD patients, while unclassified *Porphyromonadaceae* were only detected in biopsies from UC patients.

Similar findings were obtained in another study (Gophna et al. 2006), which also characterized differences in the microbiota of IBD patients using sequencing of 16S clone libraries. Members of Proteobacteria (especially, *Acinetobacter junii*) and *Bacteroidetes* (especially, *Bacteroides fragilis*) were found to be increased in CD patients. On the other hand, a significant decrease in Firmicutes, all belonging to the class Clostridia was also detected in the CD group. Most importantly in this research, the tissue-associated microbiota of UC patients was found to be broadly similar to that of healthy individuals and significantly different from that of CD patients. This relative similarity between the microbial composition of healthy individuals and individuals with UC was also found in a recent pyro-sequencing-based study (Willing et al. 2010). In addition, individuals with ileal CD differed from individuals with CD that predominantly involved the colon. In this context, high prevalence of adherent-invasive *Escherichia coli* was found to be associated with ileal mucosa in CD (Darfeuille-Michaud et al. 2004).

These findings demonstrate the complexity of IBD, and imply that there are distinct bacterial etiologies underlying the pathogenesis of CD and UC.

16.6 The Interactions Between the Gut Microbiota and the Immune System: The Key for Understanding IBD

Recent evidence suggests that the composition of the bacterial communities in the gut may be intimately linked to the proper functioning of the immune system (Round and Mazmanian 2009). Furthermore, it can be expected that disturbances in the microbiota composition, as occur in IBD, result in dysregulation of the adaptive immune cells. In this respect, direct evidence for anti-inflammatory activity by *Faecalibacterium prausnitzii*, which was under-represented in people with a higher incidence of postoperative recurrence of ileal CD, has been demonstrated (Sokol et al. 2008b). In general, low counts of this representative species of Clostridia (belonging to the *Clostridium leptum* group) have been documented both in CD and in UC (Sokol et al. 2009). More recently, it was demonstrated in mice that *Clostridium* species belonging to clusters IV (the phylogenetic cluster that includes *F. prausnitzii*) and XIVa are outstanding inducers of regulatory T cells, important for suppression of inflammation, in the colon (Atarashi et al. 2011). Thus, it is possible that indigenous *Clostridium*-dependent induction of regulatory T cells is required for maintaining immune homeostasis in mice and humans. Future research that will focus on identification of the specific mechanisms or metabolites that mediate this *Clostridium*-host crosstalk, to which butyrate may contribute, would obviously advance our understanding of the interactions between the gut microbiota and the immune system. Furthermore, these studies can potentially aid in the search for novel and better therapeutics for IBD.

Taken together, it seems that dysbiosis of the gut microbiota in genetically susceptible individuals may indeed play a central role in the initiation of IBD. It was shown repeatedly that bacterial profiles associated with gut biopsies from IBD patients are not different between inflamed and non-inflamed mucosa (Bibiloni et al. 2006; Gophna et al. 2006; Vasquez et al. 2007). This suggests that the observed compositional differences in the microbiota of IBD patients are not a result of inflammation. A recent study, further strengthening the dysbiosis hypothesis, describes subclinical dysbiosis in asymptomatic relatives of patients with CD (Joossens et al. 2011). The authors suggest that this dysbiosis might be an intermediate step toward a disease-associated dysbiosis.

To conclude, the gut microbiota clearly plays an important role in the etiology of IBD. Although direct causality has not been proven, there is a large body of evidence supporting the concept of bacterial dysbiosis as a major factor in the initiation of CD. The role of the microbiota in UC, in comparison, remains to be elucidated.

16.7 Unresolved Questions and Future Research

Further research is required in order to better characterize the bacterial dysbiosis, if such exists, in UC patients, as the large body of evidence in IBD in that context was described for CD. More studies, which will investigate the mucosal microbiota in the colon of UC patients, especially those who are newly diagnosed and untreated, should shed some light on the bacterial involvement in this disease.

Additional research should address the effects of the different treatments for IBD patients, especially antibiotics and probiotics, as well as nutritional habits on the composition of the gut microbiota, and should try and correlate these parameters to the disease progression. There are still open questions regarding the temporal stability of the gut microbiota generally in healthy individuals, but more interestingly in IBD patients that alternate between periods of remission and inflammation.

Characterization of the host-microbe interactions, and more specifically the interactions of the altered microbiota and the genetically impaired immune system of the IBD patients, will provide better understanding of the disease pathogenesis, and should potentially uncover more specific targets for therapeutic interventions.

It is important to highlight that each of the IBD, CD, and UC, most probably represent by itself a variety of heterogeneous diseases, as implied by different genetic variants and a variety of clinical features. Future research should identify and define clinically important subsets of patients with distinct etiologies for whom the natural history of disease and response to treatment could then be better predicted.

References

Atarashi K, Tanoue T et al (2011) Induction of colonic regulatory T cells by indigenous Clostridium species. Science 331(6015):337–341

Baumgart M, Dogan B et al (2007) Culture independent analysis of ileal mucosa reveals a selective increase in invasive Escherichia coli of novel phylogeny relative to depletion of Clostridiales in Crohn's disease involving the ileum. ISME J 1(5):403–418

Bibiloni R, Mangold M et al (2006) The bacteriology of biopsies differs between newly diagnosed, untreated, Crohn's disease and ulcerative colitis patients. J Med Microbiol 55(Pt 8):1141–1149

Darfeuille-Michaud A, Boudeau J et al (2004) High prevalence of adherent-invasive Escherichia coli associated with ileal mucosa in Crohn's disease. Gastroenterology 127(2):412–421

Diakos C, Prieschl EE et al (2002) Novel mode of interference with nuclear factor of activated T-cells regulation in T-cells by the bacterial metabolite n-butyrate. J Biol Chem 277(27): 24243–24251

Franchimont D, Vermeire S et al (2004) Deficient host-bacteria interactions in inflammatory bowel disease? The toll-like receptor (TLR)-4 Asp299gly polymorphism is associated with Crohn's disease and ulcerative colitis. Gut 53(7):987–992

Frank DN, St Amand AL et al (2007) Molecular-phylogenetic characterization of microbial community imbalances in human inflammatory bowel diseases. Proc Natl Acad Sci USA 104(34): 13780–13785

Gilbert KM, Weigle WO (1993) Th1 cell anergy and blockade in G1a phase of the cell cycle. J Immunol 151(3):1245–1254

Girardin SE, Boneca IG et al (2003) Nod2 is a general sensor of peptidoglycan through muramyl dipeptide (MDP) detection. J Biol Chem 278(11):8869–8872

Gophna U, Sommerfeld K et al (2006) Differences between tissue-associated intestinal microfloras of patients with Crohn's disease and ulcerative colitis. J Clin Microbiol 44(11):4136–4141

Greenstein RJ (2003) Is Crohn's disease caused by a mycobacterium? Comparisons with leprosy, tuberculosis, and Johne's disease. Lancet Infect Dis 3(8):507–514

Hendrickson BA, Gokhale R et al (2002) Clinical aspects and pathophysiology of inflammatory bowel disease. Clin Microbiol Rev 15(1):79–94

Hviid A, Svanstrom H et al (2011) Antibiotic use and inflammatory bowel diseases in childhood. Gut 60(1):49–54

Inohara N, Ogura Y et al (2003) Host recognition of bacterial muramyl dipeptide mediated through NOD2. Implications for Crohn's disease. J Biol Chem 278(8):5509–5512

Jackson SK, DeLoose A et al (2002) The ability of antigen, but not interleukin-2, to promote n-butyrate-induced T helper 1 cell anergy is associated with increased expression and altered association patterns of cyclin-dependent kinase inhibitors. Immunology 106(4):486–495

Joossens M, Huys G et al (2011) Dysbiosis of the faecal microbiota in patients with Crohn's disease and their unaffected relatives. Gut 60:631–637

Kim SC, Tonkonogy SL et al (2005) Variable phenotypes of enterocolitis in interleukin 10-deficient mice monoassociated with two different commensal bacteria. Gastroenterology 128(4):891–906

Kim SC, Tonkonogy SL et al (2007) Dual-association of gnotobiotic IL-10−/− mice with 2 nonpathogenic commensal bacteria induces aggressive pancolitis. Inflamm Bowel Dis 13(12): 1457–1466

Leenen CH, Dieleman LA (2007) Inulin and oligofructose in chronic inflammatory bowel disease. J Nutr 137(11 Suppl):2572S–2575S

Loftus EV Jr (2004) Clinical epidemiology of inflammatory bowel disease: incidence, prevalence, and environmental influences. Gastroenterology 126(6):1504–1517

Luhrs H, Gerke T et al (2002a) Butyrate inhibits NF-kappaB activation in lamina propria macrophages of patients with ulcerative colitis. Scand J Gastroenterol 37(4):458–466

Luhrs H, Kudlich T et al (2002b) Butyrate-enhanced TNFalpha-induced apoptosis is associated with inhibition of NF-kappaB. Anticancer Res 22(3):1561–1568

Macdonald TT, Monteleone G (2005) Immunity, inflammation, and allergy in the gut. Science 307(5717):1920–1925

Macfarlane S, Furrie E et al (2004) Chemotaxonomic analysis of bacterial populations colonizing the rectal mucosa in patients with ulcerative colitis. Clin Infect Dis 38(12):1690–1699

Manichanh C, Rigottier-Gois L et al (2006) Reduced diversity of faecal microbiota in Crohn's disease revealed by a metagenomic approach. Gut 55(2):205–211

Ott SJ, Musfeldt M et al (2004) Reduction in diversity of the colonic mucosa associated bacterial microflora in patients with active inflammatory bowel disease. Gut 53(5):685–693

Quirke P (2001) Antagonist. Mycobacterium avium subspecies paratuberculosis is a cause of Crohn's disease. Gut 49(6):757–760

Roediger WE (1982) Utilization of nutrients by isolated epithelial cells of the rat colon. Gastroenterology 83(2):424–429
Round JL, Mazmanian SK (2009) The gut microbiota shapes intestinal immune responses during health and disease. Nat Rev Immunol 9(5):313–323
Sartor RB (2001) Intestinal microflora in human and experimental inflammatory bowel disease. Curr Opin Gastroenterol 17:324–330
Sartor RB (2005) Does Mycobacterium avium subspecies paratuberculosis cause Crohn's disease? Gut 54(7):896–898
Segain JP, Raingeard de la Bletiere D et al (2000) Butyrate inhibits inflammatory responses through NFkappaB inhibition: implications for Crohn's disease. Gut 47(3):397–403
Sellon RK, Tonkonogy S et al (1998) Resident enteric bacteria are necessary for development of spontaneous colitis and immune system activation in interleukin-10-deficient mice. Infect Immun 66(11):5224–5231
Sokol H, Seksik P et al (2006) Specificities of the fecal microbiota in inflammatory bowel disease. Inflamm Bowel Dis 12(2):106–111
Sokol H, Lay C et al (2008a) Analysis of bacterial bowel communities of IBD patients: what has it revealed? Inflamm Bowel Dis 14(6):858–867
Sokol H, Pigneur B et al (2008b) Faecalibacterium prausnitzii is an anti-inflammatory commensal bacterium identified by gut microbiota analysis of Crohn disease patients. Proc Natl Acad Sci USA 105(43):16731–16736
Sokol H, Seksik P et al (2009) Low counts of Faecalibacterium prausnitzii in colitis microbiota. Inflamm Bowel Dis 15(8):1183–1189
Tamboli CP, Neut C et al (2004) Dysbiosis in inflammatory bowel disease. Gut 53(1):1–4
Tannock GW (2010) The bowel microbiota and inflammatory bowel diseases. Int J Inflam 2010: 954051
Uhlig HH, Powrie F (2009) Mouse models of intestinal inflammation as tools to understand the pathogenesis of inflammatory bowel disease. Eur J Immunol 39(8):2021–2026
Vasquez N, Mangin I et al (2007) Patchy distribution of mucosal lesions in ileal Crohn's disease is not linked to differences in the dominant mucosa-associated bacteria: a study using fluorescence in situ hybridization and temporal temperature gradient gel electrophoresis. Inflamm Bowel Dis 13(6):684–692
Willing BP, Dicksved J et al (2010) A pyrosequencing study in twins shows that gastrointestinal microbial profiles vary with inflammatory bowel disease phenotypes. Gastroenterology 139(6): 1844–1854, e1841
Xavier RJ, Podolsky DK (2007) Unravelling the pathogenesis of inflammatory bowel disease. Nature 448(7152):427–434

Chapter 17
A Role for Bacteria in the Development of Autoimmunity for Type 1 Diabetes

Adriana Giongo and Eric W. Triplett

Abstract The human gut microbiome consists of hundreds of bacterial species and many billions of cells. Recent work suggests that these bacterial populations play a very important role in sustaining human health. In particular, the composition of these bacterial communities changes significantly in autoimmune subjects compared to healthy controls. Feeding antibiotics or probiotics to diabetes prone mice or rats delays or prevents diabetes. Specific gut bacteria were identified that were either positively or negatively correlated with the development of autoimmunity for type 1 diabetes in a small set of Finnish children. Further analysis of the communities from these children suggests that the gut microbiome of autoimmune children is a less healthy bacterial population compared to what is found in healthy children. Future work is needed that will inform not just the identity of these bacteria but provide insights into their function. In turn, these analyses may suggest how specific bacteria may cause a leaky gut leading to an altered immune responsiveness.

17.1 Human Microbiome

The human body is a superorganism. It forms a complex ecosystem where parasitic, commensal, and symbiotic microorganisms interact and modify their host (Lederberg 2000). Collectively called the human microbiome, these microorganisms co-evolve with the host and are considered indispensable for life under ordinary conditions (Savage 2001; Artis 2008; Ley et al. 2006, 2008). They participate in several vital metabolic processes such as extracting nutrients from the diet, production of vitamins, regulating host fat storage, metabolism of xenobiotics, resistance to

A. Giongo • E.W. Triplett (✉)
Department of Microbiology and Cell Science, Institute of Food and Agricultural Sciences, University of Florida, Gainesville, FL, USA
e-mail: EWT@ufl.edu

E. Rosenberg and U. Gophna (eds.), *Beneficial Microorganisms in Multicellular Life Forms*, 231
DOI 10.1007/978-3-642-21680-0_17, © Springer-Verlag Berlin Heidelberg 2011

tumors and cancer leading neoplasms, intestinal epithelium cell proliferation and differentiation, and in development of a mature immune system (Bäckhed et al. 2004; Turnbaugh et al. 2006; Blaut and Clavel 2007; Turnbaugh et al. 2007).

It is estimated that the human microbiota is comprised ~ tenfold more than of human cells (Kurokawa et al. 2007; Hattori and Taylor 2009) and 100-fold more genes than the human genome (Gill et al. 2006). In terms of genomes, we can see ourselves as a combination of two genomes: the human genome, genetically inherited from parents, and the human microbiome, the genome of our indigenous microbial partners (microflora), acquired from the environment after birth (Turnbaugh et al. 2006; Hooper and Gordon 2001; Zhao 2010). Although microbiota from different micro niches in human body have been characterized (Costello et al. 2009), such as skin surface (Gao et al. 2007; Down et al. 2008; Fierer et al. 2008; Grice et al. 2008; Price et al. 2009), oral, nostril, or oropharynx cavities (Kuboniwa et al. 2000; Aas et al. 2005; Jenkinson and Lamont 2005; Keijser et al. 2008; Lazarevic et al. 2009; Nasidze et al. 2009a, b; Zaura et al. 2009; Lemon et al. 2010), vaginal tract (Zhou et al. 2004; Dominguez–Bello et al. 2010; Ling et al. 2010), by far the most extensively studied habitat for the human microbiome is the gastrointestinal tract (Gill et al. 2006; Andersson et al. 2008; Bjerketorp et al. 2008; Dethlefsen et al. 2008; Duncan et al. 2008; Mitsou et al. 2008; Armougom et al. 2009; Claesson et al. 2009; Matsuda et al. 2009; Roesch et al. 2009b; Booijink et al. 2010; Cox et al. 2010; De Filippo et al. 2010; Durban et al. 2010; Gillevet et al. 2010; Giongo et al. 2010; Walker et al. 2010; Wu et al. 2010a, b). It is the largest microbial community in the human body comprising of 10^{14} bacterial cells (Savage 1977) and has a significant impact on human health (Hooper and Gordon 2001; Round and Mazmanian 2009).

Due to its structure, a mammal's gut surface is continuously exposed to toxins, potential pathogens, and commensal microorganisms (Artis 2008). Nevertheless, it must remain permeable to allow efficient nutrient absorption. In addition, beneficial gut microbiota should be preserved while pathogens are eliminated by the immune system. The intestinal microbiota are crucial to mammalian development and play a role in the maturation of the metabolic capabilities of the gut, cell proliferation, immune tolerance (Hooper and Gordon 2001; Hooper 2004; MacDonald and Gordon 2005; Mazmanian et al. 2005; Blaut and Clavel 2007; Maslowski et al. 2009; Sansonetti and Medzhitov 2009; Chung and Kasper 2010), and also body weight gain in animal models with a genetic predisposition for obesity and type 2 diabetes (Ley et al. 2005; Turnbaugh et al. 2006; Cani et al. 2008a, b; Duncan et al. 2008; Armougom et al. 2009; Wu et al. 2010b).

Several methodologies are used to characterize human microbiota. The analysis of amplified and sequenced 16S rRNA genes has become the most important single approach for the rapid identification and classification of prokaryotes. High throughput methods, such as next generation sequencing and microarray analysis have been the base for several animal and human microbiome studies (Luna et al. 2007; Andersson et al. 2008; Dethlefsen et al. 2008; Dowd et al. 2008; Fierer et al. 2008; Keijser et al. 2008; McKenna et al. 2008; Price et al. 2009; Rajilic-Stojanovic et al. 2009; Roesch et al. 2009; Biagi et al. 2010; Giongo et al. 2010).

17.2 Microbiome in Health and Disease

The gut has the largest surface area in the body and is continually exposed to microbes and antigens. The intestinal epithelium and its commensal microbiome are likely to be central to many autoimmune disorders and metabolic syndromes, such as mucosal inflammation, allergenic and autoimmune diseases including Crohn's disease, celiac disease, ulcerative colitis, rheumatoid arthritis, psoriasis, type 1 diabetes (T1D), and multiple sclerosis (Frank et al. 2007a, b; Dicksved et al. 2008; Wen et al. 2008; Chow and Mazmanian 2009; Willing et al. 2009a, b; Zhang et al. 2010).

The immune system can be modulated by the gut microbiome and these microbiome-mediated changes can influence the balance between health and disease (Singh et al. 2001; Baumgart and Carding 2007; Mazmanian et al. 2008; Hand and Belkaid 2010). An inappropriate immune response to gut bacteria contributes significantly to the pathogenesis of idiopathic inflammatory bowel disease (IBD), including Crohn's disease and ulcerative colitis. This can be induced by transfer of non-regulatory T cells into immune-compromised mice (Singh et al. 2001). An uncontrolled reaction from Th1/Th7 can drive autoimmune diseases such as multiple sclerosis, T1D, rheumatoid arthritis, and psoriasis (Chow and Mazmanian 2009). Immunomodulation by gut microbes is not limited to intestinal tissue. Immunological responses to gut microflora are observed in locations distal to the gut (Borchers et al. 2009). For instance, spleens isolated from pathogen free mice with diverse gut microflora have a larger proportion of CD4+ T cells than those derived under germ free conditions (Mazmanian et al. 2005).Moreover, the effectiveness of oral probiotic treatments for allergies strongly suggests that interactions of the immune system with gut bacteria mediate systemic immunological changes (Ruemmele et al. 2009). Mazmanian and co-workers (2008) suggested that the *Bacteroides fragilis* could prevent an intestinal inflammatory disease in animal models stimulated by *Helicobacter hepaticus*.

Socioeconomic changes affecting hygiene, lifestyle conditions, vaccination, antibiotics, and dietary habits are known to influence alterations in the human microbiota (Bäckhed et al. 2004; Dethlefsen et al. 2008; Martinez et al. 2009; Okada et al. 2010; Zhang et al. 2010). The link between the increasing incidence of allergies and the modern hygienic lifestyle encouraged the development of the "hygiene hypothesis" (Strachan 1989). It suggests that the risk of developing allergic and autoimmune diseases decreases when the subject is exposed to infectious, symbiotic, or parasitic organisms during the crucial stages of immune maturation (Strachan 2000; Cox et al. 2010). For example, sanitary conditions in the animal facilities seem to be correlated with the incidence of T1D in NOD mice (Bach 2002). Besides hygiene, the improvement in dietary habits, such as variations in gluten and fiber content, can be influenced by the microbial community and thus collaborate with health or disease status (De Filippo et al. 2010). Significant differences were found in the gut microbial composition in European children compared with rural African village children. European children possessed more

Firmicutes and less Bacteroidetes when compared with the African children. The genera *Shigella* and *Escherichia*, belonging to the Enterobacteriaceae family, were also significantly more abundant in European children (De Filippo et al. 2010).

17.3 Antibiotic Use and Diabetes in Rodent Models

Antibiotic use prevents the onset of diabetes in non-obese diabetic (NOD) mice (Schwartz et al. 2007) and BioBreeding Diabetes Prone (BB-DP) rats (Brugman et al. 2006). In addition, diabetes develops spontaneously in NOD mice housed in a germ-free environment (Bach 2002; Suzuki et al. 1987; Wicker et al. 1987). The administration of Freund's adjuvant or probiotics, particularly *Lactobacillus* strains, can prevent or delay diabetes in NOD mice or BB-DP rats (Alyanakian et al. 2006; Calcinaro et al. 2005; Matsuzaki et al. 1987; McInerney et al. 1991; Sadelain et al. 1990a, b; Yadav et al. 2007; Valladares et al. 2010). Pathogen-free NOD mice lacking an adaptor protein for multiple toll-like receptors known to bind to bacterial ligands fail to develop diabetes, indicating that the interaction of the intestinal microbiota with the immune system may be involved in the development of diabetes (Wen et al. 2008).

17.4 Is There Evidence for Antibiotic Use Influencing Autoimmunity and Type 1 Diabetes in Humans?

A number of recent papers report the not so unexpected result that antibiotic consumption affects the human microbiome (Dethlefsen et al. 2008; Jernberg et al. 2010; Manichanh et al. 2010). The as yet unresolved question is whether antibiotic use affects the progression of autoimmune diseases.

However, several studies with humans have found no association between antibiotic use and T1D in children. A nationwide cohort study of all Danish children born from 1995 to 2003 was conducted. Among the 606,420 Danish children born during that period, 454 cases of T1D were identified. No association with antibiotic use and diabetes was found (Hviid and Svanstrom 2009). In a mother–child cohort with 437 children born in Finland between 1996 and 2000, no correlation was found between the use of antimicrobials by the mother before or during pregnancy and subsequent risk of becoming diabetic (Kilkkinen et al. 2006).

Nevertheless, this topic is worthy of additional study. Antibiotics may still play a role in the development of autoimmunity for type 1 diabetes in those children at high risk for the disease based on their HLA genotype. In the two studies described above, children throughout the general population were studied, not just high-risk children.

17.5 Culture Independent Analysis of Bacteria Correlated with Diabetes in Rats or in Autoimmunity for Type 1 Diabetes in Humans

As described above, antibiotics and probiotics can delay or prevent diabetes in the two murine models for the disease. These results encouraged the examination of the native microflora in Bio-Breeding diabetes resistant (BB-DR) and diabetes prone (BB-DP rats (Roesch et al. 2009b). Over 3,000 partial 16S rRNA sequences were obtained for ten BB-DP and ten BB-DR rats with stool collected at 70 days after birth. This is about the time when diabetes occurs in the BB-DP rats. Several genera and species of bacteria were discovered to be both negatively and positively correlated with the onset of diabetes in these animals. Among these, *Lactobacillus* and *Bifidobacterium* were in higher abundance in the BB-DR rats and this result was confirmed by quantitative PCR. *Bacteroides* and *Clostridium* were in higher numbers in BB-DP rats. These results are consistent with the notion that the microbiome of diabetes resistant rats contains bacteria that may have probiotic-like properties. Also, the diabetes-prone rats contained gut bacteria often found in unhealthy states such as obesity.

In addition, many groups of bacteria were identified in higher numbers in BB-DR rats that are uncultured close relatives of *Lactobacillus*. In contrast, many uncultured relatives of *Clostridium* were in higher numbers in BB-DP rats (Roesch et al. 2009b).

These results in rats suggested that bacteria related to a healthy state are in BB-DR rats while BB-DP rats contained many more bacteria typically found in unhealthy states. The next step was to determine whether this might be the case in humans. However, given that human samples are not frozen immediately upon collection, it was necessary to show the level of stability of microbial communities in human feces after storage at room temperature for 12–72h (Roesch et al. 2009a). In this work, no significant change in community composition was observed in the first 24h of storage at room temperature. Even after 72h at room temperature, the communities differed by less than 10% compared to those that were immediately frozen. Hence, the fecal samples collected by the major type 1 diabetes experiments in Europe and the United States are suitable for microbiome analysis.

While human samples were being extracted and analyzed, tools were constructed to rapidly analyze hundreds of thousands of 16S rRNA sequences. These tools include PANGEA, a pipeline for the analysis of next generation amplicons (Giongo et al. 2010a) and TaxCollector, a tool to add the full taxonomic information to every sequence analyzed (Giongo et al. 2010c).

Using PANGEA and TaxCollector, an average of 15,000 partial 16S rRNA sequences were analyzed from 24 human fecal samples (Giongo et al. 2010b). The samples were collected at three time points, approximately 4 months, 1 year, and 2 years after birth. This was a matched case-control study where the four pairs of children were matched by age and HLA genotype.

As in the rat study (Roesch et al. 2009b), specific bacteria were identified that were negatively or positively correlated with autoimmunity for type 1 diabetes (Giongo et al. 2010b). However, the bacteria that were negatively correlated with autoimmunity were different from those that were negatively correlated in the rat model. For example, three as yet uncultured Firmicutes were found in much higher numbers in controls than in cases, while the *Lactobacillus* strains simply were in very low abundance in cases and controls. However, similar to the BB-DP and BB-DR rats, several *Bacteroides* and *Clostridium* species, particularly *B. ovatus*, were more abundant in cases than in controls.

However, the most interesting differences between the case and control gut microbial communities lie in the analysis of the samples at the community level (Giongo et al. 2010b). Control communities were more diverse, had more unclassified reads, and were more similar to each other than the case communities and each of these characteristics increased over time. All of these data were interpreted to mean that the control communities were healthier than the case communities. For example, ecologists tell us that a more diverse community is a healthier community. Also, the study of pathogens has been a major focus of microbiology since the first discovery of bacteria in the mid-nineteenth century. Most human bacterial pathogens are well known and characterized. Hence, the control communities with more unclassified, unknown bacteria are more likely to contain a higher abundance of benign bacteria than the case communities. And finally, the higher mean molecular distance between any two case communities compared to the control communities suggests that the case communities are less stable than the control communities. An unstable community may be less able to adapt to changing circumstance than a stable community.

These data are intriguing but larger studies are needed to confirm these results. In addition, these 16S rRNA analyses only tell us which bacteria are present in these samples. These results do not tell us what these bacteria are doing or are capable of doing. For this, future work in metagenomic, transcriptomics, proteomics, and metabolomics will be informative.

17.6 How Can the Bacteria Influence Autoimmunity? The Leaky Gut Hypothesis

A recent review suggests that a concurrence of three events, an altered intestinal microbiota, a "leaky" intestinal mucosal barrier, and an altered intestinal immune responsiveness, all lead to the development of autoimmunity for type 1 diabetes (Vaarala et al. 2008). These same events may also lead to Crohn's disease, celiac disease, and multiple sclerosis. The initial studies demonstrating the altered intestinal microbiota in autoimmune subjects have been described above (Roesch et al. 2009b; Giongo et al. 2010b). The question is: how these changing bacterial populations actually cause a leaky gut. A protein called zonulin appears to be involved in the integrity of tight junctions in BB-DP rats (Watts et al. 2005). When prediabetic NOD mice are infected with the enteric bacterial pathogen, *Citrobacter rodentium* (Lee et al. 2010) a leaky gut is induced. However, the mechanism by which certain bacteria disrupt tight junctions in the intestinal barrier remains unclear (Lee et al. 2010).*Citrobacter rodentium* was not found in the analyses of either rat or human samples (Roesch et al. 2009b; Giongo et al. 2010b).

Nevertheless, we have identified specific bacteria in rats and humans that are correlated with diabetes or autoimmunity, respectively. These bacteria may be responsible for the disruption of the tight junctions. We have also found bacteria that are negatively correlated with disease in both systems. These bacteria may act against those bacteria that cause damage to tight junctions. Further study is needed to test these ideas as well as how gut bacteria regulate immune responsiveness in the host once the tight junctions are breached.

References

Aas JA, Paster BJ, Stokes LN, Olsen I, Dewhirst FE (2005) Defining the normal bacterial flora of the oral cavity. J Clin Microbiol 43:5721–5732

Alyanakian MA, Grela F, Aumeunier A, Chiavaroli C, Gouarin C, Bardel E, Normier G, Chatenoud L, Thieblemont N, Bach JF (2006) Transforming growth factor-B and natural killer T-cells are involved in the protective effect of a bacterial extract on type 1 diabetes. Diabetes 55:179–185

Andersson AF, Lindberg M, Jakobsson H, Backhed F, Nyren P, Engstrand L (2008) Comparative analysis of human gut microbiota by barcoded pyrosequencing. PLoS ONE 3:e2836

Armougom F, Henry M, Vialettes B, Raccah D, Raoult D (2009) Monitoring bacterial community of human gut microbiota reveals an increase in *Lactobacillus* in obese patients and methanogens in anorexic patients. PLoS ONE 4:e7125

Artis D (2008) Epithelial-cell recognition of commensal bacteria and maintenance of immune homeostasis in the gut. Nat Rev Immunol 8:411–420

Bach JF (2002) The effect of infections on susceptibility to autoimmune and allergic diseases. N Engl J Med 347:911–920

Bäckhed F, Ding H, Wang T, Hooper LV, Koh GY, Nagy A, Semenkovich CF, Gordon JI (2004) The gut microbiota as an environmental factor that regulates fat storage. Proc Natl Acad Sci USA 101:15718–15723

Baumgart DC, Carding SR (2007) Inflammatory bowel disease: cause and immunobiology. Lancet 369:1627–1640

Biagi E, Nylund L, Candela M, Ostan R, Bucci L, Pini E, Nikkila J, Monti D, Satokari R, Francheschi C, Brigidi P, De Vos W (2010) Through ageing, and beyond: gut microbiota and inflammatory status in seniors and centenarians. PLoS ONE 5:e10667

Bjerketorp J, Chiang ANT, Hjort K, Rosenquist M, Liu WT, Jansson JK (2008) Rapid lab-on-the-chip profiling of human gut bacteria. J Microbiol Meth 72:82–90

Blaut M, Clavel T (2007) Metabolic diversity of the intestinal microbiota: implications for health and disease. J Nutr 137:751–755

Booijink CCGM, El-Aidy S, Rajilic-Stojanovic M, Heilig HGHJ, Troost FJ, Smidt H, Kleerebezem M, deVos WM, Zoetendal EG (2010) High temporal and inter-individual variation detected in the human ileal microbiota. Environ Microbiol. doi:10.1111/j.1462-2920.2010.02294.x

Borchers AT, Selmi C, Meyers FJ, Keen CL, Gershwin ME (2009) Probiotics and immunity. J Gastroenterol 44:26–46
Brugman S, Klatter FA, Visser JTJ, Windeboer-Weloo ACM, Harmsen HJM, Rozing J, Bos NA (2006) Antibiotic treatment partially protects against type 1 diabetes in the Bio-Breeding diabetes-prone rat. Is the gut flora involved in the development of type 1 diabetes? Diabetologia 49:2105–2108
Calcinaro F, Dionisi S, Marinaro M, Candeloro P, Bonato V, Marzotti S, Corneli RB, Ferretti E, Gulino A, Grasso F, De Simone C, Di Mario U, Falorni A, Boirivant M, Dotta F (2005) Oral probiotic administration induces interleukin-10 production and prevents spontaneous autoimmune diabetes in the non-obese diabetic mouse. Diabetologia 48:1565–1575
Cani PD, Delzenne NM, Amar J, Burcelin R (2008a) Role of gut microflora in the development of obesity and insulin resistance following high-fat diet feeding. Pathol Biol 56:305–309
Cani PD, Bibiloni R, Knauf C, Waget A, Neyrinck AM, Delzenne NM, Burcelin R (2008b) Changes in gut microbiota control metabolic endotoxemia-induced inflammation in high-fat diet-induced obesity and diabetes in mice. Diabetes 57:1470–1481
Chow J, Mazmanian SK (2009) Getting the bugs out of the immune system: do bacterial microbiota "fix" intestinal T cell responses? Cell Host Microbe 5:8–12
Chung H, Kasper DL (2010) Microbiota-stimulated immune mechanisms to maintain gut homeostasis. Curr Opin Immunol 22:455–460
Claesson MJ, O'Sullivan O, Wang Q, Nikkila J, Marchesi JR, Smidt H, de Vos WM, Ross RP, O'Toole PW (2009) Comparative analysis of pyrosequencing and a phylogenetic microarray for exploring microbial community structures in the human distal intestine. PLoS ONE 4:e6669
Costello EK, Lauber CL, Hamady M, Fierer N, Gordon JI, Knight R (2009) Bacteria community variation in human body habitats across space and time. Science 326:1694–1697
Cox MJ, Huang YJ, Fujimura KE, Liu JT, McKean M, Boushey HA, Segal MR, Brodie EL, Cabana MD, Lynch SV (2010) Lactobacillus casei abundance is associated with profound shifts in the infant gut microbiome. PLoS ONE 5:e8745
De Filippo C, Cavalieri D, Di Paola M, Ramazzoti M, Poullet JB, Massard S, Collini S, Pieraccini G, Lionetti P (2010) Impact of diet in shaping gut microbiota revealed by a comparative study in children from Europe and rural Africa. Proc Natl Acad Sci USA 107: 14691–14696
Dethlefsen L, Huse S, Sogin ML, Relman DA (2008) The pervasive effects of an antibiotic on the human gut microbiota, as revealed by deep 16S rRNA sequencing. PLoS Biol 6:2383–2400
Dicksved J, Halfvarson J, Rosenquist M, Jarnerot G, Tysk C, Apajalahti J, Engstrand L, Jansson JK (2008) Molecular analysis of the gut microbiota of identical twins with Crohn's disease. ISME J 2:716–727
Dominguez-Bello MG, Costello EK, Contreras M, Magris M, Hidalgo G, Fierer N, Knight R (2010) Delivery mode shapes the acquisition and structure of the initial microbiota across multiple body habitats in newborn. Proc Natl Acad Sci USA 107:11971–11975
Down SE, Sun Y, Secor PR, Rhoads DD, Wolcott BM, James GA, Wolcott RD (2008) Survey of bacterial diversity in chronic wounds using pyrosequencing, DGGE, and full ribosome shotgun sequencing. BMC Microbiol 8:43
Duncan SH, Lobley GE, Holtrop G, Ince J, Johnstone AM, Louis P, Flint HJ (2008) Human colonic microbiota associated with diet, obesity and weight loss. Int J Obes 32:1720–1724
Durban A, Abellan JJ, Jimenez-Hernandez N, Ponce M, Ponce J, Sala T, D'Auria G, Latorre A, Moya A (2010) Assessing gut microbial diversity from feces and rectal mucosa. Microb Ecol. doi:10.1007/s00248-010-9738-y
Fierer N, Hamady M, Lauber CL, Knight R (2008) The influence of sex, handedness, and washing on the diversity of hand surface bacteria. Proc Natl Acad Sci USA 105:17994–17999
Frank DN, StAmand AL, Feldman RA, Boedeker EC, Harpaz N, Pace NR (2007a) Molecular-phylogenetic characterization of microbial community imbalance in human inflammatory bowel diseases. Proc Natl Acad Sci USA 104:13780–13785

Frank DN, Amand ALS, Feldman RA, Boedeker EC, Harpaz N, Pace NR (2007b) Molecular-phylogenetic characterization of microbial community imbalance in human inflammatory bowel diseases. Proc Natl Acad Sci USA 104:13780–13785

Gao Z, Tseng C, Pei Z, Blaser MJ (2007) Molecular analysis of human forearm superficial skin bacterial biota. Proc Natl Acad Sci USA 104:2927–2932

Gill SR, Pop M, Deboy RT, Eckburg PB, Turnbaugh PJ, Samuel BS, Gordon JI, Relman DA, Fraser-Liggett CM, Nelson KE (2006) Metagenomic analysis of the human distal gut microbiome. Science 312:1355–1359

Gillevet P, Sikaroodi M, Keshavarzian A, Mutlu EA (2010) Quantitative assessment of the human gut microbiome using multitag pyrosequencing. Chem Biodivers 7:1065–1075

Giongo A, Crabb DB, Davis-Richardson AG, Chauliac D, Mobberley JM, Gano KA, Mukherjee N, Roesch LFW, Walts B, Riva A, King G, Casella G, Triplett EW (2010a) PANGEA: pipeline for analysis of next generation amplicons. ISME J 4:852–861

Giongo A, Gano KA, Crabb DB, Mukherjee N, Novelo LL, Casella G, Drew JC, Simell O, Neu J, Wasserfall CH, Schatz D, Atkinson MA, Triplett EW (2010b) Toward defining the autoimmune microbiome for type 1 diabetes. ISME J J5:82–91

Giongo A, Davis-Richardson AG, Crabb DB, Triplett EW (2010c) TaxCollector: tools to modify existing 16S rRNA databases for the rapid classification at six taxonomic levels. Diversity 2:1015–1025. doi:10.3390/d2071015

Grice EA, Kong HH, Renaud G, Young AC, Comparative Sequencing Program NISC, Bouffard GG, Blakesley RW, Wolfsberg TG, Turner ML, Segre JA (2008) A diversity profile of the human skin microbiota. Genome Res 18:1043–1050

Hand T, Belkaid Y (2010) Microbial control of regulatory and effector T cell responses in the gut. Curr Opin Immunol 22:63–72

Hattori M, Taylor TD (2009) The human intestinal microbiome: a new frontier of human biology. DNA Res 16:1–12

Hooper LV (2004) Bacterial contribution to mammalian gut development. Trends Microbiol 12: 129–134

Hooper LV, Gordon JI (2001) Commensal host-bacterial relationship in the gut. Science 292: 1115–1118

Hviid A, Svanstrom H (2009) Antibiotic use and type 1 diabetes in childhood. Am J Epidemiol 169:1079–1084

Jenkinson HF, Lamont RJ (2005) Oral microbial communities in sickness and in health. Trends Microbiol 13:589–595

Jernberg C, Löfmark S, Edlund C, Jansson JK (2010) Long-term impacts of antibiotic exposure on the human intestinal microbiota. Microbiology 156:3216–3223

Keijser BKF, Zaura E, Huse SM, van der Vossen JMBM, Schuren FHJ, Montijn RC, ten Cate JM, Crielaard W (2008) Pyrosequencing analysis of the oral microflora of healthy adults. J Dent Res 87:1016–1020

Kilkkinen A, Virtanen SM, Klaukka T, Kenward MG, Salkinoja-Salonen M, Gissler M, Kaila M, Reunanen A (2006) Use of antimicrobials and risk of type 1 diabetes in a population-based mother–child cohort. Diabetologia 49:66–70

Kuboniwa M, Inaba H, Amano A (2000) Genotyping to distinguish microbial pathogenicity in periodontitis. Periodontol 54:136–159

Kurokawa K, Itoh T, Kuwahara T, Oshima K, Toh H, Toyoda A, Takami H, Morita H, Sharma VK, Srivastava TP, Taylor TD, Noguchi H, Mori H, Ogura Y, Ehrlich DS, Itoh K, Takagi T, Sakaki Y, Hayashi T, Hattori M (2007) Comparative metagenomics revealed commonly enriched gene sets in human gut microbiomes. DNA Res 14:169–181

Lazarevic V, Whiteson K, Huse S, Hernandez D, Farinelli L, Osteras M, Schrenzel J, Francois P (2009) Metagenomic study of the oral microbiota by illumina high-throughput sequencing. J Microbiol Meth 79:266–271

Lederberg J (2000) Infectious history. Science 288:287–293

Lee AS, Gibson DL, Zhang Y, Sham HP, Vallance BA, Dutz JP (2010) Gut barrier disruption by an enteric bacterial pathogen accelerates insulitis in NOD mice. Diabetologia 53:741–748

Lemon KP, Klepac-Ceraj V, Schiffer HK, Brodie EL, Lynch SV, Kolter R (2010) Comparative analyses of the bacterial microbiota of the human nostril and oropharynx. mBio 1:e10–e18

Ley RE, Backhed F, Turnbaugh P, Lozupone CA, Knight RD, Gordon JI (2005) Obesity alters gut microbial ecology. Proc Natl Acad Sci USA 102:11070–11075

Ley RE, Peterson DA, Gordon JI (2006) Ecological and evolutionary forces shaping microbial diversity in the human intestine. Cell 124:837–848

Ley RE, Hamady M, Lozupone C, Turnbaugh PJ, Ramey RR, Bircher JS, Schlegel ML, Tucker TA, Schrenzel MD, Knight R, Gordon JI (2008) Evolution of mammals and their gut microbes. Science 320:1647–1651

Ling XZ, Kong MJ, Liu F, Zhu BH, Chen YX, Wang ZY, Li JL, Nelson KE, Xia XY, Xiang C (2010) Molecular analysis of the diversity of vaginal microbiota associated with bacterial vaginosis. BMC Genomics 11:488

Luna RA, Fasciano LR, Jones SC, Boyanton BL Jr, Ton TT, Versalovic J (2007) DNA pyrosequencing-based bacterial pathogen identification in a pediatric hospital setting. J Clin Microbiol 45:2985–2992

MacDonald TT, Gordon JN (2005) Bacterial regulation of intestinal immune responses. Gastroenterol Clin N Am 34:401–412

Manichanh C, Reeder J, Gibert P, Varela E, Llopis M, Antolin M, Guigo R, Knight R, Guarner F (2010) Reshaping the gut microbiome with bacterial transplantation and antibiotic intake. Genome Res. doi:10.1101/gr.107987.110

Martinez I, Wallace G, Zhang C, Legge R, Benson AK, Carr TP, Moriyama EN, Walter J (2009) Diet-induced metabolic improvements in a hamster model of hypercholesterolemia are strongly linked to alterations of the gut microbiota. Appl Environ Microbiol 75:4175–4184

Maslowski KM, Vieira AT, Ng A, Kranich J, Sierro F, Yu D, Schilter HC, Rolph MS, Mackay F, Artis D, Xavier RJ, Teixeira MM, Mackay CR (2009) Regulation of inflammatory responses by gut microbiota and chemoattractant receptor GPR43. Nature 461:1282–1286

Matsuda K, Tsuji H, Asahara T, Matsumoto K, Takada T, Nomoto K (2009) Establishment of the analytical system for human fecal microbiota based on reverse transcription-quantitative PCR (RT-qPCR) targeting multicopy rRNA molecules. Appl Environ Microbiol 75:1961–1969

Matsuzaki T, Nagata Y, Kado S, Uchida K, Kato I, Hashimoto S, Yokokura T (1987) Prevention of onset in an insulin-dependent diabetes mellitus model, NOD mice, by oral feeding of *Lactobacillus casei*. APMIS 105:643–649

Mazmanian SK, Liu CH, Tzianabos AO, Kasper DL (2005) An immuno modulation molecule of symbiotic bacteria directs maturation of the host immune system. Cell 122:107–118

Mazmanian SK, Round JL, Kasper DL (2008) A microbial symbiosis factor prevents intestinal inflammatory disease. Nature 453:620–625

McInerney MF, Pek SB, Thomas DW (1991) Prevention of insulitis and diabetes onset by treatment with complete Freund's adjuvant in NOD mice. Diabetes 40:715–725

McKenna P, Hoffmann C, Minkah N, Aye PP, Lackner A, Liu Z, Lozupone CA, Hamady M, Knight R, Bushman FD (2008) The macaque gut microbiome in health, lentiviral infection, and chronic enterocolitis. PLoS Pathog 4:e20

Mitri J, Pittas AG (2010) Shining a light: the role of vitamin D in diabetes mellitus. Nat Rev Endocrinol 6:478–480

Mitsou EK, Kirtzalidou E, Oikonomou I, Liosis G, Kyriacou A (2008) Fecal microflora of Greek healthy neonates. Anaerobe 14:94–101

Nasidze I, Li J, Quinque D, Tang K, Stoneking M (2009a) Global diversity in the human salivary microbiome. Genome Res 19:636–643

Nasidze I, Quinque D, Jing L, Mingkun L, Tang K, Stoneking M (2009b) Comparative analysis of human saliva microbiome diversity by barcode pyrosequencing and cloning approaches. Anal Biochem 391:64–68

Okada H, Kuhn C, Feillet H, Bach JF (2010) The 'hygiene hypothesis' for autoimmune and allergic diseases: an update. Clin Exp Immunol 160:1–9

Price LB, Liu CM, Melendez JH, Frankel YM, Elgelthaler D, Aziz M, Bowers J, Rattray R, Ravel J, Kingsley C, Keim PS, Lazarus GS, Zenilman JM (2009) Community analysis of chronic wound bacteria using 16S rRNA gene-based pyrosequencing: impact of diabetes and antibiotics on chronic wound microbiota. PLoS ONE 4:e6462

Rajilic-Stojanovic M, Heilig HGHJ, Molenaar D, Kajander K, Surakka A, Smidt H, de Vos WM (2009) Development and application of the human intestinal tract chip, a phylogenetic microarray: analysis of universally conserved phylotypes in the abundant microbiota of young and elderly adults. Environ Microbiol 11:1736–1751

Roesch LFW, Casella G, Simell O, Krischer J, Wasserfall CH, Schatz D, Atkinson MA, Neu J, Triplett EW (2009a) Influence of sample storage on bacterial community diversity in fecal samples. Open Microbiol J 3:40–46

Roesch RFW, Lorca GL, Casella G, Giongo A, Naranjo A, Pionzo AM, Li N, Mai V, Wasserfall CH, Schatz D, Atkinson MA, Triplett EW (2009b) Culture-independent identification of gut bacteria correlated with the onset of diabetes in a rat model. ISME J 3:536–548

Round JL, Mazmanian SK (2009) The gut microbiota shapes intestinal immune responses during health and disease. Nat Rev 9:313–323

Ruemmele FM, Bier D, Marteau P, Rechkemmer G, Bourdet-Sicard R, Walker WA, Goulet O (2009) Clinical evidence of immunomodulatory effects of probiotic bacteria. J Pediatr Gastroenterol Nutr 48:126–141

Sadelain MW, Qin HY, Lauzon J, Singh B (1990a) Prevention of type 1 diabetes in NOD mice by adjuvant immunotherapy. Diabetes 39:583–589

Sadelain MW, Qin HY, Sumoski W, Parfery N, Singh B, Rabinovitch A (1990b) Prevention of diabetes in the BB rat by early immunotherapy using Freund's adjuvant. J Autoimmun 3: 671–80

Sansonetti PJ, Medzhitov R (2009) Learning tolerance while fighting ignorance. Cell 138:416–420

Savage DC (1977) Microbial ecology of the gastrointestinal tract. Annu Rev Microbiol 31: 107–133

Savage DC (2001) Microbial biota of the human intestine: a tribute to some pioneering scientists. Curr Issues Intest Microbiol 2:1–15

Schwartz RF, Neu J, Schatz D, Atkinson MA, Wasserfall C (2007) Comment on: Brugman S et al. (2006) Antibiotic treatment partially protects against type 1 diabetes in the Bio-Breeding diabetes-prone rat. Is the gut flora involved in the development of type 1 diabetes? Diabetologia 49:2105–2108. Diabetologia 50:220–221

Singh B, Read S, Asseman C, Malmstrom V, Mottet C, Stephens LA, Stepankova R, Tlaskalova H, Powrie F (2001) Control of intestinal inflammation by regulatory T cells. Immunol Rev 182: 190–200

Strachan DP (1989) Hay fever, hygiene, and house-hold size. Br Med J 299:1259–1260

Strachan DP (2000) Family size, infection and atopy: the first decade of the 'hygiene hypothesis'. Thorax 55:S2–10

Suzuki T, Yamada T, Fujimura T, Kawamura E, Shimizu M, Yamashita R (1987) Diabetogenic effects of lymphocyte transfusion on the NOD or NOD nude mouse. In: Rygaard J, Brunner N, Groem N, Spang-Thomsen M (eds) Immune-deficient animals in biomedical research, Copenhagen, 1985. Karger, Basel, pp 112–116

Turnbaugh PJ, Ley RE, Mahowald MA, Magrini V, Mardis ER, Gordon JI (2006) An obesity-associated gut microbiome with increased capacity for energy harvest. Nature 444:1027–1031

Turnbaugh PJ, Ley RE, Hamady M, Fraser-Liggett CM, Knight R, Gordon JI (2007) The human microbiome project. Nature 449:804–810

Vaarala O, Atkinson MA, Neu J (2008) The 'Perfect Storm' for type 1 diabetes. The complex interplay between intestinal microbiota, gut permeability, and mucosal immunity. Diabetes 57:2555–2562

Valladares R, Sankar D, Li N, Williams E, Lai KK, Abdelgeliel AS, Gonzalez CF, Wasserfall CH, Larkin J, Schatz D, Atkinson MA, Triplett EW, Neu J, Lorca GL (2010) *Lactobacillus johnsonii* N6.2 mitigates the development of type 1 diabetes in BB-DP rats. PLoS One 5:e10507

Walker AW, Ince J, Duncan SH, Webster LM, Holtrop G, Ze X, Brown D, Stares MD, Scott P, Bergerat A, Louis P, McIntosh F, Johnstone AM, Lobley GE, Parkhill J, Flint HJ (2010) Dominant and diet-responsive groups of bacteria within the human colonic microbiota. ISME J. doi:10.1038/ISMEJ.2010.118

Watts T, Berti I, Sapone A, Gerarduzzi T, Not T, Zielke R, Fasano A (2005) Role of the intestinal tight junction modulator zonulin in the pathogenesis of type I diabetes in BB diabetic-prone rats. Proc Natl Acad Sci USA 102:2916–2921

Wen L, Ley RE, Volchkov PY, Stranges PB, Avanesyan L, Stonebraker AC, Hu CY, Wong FS, Szot GL, Bluestone JA, Gordon JI, Chervonsky AV (2008) Innate immunity and intestinal microbiota in the development of Type 1 diabetes. Nature 455:1109–1113

Wicker LS, Miller BJ, Coker LZ, McNally SE, Scott S, Mullen Y, Appel MC (1987) Genetic control of diabetes and insulitis in the nonobese diabetic (NOD) mouse. J Exp Med 165: 1639–1654

Willing B, Halfvarson J, Dicksved J, Rosenquist M, Jarnerot G, Engstrand L et al (2009a) Twin studies reveal specific imbalances in the mucosa-associated microbiota of patients with ileal Crohn's disease. Inflamm Bowel Dis 15:653–660

Willing B, Dicksved J, Halfvarson J, Andersson A, Lucio M, Zeng Z et al (2009b) A pyrosequencing study in twins shows that GI microbial profiles vary with inflammatory bowel disease phenotypes. Gastroenterology. doi:10.1053/j.gastro.2010.08.049

Wu GD, Lewis JD, Hoffmann C, Chen YY, Knight R, Bittinger K, Hwang J, Chen J, Berkowsky R, Nessel L, Li H, Bushman FD (2010a) Sampling and pyrosequencing methods for characterizing bacterial communities in the human gut using 16S sequence tags. BMC Microbiol 10:206

Wu X, Ma C, Han L, Nawaz M, Gao F, Zhang X, Yu P, Zhao C, Li L, Zhou A, Wang J, Moore JE, Millar BC, Xu J (2010b) Molecular characterization of the faecal microbiota in patients with type II diabetes. Curr Microbiol 61:69–78

Yadav H, Shalini J, Sinha PR (2007) Antidiabetic effect of probiotic dahi containing *Lactobacillusacidophilus* and *Lactobacillus casei* in high fructose fed rats. Nutrition 23:62–68

Zaura E, Keijser BJF, Huse SM, Crielaard W (2009) Defining the healthy "core microbiome" of oral microbial communities. BMC Microbiol 9:259

Zhang C, Zhang M, Wang S, Han R, Cao Y, Hua W, Mao Y, Zhang X, Pang X, Wei C, Zhao G, Chen Y, Zhao L (2010) Interactions between gut microbiota, host genetics, and diet relevant to development of metabolic syndromes in mice. ISME J 4:232–241

Zhao L (2010) The tale of our other genome. Nature 465:879–880

Zhou X, Bent SJ, Schneider MG, Davis CC, Islam MR, Forney LJ (2004) Characterization of vaginal microbial communities in adult healthy women using cultivation-independent methods. Microbiology 150:2565–2573

Chapter 18
Impact of Intestinal Microbial Communities upon Health

Harry J. Flint, Sylvia H. Duncan, and Petra Louis

Abstract The microbial community of the human large intestine plays an important role in nutrition and health. Molecular analyses have revealed the diversity of gut bacteria; although some dominant species are abundant in most individuals, humans show considerable inter-individual variation in the strain composition of their intestinal microbiota. It is helpful therefore to define broad functional groups ('clubs') that share one or more characteristics of interest. This approach has been applied to the formation by microbial fermentation of short chain fatty acids that are actively absorbed by the host as energy sources. A combination of targeted molecular detection and information on cultured isolates is helping to reveal the impact of diet upon groups responsible for the fermentation of particular dietary carbohydrates and the formation of butyrate, and should provide the basis for theoretical modelling of this complex community.

18.1 Introduction

The importance of the intestinal microbiota to human health is increasingly recognised. The activities of this microbial community have wide-ranging impacts on nutrition, immune function and metabolic health (Vrieze et al. 2010; Flint et al. 2007; Backhed et al. 2005). In the healthy state, the dominant members of the gut community contribute positively to maintaining gut health, supplying energy to the gut mucosa and creating a gut environment that is hostile to pathogens. On the other hand, this community can also be the source of infection, and the gut microbiota plays a role that is poorly understood in the aetiology of gut disorders including colorectal cancer, inflammatory bowel disease and irritable bowel syndrome

H.J. Flint (✉) • S.H. Duncan • P. Louis
Microbial Ecology Group, Rowett Institute of Nutrition and Health, University of Aberdeen, Greenburn Road, Bucksburn, Aberdeen AB21 9SB, UK
e-mail: h.flint@abdn.ac.uk

E. Rosenberg and U. Gophna (eds.), *Beneficial Microorganisms in Multicellular Life Forms*, 243
DOI 10.1007/978-3-642-21680-0_18,

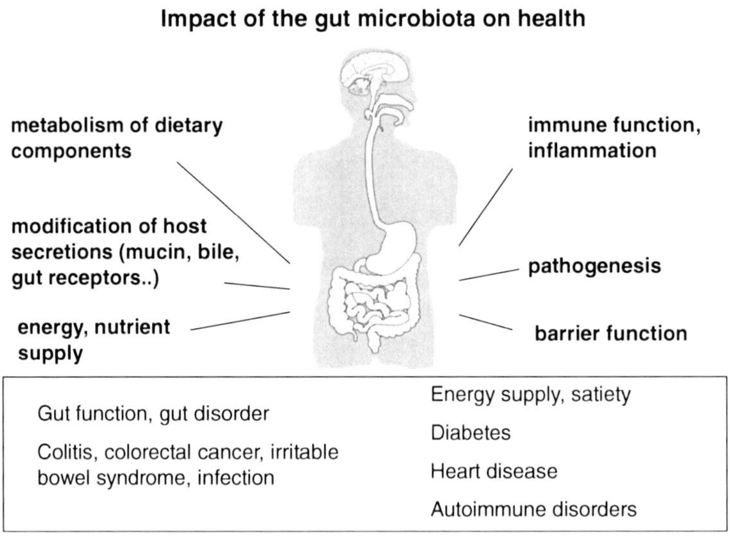

Fig. 18.1 Impacts of the human intestinal microbiota upon health

(Fig. 18.1). The greatest concentrations of bacteria ($>10^{11}$/g contents) are found in the large intestine, where the major energy sources for bacterial growth are derived from material (mainly plant fibre and resistant starches) that is non-digestible in the upper gut. It can be expected that the diet will have a major role in determining the activities and species composition of the gut microbiota, and this is the main topic to be explored here.

18.2 Culturability and Diversity of the Human Intestinal Microbiota

The mammalian intestine offers many incentives to microbial colonisation, including a constant temperature and a regular supply of nutrients. At the same time, the gut is an open system and transit is relatively fast, especially in the small intestine, which means that microorganisms require to maintain a certain minimum growth rate if they are not to be washed out. This applies even to organisms that are attached to the intestinal wall since gut epithelial cells are themselves subject to rapid turnover. We might therefore expect the culturability of gut communities to be relatively high compared with communities from the wider environment. Sequencing of directly amplified 16S rRNA genes indicates that 20–30% of phylotypes from human faecal samples correspond to cultured bacterial isolates (Suau et al. 1999; Hold et al. 2002; Eckburg et al. 2005), a much higher figure than for external environments such as soils. In-depth sequencing reveals many hundreds of phylotypes within each individual: inter-individual variation is

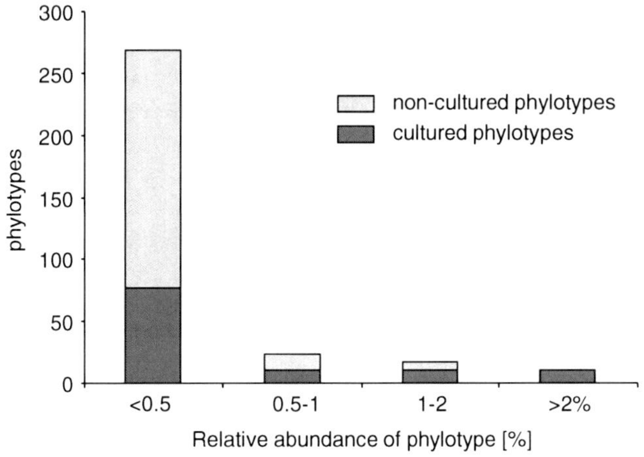

Fig. 18.2 Frequency distribution for 320 16S rRNA phylotypes identified in six obese males (mean BMI 39) (Data from Walker et al. 2011). Phylotypes (defined at a cut-off of 98% sequence identity) corresponding to cultured human gut bacteria are shown in black, while those with no close cultured relative are shown in grey

substantial and there does not appear to be a strict phylogenetic 'core' of human intestinal species, if this is taken to mean phylotypes that will be found in every healthy individual. Nevertheless, recent studies (Walker et al. 2011; Tap et al. 2009) have demonstrated the existence of certain phylotypes that are highly abundant (e.g. accounting for >1% of total bacterial 16S rRNA) and have a particularly high probability of being found among the dominant bacterial colonisers of most individuals. The majority of these common phylotypes correspond to cultured species (Walker et al. 2011) (Fig. 18.2) and show considerable overlap with the dominant species described in earlier cultural studies (Moore and Moore 1995). In general, therefore, it is the rarer phylotypes that are under-represented by culturing, suggesting that these should be regarded as 'not yet cultured' rather than 'unculturable' (Walker et al. 2011). Recent evidence also points to considerable geographic variation in gut microbiota composition that is proposed to be driven by dietary differences (de Filippo et al. 2010). This suggests that the most dominant human gut bacteria may have been described only for better studied (mainly Western) subject groups.

18.3 Responsiveness of Gut Communities to Dietary Change

There is much interest in using dietary additives (prebiotics) to select for specific bacteria within the gut community with the aim of delivering heath benefits. This assumes that community composition can be manipulated by diet, and there is good evidence for enrichment of bifidobacteria in human gut communities through supplementation with fructo-oligosaccharides or inulin (e.g. Bouhnik et al. 2004). Stimulation may not be limited to bifidobacteria, however, and it is also clear that not all individuals respond in the same way (Ramirez-Farias et al. 2009; Flint et al. 2007). Of much wider interest is the degree to which the species composition of the gut microbiota is subject to modification by normal constituents of the diet. A recent study has examined the impact of tightly controlled dietary intakes upon the faecal microbiota of overweight men (Walker et al. 2011). This showed that a diet supplemented with a type III resistant starch caused a major enrichment of certain groups, notably ruminococci related to *R. bromii*, except in those individuals whose gut communities apparently lacked these organisms at the start of the study. These population responses occurred within a few days of the dietary switch and were reversed in a similar time period on the succeeding (low RS) diet. If such responses to specific dietary components are common among the gut microbiota, then we must conclude that the species composition of the gut community is in a constant state of change driven by dietary intake. We might also expect to find major compositional differences in the gut microbiota between population groups depending upon staple food intake, as suggested by a recent study on Italian and African children (de Filippo et al. 2010).

18.4 Breakdown of Recalcitrant Dietary Substrates: Keystone Species

Much of the dietary material that remains undigested in the human small intestine is in an insoluble form, especially fragments of plant material and particles of resistant starch. Work on the rumen fermentation has shown that degradation of such substrates generally relies on a small number of specialist primary degraders (Forsberg et al. 1997). Thus only those rumen bacteria possessing cellulolytic activity are particularly effective in degrading plant cell wall material in pure culture (Morris and van Gylswyk 1980). Specialist cellulolytic bacteria might play a similar role in the human colon, even though degradation of crystalline cellulose is less extensive than in the rumen (Flint et al. 2007; Flint et al. 2008). Associations between specific groups of human colonic bacteria and particular insoluble substrates (wheat bran, particulate resistant starch and porcine mucin) have been demonstrated in a modified fermentor system (Leitch et al. 2007). 16S rRNA analysis has also shown that one group of Firmicutes bacteria, ruminococci belonging to Clostridial cluster IV, is preferentially associated with insoluble fibre in human faecal samples (Walker et al. 2008). Cellulolytic ruminococci have been isolated from the human colon and may play a key role in fibre breakdown (Robert and Bernalier–Donadille 2003). Moreover individuals lacking detectable cluster IV ruminococci within their faecal microbiota were recently shown to have much reduced resistant starch (RS) fermentation (Walker et al. 2011) suggesting that this group may also include 'keystone' species required for the degradation of RS. If such keystone species vary between individuals, this could lead to variations in the efficiency with which energy is recovered from the diet via microbial fermentation.

18.5 Relationship Between Phylogenetic Groups and Metabolic Outputs: Butyrate Production in the Human Colon

The metabolic products of the gut microbial community have a major impact on the host. In particular, short chain fatty acids are the major anions in the colon and play an important role in absorption of salt. They are also utilised as energy sources, contributing perhaps to 10% of daily diet-derived energy in humans (McNeil 1984), and interact with free fatty acid receptors (FFAR) to influence host physiology and gut hormone responses (Sleeth et al. 2010). Each acidic product, however, has quite distinct effects on the host; butyrate, for example, is taken up preferentially by the colonic epithelium as an energy source and contributes to the prevention of colorectal cancer and colitis (Hamer et al. 2008). Understanding how and why dietary intake influences the output of different microbially-produced organic acids is therefore a fundamental aspect of digestive physiology and gut health.

Pathways for production of the major acidic fermentation products including acetate, propionate, lactate, succinate and butyrate are widespread in gut anaerobes, but by no means universal. The butyrate pathway is present in many Firmicutes, and the majority of butyrate producers in the human colon appear to use butyryl CoA: acetate CoA transferase, rather than butyrate kinase, for the final step in butyrate formation (Louis et al. 2004; Louis and Flint 2009). A recent analysis of amplified butyryl CoA:acetate CoA transferase sequences revealed extensive diversity among human colonic butyrate-producing Firmicutes bacteria in ten healthy volunteers (Louis et al. 2010). Ninety percent of these sequences belonged to three phylogenetic groups previously defined by 16S rRNA sequence analysis of cultured butyrate-producers. Two of these groups (one related to *Roseburia* spp + *Eubacterium rectale* and the other to *Anaerostipes* spp. + *Eubacterium hallii*) were detected in all ten individuals. The dominant species within each group were seen to vary markedly between individuals however (Fig. 18.3) making it more fruitful to explore the microbial ecology of SCFA metabolism at the group rather than species or strain level.

Relatives of the *Roseburia*/ *Eubacterium rectale* group (Aminov et al. 2006) can be enumerated using targeted 16S rRNA primers and probes. Members of this group are flagellated and possess a characteristic chromosomal arrangement of butyrate synthetic genes (Louis and Flint 2009). Their representation within the bacterial community detected in faecal samples was shown to respond to controlled variations in dietary carbohydrate intake in overweight human volunteers (Duncan et al. 2007; Walker et al. 2011; Russell et al. 2011). This behaviour may possibly be explained by a dependence upon resistant starch as a growth substrate (Fig. 18.4a). Faecal butyrate concentrations correlated with absolute numbers of *Roseburia* + *E. rectale* in these studies (Duncan et al. 2007) while the % of butyrate among total

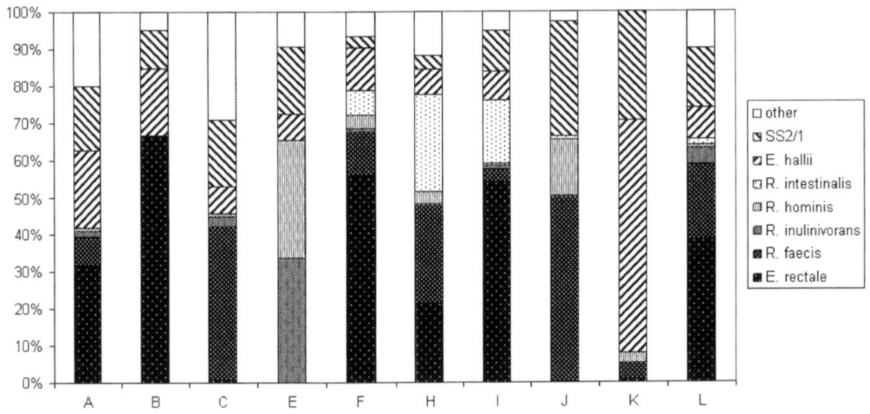

Fig. 18.3 Relative abundance of different species butyrate-producing bacteria in ten healthy human volunteers, estimated from sequence analysis of 1,718 amplified butyryl CoA: acetate CoA transferase sequences (Data from Louis et al. 2010). SS2/1 refers to a cultured, but previously un-named, phylotype (proposed name *Anaerostipes coli*). *F. prausnitzii* was underestimated by this approach due to a cloning bias, and is not shown

SCFA correlated with the % representation of *Roseburia* + *E. rectale* in the microbial community (Fig. 18.4b). This suggests that the *Roseburia* group is a major, if not the major, contributor to butyrate formation from starch. Interestingly a fourfold decrease in butyrate formation by the human colonic microbiota was seen *in vitro* in a continuous flow fermentor system upon shifting from pH 5.5 up to 6.5. This again correlated with a major decrease in the population of *Roseburia* spp. (Walker et al. 2005). This group is apparently stimulated at mildly acidic pH largely because of the inhibition of competing *Bacteroides* spp. (Duncan et al. 2009) suggesting that its favoured niche may be in the proximal colon where rapid fermentation creates mildly acidic conditions. Butyrate production from RS may therefore be amplified by the decrease in pH that accompanies active substrate fermentation in the proximal colon.

In contrast, the group related to *F. prausnitzii* has not been found to respond to dietary starch in human dietary studies, and shows poor growth on starch *in vitro*, although these bacteria may be stimulated by inulin (Ramirez-Farias et al. 2009) and possibly by other non-starch polysaccharides. *F. prausnitzii* is depleted in some forms in inflammatory bowel disease, notably ileal Crohn's disease (Sokol et al. 2008) although these shifts in the disease state have yet to be explained. A special role has been proposed for *F. prausnitzii* in maintaining gut health based on evidence for anti-inflammatory action on host cells, in addition to its ability to produce butyrate (Sokol et al. 2008).

A third abundant group consists of organisms related to *Eubacterium hallii* and *Anaerostipes* spp. that are able to metabolise lactate and acetate to produce butyrate (Duncan et al. 2004; Flint et al. 2007). This creates an alternative route for butyrate formation, via cross-feeding of fermentation products formed by other bacteria, and recent estimates suggest that up to 20% of butyrate may be formed in this way within the mixed community (Belenguer et al. 2011).

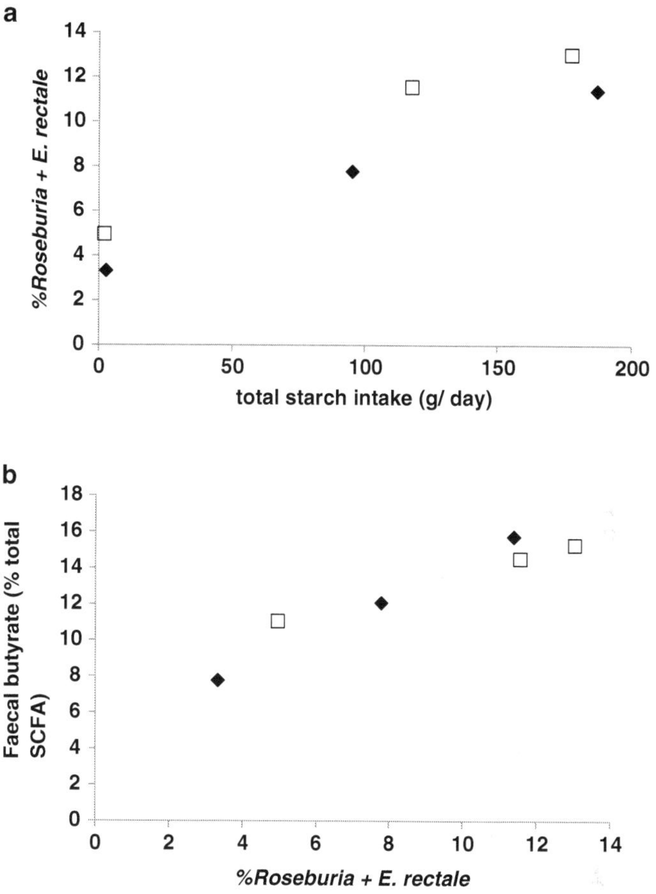

Fig. 18.4 Impact of reduced carbohydrate weight loss diets on faecal *Roseburia* + *E. rectale* populations (estimated by FISH microscopy) and molar % butyrate in obese male volunteers. Data are from two separate studies, each involving periods of weight maintenance, moderate carbohydrate and low carbohydrate intake (as reported in Duncan et al. 2007 (*diamonds*, means of 18 subjects) and Russell et al. 2011 (*squares*, means of 17 subjects)): (**a**) relationship between % *Roseburia* + *E. rectale* among total faecal bacteria and dietary starch intake; (**b**) relationship between % *Roseburia* + *E. rectale* and molar % butyrate among faecal SCFA

18.6 Modelling the Gut Community

The approach described above consists of lumping strain and species variation together by defining functional groups. We can refer to these groups as 'clubs' – in this sense a 'club' is simply 'a group of organisms that share a given functional trait or combination of traits'. Clubs are ideally suited for direct translation into theoretical models. Assuming the availability of some functional information, mainly from cultured strains, their average characteristics in terms of substrate preferences, metabolic outputs and pH tolerance can be built into equations that contribute to a simplified model of the system. This approach has been applied recently with some success using just ten clubs (functional groups) to represent the human intestinal microbiota (Kettle et al. in prep). The concept of functional clubs can be valuable as an aid to understanding and modelling highly complex microbial communities, as it removes much of the noise associated with the overwhelming extent of sequence variation at the level of species, strains and clones. At the other extreme altogether is the detailed analysis of random sequences from whole gut communities (metagenomics).

18.7 Metagenomics

Analysis of the butyryl CoA:acetate CoA transferase gene discussed above provides an example of 'targeted metagenomics', in which a specific, functionally relevant gene is amplified from the mixed community thus providing information on the diversity of a given functional group. Other examples are analysis of the cellulosomal *scaC* gene of ruminococci (Jindou et al. 2008) and the *mrcA* gene of methanogens. A practical advantage of this approach is economy of sequencing and bioinformatic effort. On the other hand, non-targeted metagenomic sequencing has now been applied on a vast scale to the human gut community, taking advantage of developments in high-throughput sequencing (Qin et al. 2010). In the long run, this has the potential to detect functional groups independent of culturing and amplification bias, and to reveal metabolic networks and associations. It is also clear that this effort will benefit from considerably more information on the genomes of cultured representatives from the human intestinal community.

18.8 Unresolved Questions and Future Research

Most of the major questions concerning the impact of the human intestinal microbiota upon health remain to be resolved. Diet clearly has an important influence on the species composition of the gut microbiota, but this appears insufficient to explain the extent of variation that is seen between individuals. For this, a better understanding is needed of the microbial ecology of early colonisation, responses to perturbation due to drugs and disease, and the impact of host physiology and defence systems. Conversely, information is still very limited on the impact of dominant groups of gut bacteria upon the host. Indeed, it is not clear whether the focus should be on trying to identify 'good' and 'bad' groups of bacteria, or rather on more, or less, beneficial states of the microbial community as a whole. Prospects appear good for continuing to link new phylogenetic groups with particular functions (e.g. substrate breakdown, formation of specific metabolites) and for exploiting high-throughput sequencing to identify community profiles that are associated with health or disease. A perennial challenge is to distinguish between bacterial 'dysbiosis' that may cause or aggravate disease, and disturbances in the microbial community that are consequences of the disease state.

Acknowledgment The authors would like to acknowledge support from the Research and Science Division of the Scottish Government.

References

Aminov RI, Walker AW, Duncan SH, Harmsen HJM, Welling GW, Flint HJ (2006) Molecular diversity, cultivation, and improved FISH detection of a dominant group of human gut bacteria related to *Roseburia* spp. or *Eubacterium rectale*. Appl Environ Microbiol 72:6371–6376

Backhed F, Ley RE, Sonnenburg JL, Gordon JI (2005) Host-bacterial mutualism in the human intestine. Science 307:1915–1920

Belenguer A, Holtrop G, Duncan SH, Anderson SE, Calder AG, Flint HJ, Lobley GE. (2011) Rates of production and utilization of lactate by microbial communities from the human colon. FEMS Microbiol Ecol 77:107–119

Bouhnik Y, Raskine L, Simoneau G, Vicaut E, Neut C, Flourie B et al (2004) The capacity of non-digestible carbohydrates to stimulate faecal bifidobacteria in healthy humans: a double blind, randomized, placebo-controlled, parallel-group, dose response relation study. Am J Clin Nutr 80:1658–1664

De Filippo C, Cavalieri D, Di Paolo M, Ramazzotti M, Poullet JB, Massart S, Collini S, Pieraccini G, Lionetti P (2010) Impact of diet in shaping gut microbiota revealed by a comparative study in children from Europe and rural Africa. Proc Natl Acad Sci USA 107:14691–14696

Duncan SH, Louis P, Flint HJ (2004) Lactate-utilizing bacteria, isolated from human feces, that produce butyrate as a major fermentation product. Appl Environ Microbiol 70:5810–5817

Duncan SH, Belenguer A, Holtrop G, Johnstone AM, Flint HJ, Lobley GE (2007) Reduced dietary intake of carbohydrates by obese subjects results in decreased concentrations of butyrate and butyrate-producing bacteria in feces. Appl Environ Microbiol 73:1073–1078

Duncan SH, Louis P, Thomson JM, Flint HJ (2009) The role of pH in determining the species composition of the human colonic microbiota. Environ Microbiol 11:2112–2122

Eckburg PB, Bernstein CN, Purdom E, Dethlefsen L, Sargent M, Gill SR (2005) Diversity of the human intestinal microbial flora. Science 308:1635–1638

Flint HJ, Duncan SH, Scott KP, Louis P (2007) Interactions and competition within the microbial community of the human colon: links between diet and health. Environ Microbiol 9:1101–1111

Flint HJ, Bayer EA, RinconMT LR, White BA (2008) Polysaccharide utilization by gut bacteria: potential for new insights from genomic analysis. Nat Rev Microbiol 6:121–131

Forsberg CW, Cheng K-J, White BA (1997) Polysaccharide degradation in the rumen and large intestine. In: Gastrointestinal microbiology, vol 1. Gastrointestinal ecosystems and fermentations. Chapman & Hall, New York, pp 319–379

Hamer HM, Jonkers D, Venema K, Vanhoutin S, Troost FJ, Brummer RJ (2008) Review article: the role of butyrate on colonic function. Aliment Pharmacol Ther 27:104–119

Hold GL, Pryde SE, Russell VJ, Furrie E, Flint HJ (2002) Assessment of microbial diversity in human colonic samples by 16S rDNA sequence analysis. FEMS Microbiol Ecol 39:33–39

Jindou S, Brulc JM, Levy-Assaraf M, Rincon MT, Flint HJ, Berg ME, Wilson MK, White BA, Bayer EA, Lamed R, Borovok I (2008) Cellulosome gene cluster analysis for gauging the diversity of the ruminal cellulolytic bacterium *Ruminococcus flavefaciens*. FEMS Microbiol Lett 285:188–194

Kettle H, Louis P, Flint HJ, Holtrop G Modelling the emergent dynamics of microbial communities in the human colon (in preparation)

Leitch ECM, Walker AW, Duncan SH, Holtrop G, Flint HJ (2007) Selective colonization of insoluble substrates by human colonic bacteria. Environ Microbiol 72:667–679

Louis P, Flint HJ (2009) Diversity, metabolism and microbial ecology of butyrate-producing bacteria from the human large intestine. FEMS Microbiol Lett 294:1–8

Louis P, Duncan SH, McCrae SI, Millar J, Jackson MS, Flint HJ (2004) Restricted distribution of the butyrate kinase pathway among butyrate-producing bacteria from the human colon. J Bacteriol 186:2099–2106

Louis P, Young P, Holtrop G, Flint HJ (2010) Diversity of human colonic butyrate-producing bacteria revealed by analysis of the butyryl-CoA:acetate CoA transferase gene. Environ Microbiol 12:304–314

McNeil NI (1984) The contribution of the large intestine to energy supplies in man. Am J Clin Nutr 39:338–342

Moore WEC, Moore LH (1995) Intestinal floras of populations that have a high risk of colon cancer. Appl Environ Microbiol 61:3202–3207

Morris EJ, van Gylswyk NO (1980) Comparison of the action of rumen bacteria on cell walls from *Eragrostis tef*. J Agric Sci Camb 95:313–323

Qin JJ, Li RQ, Raes J et al (2010) A human gut microbial gene catalogue established by metagenome sequencing. Nature 464:59–70

Ramirez-Farias C, Slezak K, Fuller Z, Duncan A, Holtrop G, Louis P (2009) Effect of inulin on the human gut microbiota: stimulation of *Bifidobacterium adolescentis* and *Faecalibacterium prausnitzii*. Br J Nutr 101:541–550

Robert C, Bernalier-Donadille A (2003) The cellulolytic microflora of the human colon: evidence of microcrystalline cellulose-degrading bacteria in methane-excreting subjects. FEMS Microbiol Ecol 46:81–89

Russell WR, Gratz S, Duncan SH, Holtrop G, Ince J, Scobbie L, Duncan G, Johntsone AM, Lobley GE, Wallace RJ, Duthie GG, Flint HJ (2011) High protein, reduced carbohydrate diets promote metabolite profiles likely to be detrimental to colonic health. Am J Clin Nutr 93:1062–1072

Sleeth ML et al (2010) Free fatty acid receptor 2 and nutrient sensing: a proposed role for fibre, fermentable carbohydrates and short chain fatty acids in appetite regulation. Nutr Res Rev 23: 135–145

Sokol H, Pigneur B, Watterlot L, Lakhdari O, Bermúdez-Humarán LG, Gratadoux J et al (2008) *Faecalibacterium prausnitzii* is an anti-inflammatory commensal bacterium identified by gut microbiota analysis of Crohn's disease patients. Proc Natl Acad Sci USA 105:16731–16736

Suau A, Bonnet R, Sutren M, Godon JJ, Gibson GR, Collins MD et al (1999) Direct analysis of genes encoding 16S rRNA from complex communities reveals many novel molecular species within the human gut. Appl Environ Microbiol 24:4799–4807

Tap J, Mondot S, Levenez F, Pelletier E, Caron C, Furet JP et al (2009) Towards the human intestinal microbiota phylogenetic core. Environ Microbiol 11:2574–2584

Vrieze A et al (2010) The environment within: how gut microbiota may influence metabolism and body composition. Diabetologia 53:606–613

Walker AW, Duncan SH, Leitch ECM, Child MW, Flint HJ (2005) pH and peptide supply can radically alter bacterial populations and short-chain fatty acid ratios within microbial communities from the human colon. Appl Environ Microbiol 71:3692–3700

Walker AW, Duncan SH, Harmsen HJM, Holtrop G, Welling GW, Flint HJ (2008) The species composition of the human intestinal microbiota differs between particle-associated and liquid phase communities. Environ Microbiol 10:3275–3283

Walker AW, Ince J, Duncan SH, Webster LM, Holtrop G, Ze X, Brown D, Stares MD, Scott P, Bergerat A, Louis P, McIntosh F, Johnstone AM, Lobley GE, Parkhill J, Flint HJ (2011) Dominant and diet-responsive groups of bacteria within the human colonic microbiota. ISME J 5:220–230

Chapter 19
Commensalism Versus Virulence

Dvora Biran, Anat Parket, and Eliora Z. Ron

Abstract Recent genomic research tools enable a detailed genetic and physiological analysis of virulent bacteria and provide a deeper understanding of the genes which are involved in pathogenic processes. The results obtained from such experiments indicate that the distinction between commensals and virulent bacteria is complicated and cannot be performed on the basis of genomic data. Using *Escherichia coli* as a model system, we show that commensalism and virulence cannot be predicted on the basis of genomic data alone. Rather, virulence is determined by the location of the bacteria in the host, by the pattern of their gene expression and by the genetic make-up of the host with whom they interact.

19.1 Introduction

Escherichia coli is a major inhabitant of the intestinal tract of animals. The human intestine is colonized with *E. coli* in the first 40 h after birth. *E. coli* strains are defined by serology of O antigens (of the LPS = lipopolysaccharide fraction of the outer membrane), K (capsule) antigens, and H (flagellar) antigens. There are more than 60,000 known strains.

For a long time, these bacteria were considered commensals, but it is now clear that many *E. coli* strains are virulent. Pathogenic *E. coli* strains can be divided into two groups, in terms of the disease they cause and the virulence factors they express:

- Intestinal *E. coli* strains – they cause diarrhea with varying degrees of severity (Campos et al. 2004; Eckmann 2002; Frankel and Phillips 2008; Garmendia et al. 2005; Mellies et al. 2007; Ochoa et al. 2008). They are a major factor of

D. Biran • A. Parket • E.Z. Ron (✉)
Tel Aviv University, Tel Aviv, Israel
e-mail: eliora@post.tau.ac.il

E. Rosenberg and U. Gophna (eds.), *Beneficial Microorganisms in Multicellular Life Forms*, 253
DOI 10.1007/978-3-642-21680-0_19,

infant mortality in developing countries. Adults usually encounter this disease as "traveler's diarrhea" which is acute but short. Exceptions are strains such as these of serotype O157 which cause infection that cause mortality even in adults, usually due to kidney failure (Nauschuetz 1998; Peacock et al. 2001).

- Extraintestinal pathogenic *E. coli* strains (ExPEC) – these cause a variety of infections, all of which are extraintestinal (Johnson and Russo 2002a, b; Orskov and Orskov 1985; Smith et al. 2007). These include:
  - Urinary tract infections (UTI) (pyelonephritis, kidney failure, productivity loss). UTIs are responsible for > seven million patient visits and one million hospital admissions (due to complications) per year in the United States only. 80–90% of the cases are caused by *E. coli*.
  - Neonatal meningitis: bacterial meningitis. 0.25 per 1,000 live births in industrialized countries (2.66 per 1,000 in developing countries) ~30% caused by *E. coli*, ~10% mortality.
  - Intra-abdominal infections, respiratory tract infections, wound and surgical infections.
  - Septicemia. Colisepticemia is the major cause of mortality from community and hospital-acquired infections (more than 80%) and the main cause of mortality in immuno-suppressed patients (HIV, chemotherapy, and old age). Colisepticemia is an emerging disease –83% increase 1980–1992, over 40% of the bacteremia cases in community acquired infections.

ExPEC strains were not considered an important health hazard but this perception changed in the last century. The discovery of antibiotics almost eliminated pathogens such as Yersinia, Brucella, and Vibrio cholera that are usually sensitive to antibiotics. These were replaced by the less virulent but relatively antibiotic resistant bacteria such as *E. coli*, *Salmonella*, and *Klebsiella*.

19.2 The Rise of ExPEC

E. coli is very abundant and so are its virulent strains. These bacteria acquire resistance to antibiotics very rapidly, especially because they carry plasmids with many genes coding for antibiotics resistance. These plasmids are often conjugative and can be transferred from one bacterium to another. Thus, every environment in which antibiotics are used is a potential incubation place for virulent, antibiotic resistant ExPEC strains. These days more than 70% of the bacteria that cause hospital-acquired infections are resistant to at least one of the antibiotics used to treat them. People infected with antibiotic-resistant organisms have longer hospital stays and require treatment with second or third choice medicines that are less effective, more toxic, and more expensive (Amabile-Cuevas 2011; Bader et al. 2011; Oteo et al. 2011; Paterson 2002; Petrosillo et al. 2011; Pitout and Laupland 2008; Vicente et al. 2006).

In the last 60 years, the use of antibiotics became widespread and is correlated with an increase in life expectancy. Therefore, there is an increase in the percentage of old people in the population and they spend much of their time in institutions and in hospitals, where use of antibiotics is common.

This fact, together with the weakening of the immune system results in susceptibility to infections. This infection is usually caused by bacteria which are antibiotic resistant. The situation is grave and today hospital-acquired infections are a leading cause of illness and death in industrialized countries (Apisarnthanarak et al. 2008; Cheong et al. 2007; George and Manges ; Jackson et al. 2005; Johnson et al. 2006a, b; Laupland et al. 2007, 2008; Marschall et al. 2008; Mylotte 2005; Wyllie et al. 2007).

ExPEC strains are a major cause of community and hospital acquired infections. ExPEC strains include a very large number of serotypes which do not cross react. Therefore, it is not possible to develop a "simple" vaccine (inactivated bacteria etc.), and because ExPEC strains are highly resistant to antibiotics, they are difficult to treat. Hence, it is essential to understand what the genetic determinants are which enable ExPEC strains to cause infection and the properties that distinguish them from commensals.

19.3 Virulence Factors of ExPEC

Genomic data is now available for a large number of ExPEC strains, making it possible to compare genomes and identify genes which are present in virulent strains and absent from avirulent strains. Of high interest are bacteria that cause septicemia and can spread in the blood system and infect internal organs (Angus and Wax 2001; Apisarnthanarak et al. 2008; Kim et al. 2002; Laupland et al. 2008; Mylotte et al. 2002; Mylotte 2005; Paterson 2002; Vazquez et al. 1992).

The genomic studies point out two important facts:

1. Septicemic strains have to contain at least one gene for the following functions:

    a. Ability to adhere to the host tissues
    b. Ability to bind iron with high affinity – essential as there is no free iron in the blood and tissues
    c. Ability to survive in serum

    It should be noted that many of these virulent genes are usually located on plasmids. Important and conserved is the ColV plasmid, which carries many of the genes that contribute to virulence (Binns et al. 1979; Fernandez–Beros et al. 1990; Gophna et al. 2003; Johnson et al. 2006a, b, 2008; Nolan et al. 2003; Perez–Casal and Crosa 1984; Smith and Huggins 1976; Tivendale et al. 2004; Valvano and Crosa 1984; Warner et al. 1981; Waters and Crosa 1991; Zgur–Bertok et al. 1990).

2. There is usually more than one genetic system coding for each of these functions. For example – ExPEC strains code for several iron-binding systems:

the Aerobactin iron uptake system, the *sit* system, the salmochelin system, and the yersiniabactin of the Shigella high pathogenicity island. These iron binding systems are additional to the enterobactin iron binding system which is also present in avirulent strains such as *E. coli* K-12 (Babai et al. 1997; Blum-Oehler et al. 2000; Carniel 2001; Hacker et al. 2003; Oelschlaeger et al. 2002a).

3. There is a high degree of variability in virulence factors of ExPEC strains. It appears that there is a pool of virulence-related genes which are present in various combinations in the different strains (Brzuszkiewicz et al. 2009; Mokady et al. 2005; Ron 2006). The genes involved in virulence are often located on PAIs = pathogenicity islands and there are extensive evidence for the occurrence of lateral gene transfer of these genes and these PAIs (Bach et al. 2000; Bingen-Bidois et al. 2002; Blum-Oehler et al. 2000; Blum et al. 1994; Brzuszkiewicz et al. 2006, 2009; Buchrieser et al. 1999; Carniel et al. 1996; Dezfulian et al. 2004; Dobrindt et al. 1998, 2000, 2002, 2003; Dozois and Curtiss 1999; Finlay and Falkow 1997; Hacker et al. 1992, 1997, 2003; Hacker and Kaper 1999, 2000; Houdouin et al. 2002; Johnson et al. 2001; Morschhauser et al. 1994; Oelschlaeger et al. 2002b; Rakin et al. 1999; Redford and Welch 2002; Ritter et al. 1995, 1997; Schneider et al. 2004; Schubert et al. 1999).

19.4 Virulent Versus Commensal

Genomic analyses of several avirulent *E. coli* strains, such as Nissle and K-12, indicate that several virulence factors are present in the nonpathogenic strains. One good example is the genetic system coding for curli fibers.

Curli are thin coiled fibers have a high affinity binding for several host proteins-Plasminogen and plasminogen activator, MHC Class I molecules, Laminin, and Fibronectin (Olsen et al. 1989, 1998). Isolates from human sepsis constitutively express high levels of curli and O157:H7 isolates with increased curli expression are more invasive to cells and more virulent in mice (Uhlich et al. 2002). It was also shown that curli mutants of avian septicemic *E. coli* serotype O78 are attenuated *in vivo* (Hammar et al. 1995; La Ragione et al. 1999, 2000; Uhlich et al. 2002). Curli fibers were also shown to contribute to internalization of septicemic ExPEC strains by epithelial cells (Gophna et al. 2001). All these facts indicate that curli fibers are associated with virulence. However, the gene cluster coding for curli is present in nonpathogenic strains, as well as in the pathogens.

This apparent paradox could be explained by analyses of the expression of the operon coding for curli. Thus, in the avirulent K-12 strains, curli fibers are expressed only at low temperatures and low ionic strength, while in a virulent O78 strain, curli fibers are expressed at a higher level and in host conditions – high temperature and high osmolarity (Gophna et al. 2001, 2002).

These data suggest that the difference between virulent and commensal strains can be due to different expression patterns of virulence-associated genes, rather than their presence or absence.

19.5 Functional Genomics of Virulent and Commensal Strains

Functional genomic studies provide further support to the idea that the difference between virulence and commensalism is due to difference in genes expression and not merely to the presence or absence of virulence associated genes.

One such study analyzed the genetics of virulence-associated genes in *E. coli* strain Nissle 1917. This strain is a probiotic strain used to reduce intestinal *E. coli* infections (Altenhoefer et al. 2004). As the Nissle strain is closely related to CF073 – a strain causing urinary tract infections – its genetic composition could be compared to the virulent strain. This comparison indicates that Nissle contains a large number of genes associated with virulence. However, many of these genes contain mutations, deletions, and stop codons. Moreover, the virulence-associated genes are often not expressed under host conditions (Homburg et al. 2007).

In a recent paper, the role of the intestinal microbiota and the adaptive immune system in preventing translocation of probiotics (e.g., *Escherichia coli* Nissle) was examined (Gronbach et al. 2011). The data indicated that *E. coli* strain Nissle 1917 could cause mortality, and that the rate was significantly higher in the germ-free raised mice. The results suggested that if both the microbiota and adaptive immunity are defective, translocation across the intestinal epithelium and dissemination of the probiotic *E. coli* strain Nissle 1917 may occur and have potentially severe adverse effects. Further experiments are required to identify the molecular factors that promote probiotic functions, fitness, and facultative pathogenicity.

19.6 Virulence or Commensalism Also Depends on Location of the Bacteria

E. coli strains can colonize the intestinal epithelial as well as surfaces of the bladder and upper respiratory tract. As already stated, the virulent strains can cause intestinal infections or extra-intestinal infections. To cause intestinal infections, bacteria have to adhere to the intestinal epithelium and produce enterotoxins. If such a strain enters the blood stream, it does not survive because it does not have the serum survival factors and therefore does not cause disease. On the other hand, ExPEC strains survive in serum and cause septicemia, but if they enter the intestine, they do not produce toxins and are avirulent.

This fact is important for the understanding of the epidemiology of ExPEC strains. Because they do not cause disease when located in the intestine, they can be carried by healthy hosts and their presence is not detected. Therefore, they are very abundant in the environment and are harmless. However, once they get into a bladder they will cause UTI, and if they get into the blood stream they will survive in serum and cause septicemia and bacteremia.

19.7 Evolution of Bacteria in the Host

In some patients, *E. coli* strains establish significant bacteriuria without causing symptoms of urinary tract infection = Asymptomatic Bacteriuria (ABU) (Emans et al. 1979; Tencer 1988; Wullt et al. 2003). ABU isolates from different hosts are genotypically and phenotypically heterogeneous (Roos et al. 2006b; Vranes et al. 2003) . Some of these isolates are related to UPEC strains by overall genotype. Although these isolates do not cause a disease in their hosts, all of them carry many typical UPEC virulence genes. However, these genes are frequently not expressed due to genome reduction by deletions or accumulated point mutations (Hancock and Klemm 2007; Mabbett et al. 2009; Roos et al. 2006a; Zdziarski et al. 2008).

Recently, ABU strains were used for a novel experimental therapy against UTI (Sunden et al. 2006). Recurrent UTI is common in girls and young women, who have to be treated with antibiotics for very long time periods. The disadvantages of prolonged antibiotic treatments are obvious, especially as they lead to selection of antibiotics resistant bacteria. An alternative approach was to use therapeutic innoculation with ABU strains that were supposed to act as probiotics and prevent infection with virulent UPEC strains.

In the experiment, a single ABU strain – *E. coli* 83972 – was deliberately colonized in several patients as therapeutic bladder colonization (Zdziarski et al. 2010). This strain was genome sequenced before inoculation and samples were taken from the patients at time intervals after the inoculation. The bacteria evolved rapidly in the bladder, but there were significant differences between the patients. Overall 34 mutations were identified, which affect metabolic and virulence -related genes. Transcriptome and proteome analyses proved that these genome changes altered bacterial gene expression. There was a unique adaptation pattern in each patient, which was repeated after consecutive inoculations, indicating that adaptive bacterial evolution is driven by individual host environments. Altogether, there was an ongoing loss of gene function, which supports the hypothesis that growth following bladder colonization favors evolution towards commensalism rather than virulence.

19.8 Unresolved Questions and Future Research

What makes a pathogen and what are the differences between pathogens and commensals? This complex question is far from being resolved. When the genes involved in pathogenesis are identified, it is clear that many of them are present in commensals. Thus, the difference is not only in the composition of the genome. Clearly, a major difference between pathogens and commensals is in the expression of genes, mainly genes which are important for virulence. Therefore, it appears that in ExPEC strains, commensalism involves genetic changes in virulence genes, as well as in the regulatory elements that control their expression.

Following are three of the many unresolved questions:

1. What are the forces that drive evolution from virulence to commensalism?

Several experimental approaches indicate that there is an evolutionary process which drives for the loss of virulence. What are the forces that drive these processes? Are there conditions in which the opposite trend exists – i.e. a drive towards virulence?

2. What are the host factors that determine the susceptibility to the bacteria?

Infections caused by ExPEC strains can be maintained and controlled by many hosts, but some hosts are especially sensitive. One clear case is this of UTI, where most girls/women are not infected but others have recurrent infections. The problem is more complicated for severe diseases, such as septicemia.

3. What are the host factors that affect the evolution of the bacteria?

It was shown that the host drives the evolution of colonizing bacteria – ABU. Which host factors are important in driving the evolution, and which host factors are the ones that vary from one host to another?

As we obtain more data about functional genomics of ExPEC strains, in vitro and in vivo, we should be able to understand the pathogenesis of these bacteria and their interactions with the hosts. In order to understand the role of the host in pathogenesis and evolution of the bacteria, it is essential to learn more about the molecular response of the host to the pathogen and the host-specific systems that function in the recognition of and response to the infecting ExPEC.

Understanding the basis of virulence versus commensalism will also help develop advance methods for prevention and treatment for ExPEC, which are now a significant medical threat due to their widespread antibiotics resistance.

References

Altenhoefer A, Oswald S, Sonnenborn U, Enders C, Schulze J, Hacker J, Oelschlaeger TA (2004) The probiotic *Escherichia coli* strain Nissle 1917 interferes with invasion of human intestinal epithelial cells by different enteroinvasive bacterial pathogens. FEMS Immunol Med Microbiol 40:223–229

Amabile-Cuevas C (2011) Antibiotic resistance in Mexico: a brief overview of the current status and its cause. J Infect Dev Ctries 4:126–131

Angus DC, Wax RS (2001) Epidemiology of sepsis: an update. Crit Care Med 29:S109–S116

Apisarnthanarak A, Kiratisin P, Mundy LM (2008) Predictors of mortality from community-onset bloodstream infections due to extended-spectrum beta-lactamase-producing *Escherichia coli* and *Klebsiella pneumoniae*. Infect Control Hosp Epidemiol 29:671–674

Babai R, Blum-Oehler G, Stern BE, Hacker J, Ron EZ (1997) Virulence patterns from septicemic *Escherichia coli* O78 strains. FEMS Microbiol Lett 149:99–105

Bach S, de Almeida A, Carniel E (2000) The Yersinia high-pathogenicity island is present in different members of the family Enterobacteriaceae. FEMS Microbiol Lett 183:289–294

Bader MS, Hawboldt J, Brooks A (2011) Management of complicated urinary tract infections in the era of antimicrobial resistance. Postgrad Med 122:7–15

Bingen-Bidois M, Clermont O, Bonacorsi S, Terki M, Brahimi N, Loukil C, Barraud D, Bingen E (2002) Phylogenetic analysis and prevalence of urosepsis strains of *Escherichia coli* bearing pathogenicity island-like domains. Infect Immun 70:3216–3226

Binns MM, Davies DL, Hardy KG (1979) Cloned fragments of the plasmid ColV, I-K94 specifying virulence and serum resistance. Nature 279:778–781

Blum G, Ott M, Lischewski A, Ritter A, Imrich H, Tschape H, Hacker J (1994) Excision of large DNA regions termed pathogenicity islands from tRNA-specific loci in the chromosome of an *Escherichia coli* wild-type pathogen. Infect Immun 62:606–614

Blum-Oehler G, Dobrindt U, Janke B, Nagy G, Piechaczek K, Hacker J (2000) Pathogenicity islands of uropathogenic *E. coli* and evolution of virulence. Adv Exp Med Biol 485:25–32

Brzuszkiewicz E, Bruggemann H, Liesegang H, Emmerth M, Olschlager T, Nagy G, Albermann K, Wagner C, Buchrieser C, Emody L, Gottschalk G, Hacker J, Dobrindt U (2006) How to become a uropathogen: comparative genomic analysis of extraintestinal pathogenic *Escherichia coli* strains. Proc Natl Acad Sci USA 103:12879–12884

Brzuszkiewicz E, Gottschalk G, Ron E, Hacker J, Dobrindt U (2009) Adaptation of pathogenic *E. coli* to various niches: genome flexibility is the key. Genome Dyn 6:110–125

Buchrieser C, Rusniok C, Frangeul L, Couve E, Billault A, Kunst F, Carniel E, Glaser P (1999) The 102-kilobase *pgm* locus of *Yersinia pestis*: sequence analysis and comparison of selected regions among different *Yersinia pestis* and *Yersinia pseudotuberculosis* strains. Infect Immun 67:4851–4861

Campos LC, Franzolin MR, Trabulsi LR (2004) Diarrheagenic *Escherichia coli* categories among the traditional enteropathogenic *E. coli* O serogroups – a review. Mem Inst Oswaldo Cruz 99: 545–552

Carniel E (2001) The Yersinia high-pathogenicity island: an iron-uptake island. Microbes Infect 3: 561–569

Carniel E, Guilvout I, Prentice M (1996) Characterization of a large chromosomal "high-pathogenicity island" in biotype 1B *Yersinia enterocolitica*. J Bacteriol 178:6743–6751

Cheong HS (2007) Clinical significance of healthcare-associated infections in community-onset *Escherichia coli* bacteraemia. J Antimicrob Chemother 60:1355–1360

Dezfulian H, Tremblay D, Harel J (2004) Molecular characterization of extraintestinal pathogenic *Escherichia coli* (ExPEC) pathogenicity islands in F165-positive *E. coli* strain from a diseased animal. FEMS Microbiol Lett 238:321–332

Dobrindt U, Cohen PS, Utley M, Muhldorfer I, Hacker J (1998) The *leuX*-encoded tRNA5(Leu) but not the pathogenicity islands I and II influence the survival of the uropathogenic *Escherichia coli* strain 536 in CD-1 mouse bladder mucus in the stationary phase. FEMS Microbiol Lett 162:135–141

Dobrindt U, Janke B, Piechaczek K, Nagy G, Ziebuhr W, Fischer G, Schierhorn A, Hecker M, Blum-Oehler G, Hacker J (2000) Toxin genes on pathogenicity islands: impact for microbial evolution. Int J Med Microbiol 290:307–311

Dobrindt U, Blum-Oehler G, Nagy G, Schneider G, Johann A, Gottschalk G, Hacker J (2002) Genetic structure and distribution of four pathogenicity islands (PAI I(536) to PAI IV(536)) of uropathogenic *Escherichia coli* strain 536. Infect Immun 70:6365–6372

Dobrindt U, Agerer F, Michaelis K, Janka A, Buchrieser C, Samuelson M, Svanborg C, Gottschalk G, Karch H, Hacker J (2003) Analysis of genome plasticity in pathogenic and commensal *Escherichia coli* isolates by use of DNA arrays. J Bacteriol 185:1831–1840

Dozois CM, Curtiss R 3rd (1999) Pathogenic diversity of *Escherichia coli* and the emergence of 'exotic' islands in the gene stream. Vet Res 30:157–179

Eckmann L (2002) Small bowel infections. Curr Opin Gastroenterol 18:197–202

Emans SJ, Grace E, Masland RP Jr (1979) Asymptomatic bacteriuria in adolescent girls: I. Epidemiology. Pediatrics 64:433–437

Fernandez-Beros ME, Kissel V, Lior H, Cabello FC (1990) Virulence-related genes in ColV plasmids of *Escherichia coli* isolated from human blood and intestines. J Clin Microbiol 28: 742–746

Finlay BB, Falkow S (1997) Common themes in microbial pathogenicity revisited. Microbiol Mol Biol Rev 61:136–169

Frankel G, Phillips AD (2008) Attaching effacing *Escherichia coli* and paradigms of Tir-triggered actin polymerization: getting off the pedestal. Cell Microbiol 10:549–556

Garmendia J, Frankel G, Crepin VF (2005) Enteropathogenic and enterohemorrhagic *Escherichia coli* infections: translocation, translocation, translocation. Infect Immun 73:2573–2585

Gophna U, Barlev M, Seijffers R, Oelschlager TA, Hacker J, Ron EZ (2001) Curli fibers mediate internalization of *Escherichia coli* by eukaryotic cells. Infect Immun 69:2659–2665

Gophna U, Oelschlaeger TA, Hacker J, Ron EZ (2002) Role of fibronectin in curli-mediated internalization. FEMS Microbiol Lett 212:55–58

Gophna U, Parket A, Hacker J, Ron EZ (2003) A novel ColV plasmid encoding type IV pili. Microbiology 149:177–184

Gronbach K, Eberle U, Muller M, Olschlager TA, Dobrindt U, Leithauser F, Niess JH, Doring G, Reimann J, Autenrieth IB, Frick JS (2011) Safety of probiotic *Escherichia coli* strain Nissle 1917 depends on intestinal microbiota and adaptive immunity of the host. Infect Immun 78: 3036–3046

Hacker J, Kaper JB (1999) The concept of pathogenicity islands. In: Kaper JB, Hacker J (eds) Pathogenicity islands and other mobile virulence elements. ASM-Press, Washington, DC, pp 1–11

Hacker J, Kaper JB (2000) Pathogenicity islands and the evolution of microbes. Annu Rev Microbiol 54:641–679

Hacker J, Ott M, Blum G, Marre R, Heesemann J, Tschape H, Goebel W (1992) Genetics of *Escherichia coli* uropathogenicity: analysis of the O6:K15:H31 isolate 536. Int J Med Microbiol Virol Parasitol Infect Dis 276:165–175

Hacker J, Blum-Oehler G, Muhldorfer I, Tschape H (1997) Pathogenicity islands of virulent bacteria: structure, function, and impact on microbial evolution. Mol Microbiol 23:1089–1097

Hacker J, Blum-Oehler G, Hochhut B, Dobrindt U (2003) The molecular basis of infectious diseases: pathogenicity islands and other mobile genetic elements – a review. Acta Microbiol Immunol Hung 50:321–330

Hammar M, Arnqvist A, Bian Z, Olsen A, Normark S (1995) Expression of two csg operons is required for production of fibronectin and congo red-binding curli polymers in *Escherichia coli* K-12. Mol Microbiol 18:661–670

Hancock V, Klemm P (2007) Global gene expression profiling of asymptomatic bacteriuria *Escherichia coli* during biofilm growth in human urine. Infect Immun 75:966–976

Homburg S, Oswald E, Hacker J, Dobrindt U (2007) Expression analysis of the colibactin gene cluster coding for a novel polyketide in *Escherichia coli*. FEMS Microbiol Lett 275:255–262

Houdouin V, Bonacorsi S, Brahimi N, Clermont O, Nassif X, Bingen E (2002) A uropathogenicity island contributes to the pathogenicity of *Escherichia coli* strains that cause neonatal meningitis. Infect Immun 70:5865–5869

Johnson JR, Russo TA (2002a) Extraintestinal pathogenic *Escherichia coli*: "the other bad E coli". J Lab Clin Med 139:155–162

Johnson JR, Russo TA (2002b) Uropathogenic *Escherichia coli* as agents of diverse non-urinary tract extraintestinal infections. J Infect Dis 186:859–864

Johnson JR, O'Bryan TT, Kuskowski M, Maslow JN (2001) Ongoing horizontal and vertical transmission of virulence genes and papA alleles among *Escherichia coli* blood isolates from patients with diverse-source bacteremia. Infect Immun 69:5363–5374

Jackson LA, Benson P, Neuzil KM, Grandjean M, Marino JL (2005) Burden of community-onset *Escherichia coli* bacteremia in seniors. J Infect Dis 191:1523–1529

Johnson TJ, Johnson SJ, Nolan LK (2006a) Complete DNA sequence of a ColBM plasmid from avian pathogenic *Escherichia coli* suggests that it evolved from closely related ColV virulence plasmids. J Bacteriol 188:5975–5983

Johnson TJ, Siek KE, Johnson SJ, Nolan LK (2006b) DNA sequence of a ColV plasmid and prevalence of selected plasmid-encoded virulence genes among avian *Escherichia coli* strains. J Bacteriol 188:745–758

Johnson TJ, Wannemuehler YM, Nolan LK (2008) Evolution of the iss gene in *Escherichia coli*. Appl Environ Microbiol 74:2360–2369

Kim YK, Pai H, Lee HJ, Park SE, Choi EH, Kim J, Kim JH, Kim EC (2002) Bloodstream infections by extended-spectrum beta-lactamase producing *Escherichia coli* and Klebsiella pneumoniae in children: epidemiology and clinical outcome. Antimicrob Agents Chemother 46:1481–1491

La Ragione RM, Collighan RJ, Woodward MJ (1999) Non-curliation of *Escherichia coli* O78:K80 isolates associated with IS1 insertion in csgB and reduced persistence in poultry infection. FEMS Microbiol Lett 175:247–253

La Ragione RM, Cooley WA, Woodward MJ (2000) The role of fimbriae and flagella in the adherence of avian strains of *Escherichia coli* O78:K80 to tissue culture cells and tracheal and gut explants. J Med Microbiol 49:327–338

Laupland KB (2007) Burden of community-onset bloodstream infection: a population-based assessment. Epidemiol Infect 135:1037–1042

Laupland KB, Gregson DB, Church DL, Ross T, Pitout JD (2008) Incidence, risk factors, and outcomes of *Escherichia coli* bloodstream infections in a large Canadian region. Clin Microbiol Infect 14:1041–1047

Mabbett AN, Ulett GC, Watts RE, Tree JJ, Totsika M, Ong CL, Wood JM, Monaghan W, Looke DF, Nimmo GR, Svanborg C, Schembri MA (2009) Virulence properties of asymptomatic bacteriuria *Escherichia coli*. Int J Med Microbiol 299:53–63

Marschall J, Agniel D, Fraser VJ, Doherty J, Warren DK (2008) Gram-negative bacteraemia in non-ICU patients: factors associated with inadequate antibiotic therapy and impact on outcomes. J Antimicrob Chemother 61:1376–1383

Mellies JL, Barron AM, Carmona AM (2007) Enteropathogenic and enterohemorrhagic *Escherichia coli* virulence gene regulation. Infect Immun 75:4199–4210

Mokady D, Gophna U, Ron EZ (2005) Extensive gene diversity in septicemic *Escherichia coli* strains. J Clin Microbiol 43:66–73

Morschhauser J, Vetter V, Emody L, Hacker J (1994) Adhesin regulatory genes within large, unstable DNA regions of pathogenic *Escherichia coli*: cross-talk between different adhesin gene clusters. Mol Microbiol 11:555–566

Mylotte JM (2005) Nursing home-acquired bloodstream infection. Infect Control Hosp Epidemiol 26:833–837

Mylotte JM, Tayara A, Goodnough S (2002) Epidemiology of bloodstream infection in nursing home residents: evaluation in a large cohort from multiple homes. Clin Infect Dis 35: 1484–1490

Nauschuetz W (1998) Emerging foodborne pathogens: enterohemorrhagic *Escherichia coli*. Clin Lab Sci 11:298–304

Nolan LK, Horne SM, Giddings CW, Foley SL, Johnson TJ, Lynne AM, Skyberg J (2003) Resistance to serum complement, iss, and virulence of avian *Escherichia coli*. Vet Res Commun 27:101–110

Ochoa TJ, Barletta F, Contreras C, Mercado E (2008) New insights into the epidemiology of enteropathogenic *Escherichia coli* infection. Trans R Soc Trop Med Hyg 102:852–856

Oelschlaeger TA, Dobrindt U, Hacker J (2002a) Virulence factors of uropathogens. Curr Opin Urol 12:33–38

Oelschlaeger TA, Dobrindt U, Hacker J (2002b) Pathogenicity islands of uropathogenic *E. coli* and the evolution of virulence. Int J Antimicrob Agents 19:517–521

Olsen A, Jonsson A, Normark S (1989) Fibronectin binding mediated by a novel class of surface organelles on *Escherichia coli*. Nature 338:652–655

Olsen A, Wick MJ, Morgelin M, Bjorck L (1998) Curli, fibrous surface proteins of *Escherichia coli*, interact with major histocompatibility complex class I molecules. Infect Immun 66: 944–949

Orskov I, Orskov F (1985) *Escherichia coli* in extra-intestinal infections. J Hyg (Lond) 95: 551–575

Oteo J, Perez-Vazquez M, Campos J (2011) Extended-spectrum [beta]-lactamase producing *Escherichia coli*: changing epidemiology and clinical impact. Curr Opin Infect Dis 23:320–326

Paterson DL (2002) Serious infections caused by enteric gram-negative bacilli – mechanisms of antibiotic resistance and implications for therapy of gram-negative sepsis in the transplanted patient. Semin Respir Infect 17:260–264

Peacock E, Jacob VW, Fallone SM (2001) *Escherichia coli* O157:H7: etiology, clinical features, complications, and treatment. Nephrol Nurs J 28:547–550, 553–545; quiz 556–547

Perez-Casal JF, Crosa JH (1984) Aerobactin iron uptake sequences in plasmid ColV-K30 are flanked by inverted IS1-like elements and replication regions. J Bacteriol 160:256–265

Petrosillo N, Capone A, Di Bella S, Taglietti F (2011) Management of antibiotic resistance in the intensive care unit setting. Expert Rev Anti Infect Ther 8:289–302

Pitout JD, Laupland KB (2008) Extended-spectrum beta-lactamase-producing Enterobacteriaceae: an emerging public-health concern. Lancet Infect Dis 8:159–166

Rakin A, Noelting C, Schubert S, Heesemann J (1999) Common and specific characteristics of the high-pathogenicity island of Yersinia enterocolitica. Infect Immun 67:5265–5274

Redford P, Welch RA (2002) Extraintestinal *Escherichia coli* as a model system for the study of pathogenicity islands. Curr Top Microbiol Immunol 264:15–30

Ritter A, Blum G, Emody L, Kerenyi M, Bock A, Neuhierl B, Rabsch W, Scheutz F, Hacker J (1995) tRNA genes and pathogenicity islands: influence on virulence and metabolic properties of uropathogenic *Escherichia coli*. Mol Microbiol 17:109–121

Ritter A, Gally DL, Olsen PB, Dobrindt U, Friedrich A, Klemm P, Hacker J (1997) The Pai-associated leuX specific tRNA 5(Leu) affects type 1 fimbriation in pathogenic *Escherichia coli* by control of FimB recombinase expression. Mol Microbiol 25:871–882

Ron EZ (2006) Host specificity of septicemic *Escherichia coli*: human and avian pathogens. Curr Opin Microbiol 9:28–32

Roos V, Schembri MA, Ulett GC, Klemm P (2006a) Asymptomatic bacteriuria *Escherichia coli* strain 83972 carries mutations in the *foc* locus and is unable to express F1C fimbriae. Microbiology 152:1799–1806

Roos V, Ulett GC, Schembri MA, Klemm P (2006b) The asymptomatic bacteriuria *Escherichia coli* strain 83972 outcompetes uropathogenic *E. coli* strains in human urine. Infect Immun 74:615–624

Schneider G, Dobrindt U, Bruggemann H, Nagy G, Janke B, Blum-Oehler G, Buchrieser C, Gottschalk G, Emody L, Hacker J (2004) The pathogenicity island-associated K15 capsule determinant exhibits a novel genetic structure and correlates with virulence in uropathogenic *Escherichia coli* strain 536. Infect Immun 72:5993–6001

Schubert S, Rakin A, Fischer D, Sorsa J, Heesemann J (1999) Characterization of the integration site of Yersinia high-pathogenicity island in *Escherichia coli*. FEMS Microbiol Lett 179:409–414

Smith HW, Huggins MB (1976) Further observations on the association of the colicine V plasmid of *Escherichia coli* with pathogenicity and with survival in the alimentary tract. J Gen Microbiol 92:335–350

Smith JL, Fratamico PM, Gunther NW (2007) Extraintestinal pathogenic *Escherichia coli*. Foodborne Pathog Dis 4:134–163

Sunden F, Hakansson L, Ljunggren E, Wullt B (2006) Bacterial interference – is deliberate colonization with Escherichia coli 83972 an alternative treatment for patients with recurrent urinary tract infection? Int J Antimicrob Agents 28(Suppl 1):S26–S29

Tencer J (1988) Asymptomatic bacteriuria – a long-term study. Scand J Urol Nephrol 22:31–34

Tivendale KA, Allen JL, Ginns CA, Crabb BS, Browning GF (2004) Association of iss and iucA, but not tsh, with plasmid-mediated virulence of avian pathogenic *Escherichia coli*. Infect Immun 72:6554–6560

Uhlich GA, Keen JE, Elder RO (2002) Variations in the csgD promoter of *Escherichia coli* O157: H7 associated with increased virulence in mice and increased invasion of HEp-2 cells. Infect Immun 70:395–399

Valvano MA, Crosa JH (1984) Aerobactin iron transport genes commonly encoded by certain ColV plasmids occur in the chromosome of a human invasive strain of *Escherichia coli* K1. Infect Immun 46:159–167

Vazquez F, Mendoza MC, Viejo G, Mendez FJ (1992) Survey of *Escherichia coli* septicemia over a six-year period. Eur J Clin Microbiol Infect Dis 11:110–117

Vicente M, Hodgson J, Massidda O, Tonjum T, Henriques-Normark B, Ron EZ (2006) The fallacies of hope: will we discover new antibiotics to combat pathogenic bacteria in time? FEMS Microbiol Rev 30:841–852

Vranes J, Kruzic V, Sterk-Kuzmanovic N, Schonwald S (2003) Virulence characteristics of *Escherichia coli* strains causing asymptomatic bacteriuria. Infection 31:216–220

Warner PJ, Williams PH, Bindereif A, Neilands JB (1981) ColV plasmid-specific aerobactin synthesis by invasive strains of *Escherichia coli*. Infect Immun 33:540–545

Waters VL, Crosa JH (1991) Colicin V virulence plasmids. Microbiol Rev 55:437–450

Wullt B, Bergsten G, Fischer H, Godaly G, Karpman D, Leijonhufvud I, Lundstedt AC, Samuelsson P, Samuelsson M, Svensson ML, Svanborg C (2003) The host response to urinary tract infection. Infect Dis Clin North Am 17:279–301

Wyllie DH, Walker AS, Peto TE, Crook DW (2007) Hospital exposure in a UK population, and its association with bacteraemia. J Hosp Infect 67:301–307

Zdziarski J, Svanborg C, Wullt B, Hacker J, Dobrindt U (2008) Molecular basis of commensalism in the urinary tract: low virulence or virulence attenuation? Infect Immun 76:695–703

Zdziarski J, Brzuszkiewicz E, Wullt B, Liesegang H, Biran D, Voigt B, Groenberg-Hernandez J, Ragnarsdottir B, Hecker M, Ron EZ, Daniel R, Gottschalk G, Hacker J, Svanborg C, Dobrindt U (2010) Host imprints on bacterial genomes – rapid, divergent evolution in individual patients. PLoS Pathog 6:e1001078

Zgur-Bertok D, Modric E, Grabnar M (1990) Aerobactin uptake system, ColV production, and drug resistance encoded by a plasmid from an urinary tract infection *Escherichia coli* strain of human origin. Can J Microbiol 36:297–299

Chapter 20
Prebiotics: Modulators of the Human Gut Microflora

Uri Lesmes

Abstract Prebiotics are cost-effective and efficient tools to promote the growth and/or activity of certain bacteria in the indigenous flora of the human gastrointestinal tract to beneficially affect host health and well-being. Human studies have established inulin, fructo-oligosaccharides, galacto-oligosaccharides and lactulose as prebiotics. Their mechanism/s of action have been elucidated in part, mainly through various in vivo and in vitro methods outlined herein. Prebiotic efficacy has been demonstrated in healthy adults, pregnant and lactating women, elderly, infants, and various disease states including colon cancer. Prebiotics maintain a vivid field of research now using state of the art instruments, in vitro gastrointestinal models, advanced computerization tools and comprehensive approaches, e.g., metabolomics and metagenomics. It is not unlikely that future innovations and discoveries will expand and establish the use of prebiotics as modulators of the human gut microbiota and ultimately help support modern life.

20.1 Prebiotics: Definition and Major Established Prebiotics

The human large intestine hosts a myriad of organisms that exist in a dynamic but stable ecosystem which affects humans in health and disease (Guarner and Malagelada 2003; O'Hara and Shanahan 2006). It is also becoming more evident that the human large intestine and its microbiome is a specialized digestive organ that also plays a role in obesity and the well-being of infants and elderly people (Delzenne and Cani 2010; Duncan et al. 2008; O'Hara and Shanahan 2006; Parracho et al. 2007; Tuohy 2007; Tuohy et al. 2009). Consequently, three different dietary strategies have been implemented and studied for their potential to beneficially modulate the gut flora composition and/or activity towards a healthier one.

U. Lesmes (✉)
Department of Food Science and Biotechnology, Technion University, Haifa, Israel
e-mail: lesmesu@tx.technion.ac.il

E. Rosenberg and U. Gophna (eds.), *Beneficial Microorganisms in Multicellular Life Forms*,
DOI 10.1007/978-3-642-21680-0_20,

The first strategy seeks to fortify gut flora through the consumption of exogenous live bacterial feeds or probiotic bacteria, e.g., *L. Acidophilus* in dairy products. The second more adaptive approach seeks to stimulate the growth and/or activity of endogenous bacterial species found in the host gut flora. The third synergistic approach seeks to combine the potential of the two previous strategies through combined consumption of efficacious probiotic bacteria with ingredients that stimulate endogenous gut flora species to ultimately benefit host health.

Amongst the plethora of studies seeking to promote host health through nutrition, it is no surprise that many studies focus on understanding how the composition and activity of the gut flora can be modulated (Delzenne and Cani 2010; Fooks et al. 1999; Gibson 2008; Macfarlane et al. 2006; Tuohy et al. 2003). In this respect, much attention has been drawn to certain carbohydrates which have been found to evade human digestion in the upper gastrointestinal tract and are selectively fermented in the colon. This led Gibson and Roberfroid (Gibson and Roberfroid 1995) to coin the term prebiotics used to describe any indigestible food ingredient that beneficially affects the host by selectively stimulating the growth and/or activity of selected bacteria in the colon and consequently improves host health. This definition has been revisited and updated into its current form as "a selectively fermented ingredient that allows specific changes, both in the composition and/or activity in the gastrointestinal microflora that confers benefits upon host well-being and health" (Gibson et al. 2004; Roberfroid 2007; Schrezenmeir and de Vrese 2001).

Inspired by the selective effects of human milk derived oligosaccharides on gut microflora, various studies have focused on the development and fabrication of prebiotic carbohydrates. Several carbohydrates are currently marketed as prebiotics, however only inulin, fructo-oligosaccharides (FOS), galacto-oligosaccharides (GOS) and lactulose are supported by high quality data both from in vitro, in vivo and human trials (Dominguez-Vergara et al. 2009; Gibson 2004, 2008; Macfarlane et al. 2006; Rastall 2010; Sabater-Molina et al. 2009; Torres et al. 2010; Tuohy et al. 2003, 2005). Specifically, human trials have established that dietary consumption of 5–20 g/day stimulates the growth of *Bifidobacterium* and *Lactobacillus* and promotes the health and well-being of infants, adults, pregnant and lactating women, as well as the elderly (Champ and Hoebler 2009; Macfarlane et al. 2006; Parracho et al. 2007; Tuohy 2007).

20.2 Physicochemical Properties of Established Prebiotics

20.2.1 Inulin and Fructo-oligosaccharides

Under the general term fructans, one can classify three established prebiotic carbohydrates, inulin, fructo-oligosaccharides (FOS), and short chain fructo-oligosaccharides (scFOS) (Rastall 2010; Sabater-Molina, et al. 2009; Tuohy et al. 2005). As will be discussed later, these fructans have been established as prebiotics,

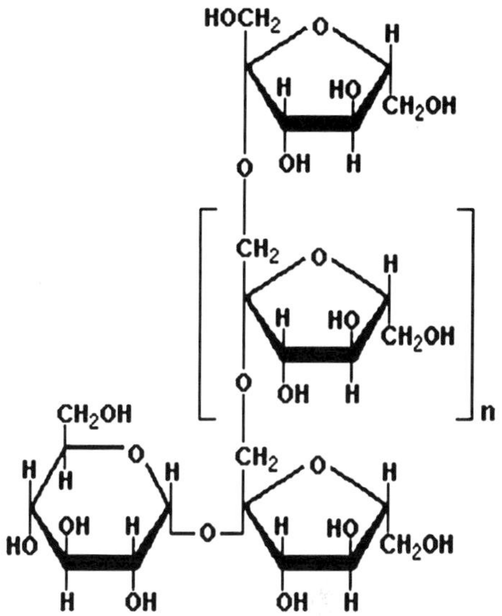

Fig. 20.1 Schematic structure of prebiotic fructans made of b-linked fructose units and terminated by a β-linked glucose unit. Inulin is defined to have 2–60 fructose units, FOS to have up to 20 fructose units

based on sound scientific evidence gathered in both in vitro and in human trials (Macfarlane et al. 2006; Rastall 2010; Sabater-Molina et al. 2009; Tuohy 2007; Tuohy et al. 2005).

Overall inulin, FOS, and scFOS are structurally related polysaccharides made up of fructose chains linked through β-1-2 glycosidic linkages and terminated by a β-2-1 linked glucose unit (as illustrated in Fig. 20.1). These fructan preparations differ in their physicochemical properties and production methods and to date are the most extensively studied prebiotics. Inulin is a naturally derived product commercially produced from chicory through enzymatic hydrolysis using inulinase and defined as the polysaccharide fraction to have a degree of polymerization of 2–60 fructose units. Inulin also naturally occurs to varying extent in onions, asparagus, garlic, artichoke, Jerusalem artichoke, and leek (Rastall 2010; Tuohy et al. 2005). Similarly, oligofructose, commonly referred to as FOS, is an enzymatic hydrolysis product produced from inulin, and may or may not carry the β-2-1 glucose terminal. It is defined as an oligosaccharide fraction which has a maximal degree of polymerization of 20, while most common commercial products have an average degree of polymerization of nine. In contrast, short chain fructo-oligosaccharides (scFOS) are synthesized through natural fermentation to yield mixtures of fructosyl chains with a maximum of five fructose units per chain. Such products are commonly mixtures comprised mainly of 1-kestose (Glu-2-Fru), nistose (Glu-3-Fru) and 1-fructosyl nistose (Glu-4-Fru). Although several methodologies, including advanced mass spectrometry (MS) and nuclear magnetic resonance (NMR) have been used to characterize these polysaccharide mixtures, their fabrication and exact identification continues to challenge manufacturers and scientists. Nutritionally, it has been found that some energy can be salvaged from fructans and their caloric value has been found to be 1.5–2.0 kcal/g, in part due to residual monosaccharides found in most preparations (Flamm et al. 2001).

Ultimately, the β-glycosidic linkages between the saccharide units of inulin and FOS render them resistant to hydrolysis by the human digestive enzymes which mainly target α-glycosidic linkages. As a consequence, these carbohydrates are expected to evade digestion in the upper gastrointestinal tract, i.e., the mouth, stomach, and small intestine, and are therefore classified as "nondigestible" carbohydrates or dietary fibers with a caloric value of 1.5–2.0 kcal/g (Flamm et al. 2001). Both in vitro and in vivo data support this notion with independent human studies demonstrating that almost 90% of ingested inulin and oligofructose escape digestion in the upper GI (Knudsen and Hessov 1995; Nilsson et al. 1988). Thus, the majority of ingested inulin and FOS are believed to reach the colon where they become accessible for fermentation by selected fractions of the gut flora.

20.2.2 *Galacto-oligosaccharides*

Oligosaccharide preparations in which galactose units make the significant portion of the carbohydrate are termed galacto-oligosaccharides (GOS). This term is now commonly used to describe oligosaccharide mixtures resulting from the application of glycosyltransferases or glycoside hydrolases on lactose syrups (Rastall 2010; Torres et al. 2010; Tuohy et al. 2005). In the past, such mixtures were also termed transgalactosylated oligosaccharides (TOS) (Rastall 2010). Commercially, GOS are oligosaccharide mixtures of the form Glu α-1-4[β-Gal-1-6]$_n$, where n can be between 2–5, and some disaccharides of galactose are produced using primarily β-galactosidases (Rastall 2010; Torres et al. 2010). They are manufactured in Japan, Europe, and the US as either clear syrups or white powders; however, their purity and clear identification vary considerably due to the use of various enzymes. This variability is also pronounced in different glycosidic linkages with β-1-6, β-1-3, and β-1-4 found to be dominant in different commercially available products (Torres et al. 2010).

As in inulin-type fructans, the β-glycosidic linkages between the saccharide units prevent GOS from being hydrolyzed by the human digestive enzymes secreted in the upper gastrointestinal tract. This has been demonstrated in several in vitro and in vivo experiments, thus it is commonly accepted that over 90% of GOS pass into the colon (Van Loo et al. 1999). In light of that and the presence of some residual sugars in GOS products, they are identified as dietary fibers with an estimated low caloric value of 1–2 kcal/g which are fermentable by the gut flora. Moreover, selective fermentation of GOS by "beneficial" bacteria on the skin has even led to incorporation of GOS in cosmetic formulations (Krutmann 2009).

20.2.3 *Lactulose*

Lactulose is a synthetic disaccharide derived from lactose and made up of galactose and fructose linked in a β-1-4 glycosidic bond. It is commonly administered at doses of over 20 g/day to chronically constipated patients to increase fecal output. It is well established that this disaccharide escapes digestion in the upper gastrointestinal tract. Human trials have consistently found that when consumed at subtherapeutic doses, lactulose reaches the colon and exerts prebiotic effects (Rastall 2010; Tuohy et al. 2005). Although this substance is an established prebiotic, it is still confined to applications as a therapeutic agent and its use in the food sectors stagnates.

20.3 Mechanisms of Action of Prebiotics

Many studies and reviews describe the fabrication, application, and beneficial effects of prebiotics (Macfarlane et al. 2006; Rastall 2010; Sabater-Molina et al. 2009; Sajilata et al. 2006; Torres et al. 2010; Tuohy 2007; Tuohy et al. 2005) . However, the complex nature of the colon microbiome, the interplay between the various bacteria and their interactions with the host cells account in part for the relatively limited understanding of the mechanisms of action (MoA) by which prebiotics confer their beneficial effects. Currently, three different mechanisms of action have been postulated and observed experimentally (Fig. 20.2).

The first paradigm is that the prebiotic action is through specific cell-associated glycosidases, which allow only certain bacteria to liberate monosaccharides from prebiotics. Consequently, such bacteria can be selectively enriched through a diet containing prebiotics. This has been demonstrated to be the case for *Bifidobacterium Infantis* strains, in which cell-associated β-fructofuranosidases

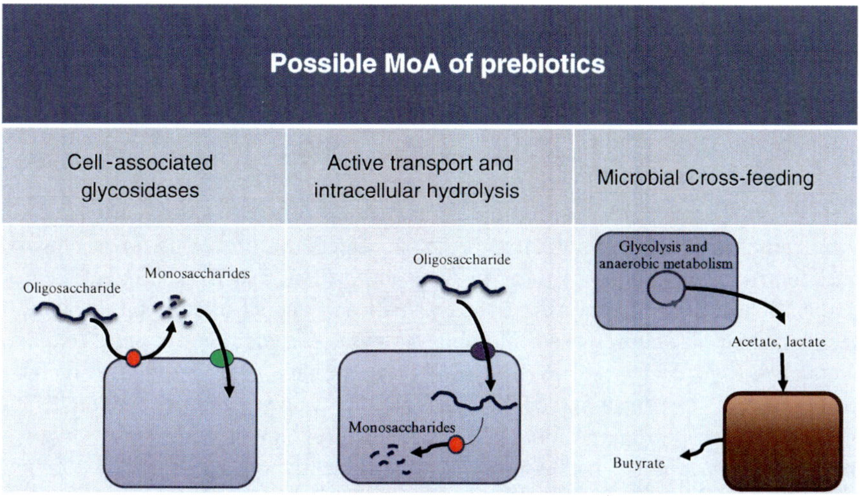

Fig. 20.2 Possible mechanisms of action (MoA) by which prebiotics can selectively induce the growth of and/or activity of specific bacteria in the colon

have been identified and isolated (Perrin et al. 2001). The second paradigm is that prebiotics selectively induce the growth of bacteria possessing an active transport system that enables them to internalize oligosaccharides for intracellular hydrolysis and utilization. Evidence supporting this notion has been gained from studies of the growth of *Lactobacillus Paracasei* and *Bifidobacterium Lactis*, which have been shown to be preferential when grown on tri and tetrasaccharide fractions of FOS or GOS (Gopal et al. 2001; Kaplan and Hutkins 2000, 2003). The third paradigm is that prebiotics exert beneficial effects not necessarily through an ecological shift but rather through changes in bacterial metabolism which can at times occur through cross-feeding between human gut microbes. This has been shown to be the case for some butyrate producing anaerobes which can utilize acetate and lactate produced by Bifidobacteria (Belenguer et al. 2006; Duncan et al. 2004; Louis and Flint 2009).

Overall, research into the prebiotic effects of orally consumed ingredients can be generally classified into in vivo and in vitro methods. Due to the need for closely controlled experimental settings, many studies have focused on the development and application of in vitro methods which range from pure cultures to highly advanced multistage continuous fermentation systems that closely mimic physiological conditions. Despite the many experimental advantages of such systems, health claims regarding prebiotic effects must be supported by in vivo experiments and more than one well-designed human trial. Consequently, one can identify both in vivo and in vitro methods for the evaluation of prebiotic effects; however, it is also imperative to understand that all these platforms are also limited by the ability to adequately analyze the bacterial and chemical composition of the samples generated by the experimental subjects/models. In this respect, the past decade has been accompanied by a significant leap in researchers' capability to analyze complex samples through comprehensive chemometric/metabolomic analyses and state-of-the-art microbiology DNA-based techniques, e.g., 16 s DNA probes, tRFLP, and metagenomics.

20.4 In Vivo Methods

Ultimately, health claims regarding prebiotic effects must rely on comprehensive well-controlled human trials. Due to the complexity of such experiments, early studies of prebiotic effects were held in small and uncontrolled populations and heavily relied on culture-based microbial analyses. To date, in vivo studies have evolved in their design and experimental subjects have been reported to include healthy human volunteers, hospital patients, ileostomy patients, sudden death victims, and various animal models. Experimentally, human studies have evolved to robust study designs which combine double-blind and placebo-controlled design with advanced microbial analyses, e.g., bacterial enumerations using 16s DNA probes in fluorescence in situ hybridizations (FISH). Despite their importance, in vivo experiments are limited by financial and ethical constraints, recruitment of subjects, access to properly trained researchers, and adequate facilities. For this and other reasons, animal models have also been used to evaluate prebiotic effects. Such studies carefully balance experimental expenditures and study design with the scientific need to closely mimic the human GI tract. They offer researchers "subjects" whose diet can be tightly controlled while allowing direct access to intestinal contents as well as tissues and organs at autopsy. Amongst various animal models, many studies use germ free animals dosed with fecal suspensions obtained from human donors. As a result, the germ-free animals become human flora-associated animals (HFA animals) which are considered to be good and reliable in vivo models for the study of human gut microbiology. For example, human flora-associated rats have been used to study the prebiotic effects of resistant starch (Silvi et al. 1999). However, data generated from animal models do not always coincide with human trials or in vitro studies of human fecal samples, as has been shown for resistant starch type III (Lesmes et al. 2008; Shimoni 2005).

20.5 In Vitro Methods

Many in vitro experimental models have been developed to simulate various aspects of the human gastrointestinal tract (Bu et al. 2010; Hur et al. 2011; Kong and Singh 2010; Macfarlane and Macfarlane 2007; Macfarlane et al. 1998; Minekus et al. 1995; Yoo and Chen 2006). Overall, these in vitro systems seek to closely mimic the conditions of an organ or a set of organs along the GI tract under closely controlled settings. Thus, they offer researchers a splendid controlled experimental system that is relatively inexpensive, easy to set up, high throughput and has limited ethical considerations. In general, in vitro models offer great possibilities for evaluating and comparing prebiotic effects as well as for mechanistic studies looking into GI flora responsiveness to indigestible food ingredients and xenobiotics.

Based on an engineering approach, various researchers have developed standard anaerobic fecal cultures into a verity of digestion models based on chemical reactors and/or a series of reactors. One of the first in vitro GI models described in literature is the multistage simulator of the human intestinal microbial ecosystem (SHIME) (Molly et al. 1994). This computer controlled model comprises five serially connected reactors simulating the duodenum/jejunum, ileum, cecum/proximal colon, transverse colon, and distal colon. This model was later expanded to a more comprehensive one through the addition of a sixth reactor simulating the human stomach (De Boever et al. 2000; Gmeiner et al. 2000). This comprehensive anaerobic model allows operators to control various parameters of physiological relevance, including gastric and pancreatic secretions, pH, transit times, feed rate, and composition, as well as sample different loci along the GI on a regular basis. Mimicking of the colon flora is achieved through inoculation with a 20% fecal suspension. Thus, the SHIME model facilitates mechanistic studies of the GI microbial community in a validated and relatively cheap model which overcomes the high costs and ethical issues associated with in vivo studies. Although this comprehensive model carefully mimics the main features of the human GI, one should bear in mind that it is an over-simplified model which does not account for GI peristaltic movement, luminal interactions with the host cells, and omits the absorption of metabolites and fluids.

Another well-known in vitro GI model was developed in the Netherlands (Minekus et al. 1995). TNO's intestinal model (TIM) is actually two separate models. TIM-1 is a series of four computer controlled chambers simulating the upper GI, i.e., the stomach, duodenum, jejunum, and ileum, while TIM-2 models the large intestine. Each chamber of the TIM model is composed of a flexible chamber jacketed by a glass water jacket in which changes in the water pressure enable simulating the peristaltic movements of the GI. The hollow fiber construct of the jejunal and ileal compartments of the TIM-1 enable active removal of water and digested materials which mimic absorption. TIM-2 compartments are inoculated with fecal suspensions and the hollow fiber construct of the system also enables controlled removal of water and small metabolites. Overall, the dynamic TIM model closely reproduces in vivo conditions found in monogastric animals and man. Similarly, simple glass reactors have been used for in batch and three stage continuous fermentation systems to closely simulate the lower GI, i.e., the proximal, transverse, and distal colon. These relatively cheap and simple to operate fermentation systems have been successfully used to study gut microbiology and validated against sudden death victims (Macfarlane et al. 1998).

Overall, in vivo methods and particularly human feeding trials are essential for establishing health claims regarding prebiotic effects on the human GI microbiota. However, these methods are limited by financial, ethical, and practical considerations. In vitro GI models are over-simplistic models that fail to account for various relevant parameters such as luminal mixing patterns and shear forces encountered by ingested bolus. This issue has been recently addressed in a novel human gastric model (Kong and Singh 2010). However, other significant parameters such as the role of mucosal flora, impact of immune components, feed-back mechanisms, and specific hormonal controls, e.g., control of gastric emptying through satiety feed-back mechanisms, still remain unanswered by in vitro models. These and other limitations explain the low predictive value of such systems; however, they do not deter from the high potency, relatively low costs, and ease of use of these in vitro models which make them excellent tools for the study of luminal biochemistry and microbiology.

20.6 Evidence on Beneficial Effects of Prebiotics

To date, various human trials have demonstrated that established prebiotics have various beneficial effects, with successful modulation of gut microflora toward a more healthy one, with the most common effect being increased levels of Bifidobacteria and to a lesser extent Lactobacilli, as reviewed by others

(Gibson 2008; Macfarlane et al. 2006; Rastall 2010; Sabater-Molina et al. 2009; Torres et al. 2010; Tuohy 2007; Tuohy et al. 2005).

20.6.1 Prebiotic Effects in Healthy Adults

Numerous studies have demonstrated different beneficial effects of fructans (Inulin and FOS) on health and well-being of healthy adult subjects (Gibson 2008; Rastall 2010; Sabater-Molina et al. 2009; Tuohy et al. 2005). Experimental data pooled from various independent studies of healthy human volunteers consistently show increases in Bifidobacteria levels in the gut microflora. Although there is no recommended daily intake or allowance (RDI or RDA) for these prebiotics, dose-response studies show consumption of 5–10 g/day are well tolerated and exert a beneficial bifidogenic effect on healthy adults. Similar doses of GOS and lactulose have been noted to exert a bifidogenic effect in vitro and in human trails (Bouhnik et al. 1997; Salminen and Salminen 1997; Tuohy et al. 2005).

As to bowel habits, constipation, and laxative effects, studies have shown various results in which fecal output is unchanged at daily intake of up to 15 g/day, while slightly increased at doses of 15 g/day or higher (Macfarlane et al. 2006). Furthermore, there is some evidence that in constipated subjects, Inulin may increase bowel habits, that is, the frequency of bowel discharge but not fecal output (Den Hond et al. 2000). Lactulose is clinically prescribed at 20 g/day to increase fecal output of chronically constipated patients, however having a bifidogenic effect on healthy adults at lower doses (Salminen and Salminen 1997). Accordingly, inulin, FOS, GOS, and lactulose can be considered mild laxatives with adverse effects observed only at doses exceeding 20 g/day. Additionally, various studies have linked the consumption of established prebiotics to include protection against enteric infections, improved mineral absorption, immunomodulation, symptomatic relief of inflammatory GI conditions, production of short chain fatty acids, particularly butyrate, and reduced risk of colon cancer (Gibson 2004, 2008; Macfarlane et al. 2006; Sabater-Molina et al. 2009; Tuohy et al. 2005). Prebiotics have also been studied in the context of women's life cycle, particularly during pregnancy and lactation. In general, these periods of specific nutritional requirements are accompanied by specific physiological and behavioral changes which affect not only the mother but also the fetus or breast fed infant. These life periods are sometimes accompanied by irregular gastrointestinal activity which is beneficially affected by the consumption of dietary fibers and prebiotics (Champ and Hoebler 2009). Moreover, modulation of gut microflora has been suggested to be a possible way to tackle gestational weight gain and postpartum weight retention (Duncan et al. 2008). As to the ability of prebiotics effects to be extended to the fetus or breast fed baby, a recent double-blind placebo-controlled human study demonstrated prebiotic manipulation of the mother's gut microflora is not transmissible to her infant offspring (Shadid et al. 2007).

20.6.2 Infants and the Elderly

The interest in prebiotics as supplements for infant formulas stems from various beneficial effects attributed in part to the 130 different human milk oligosaccharides (HMO) (Sabharwal et al. 1991). Due to the complex composition and structure of HMOs, the food industry has sought to augment infant formulas with prebiotics in an attempt to mimic human milk and beneficially affect infant health (Parracho et al. 2007; Rivero-Urgelland Santamaria-Orleans 2001). This practical application of prebiotics was evaluated in double-blind, randomized, controlled studies in 90 full term infants (Moro et al. 2002, 2003). These studies showed that 4 g/L or 8 g/L of FOS, GOS, or their combination led to a significant reduction in fecal pH and an increase in Bifidobacteria and Lactobacilli after 28 days feeding. According to this and other studies reviewed elsewhere (Parracho et al. 2007), Inulin, FOS, and GOS are now supplemented into various infant formulas for their prebiotic effects.

On the other side of the human life cycle, it is now well established that the gut microbiota changes at old age, with an increased number of bacterial groups represented in the predominant elderly gut microbiota (Tiihonen et al. 2010). This shift in the gut microflora is concomitant with changes in immune function, diet, and lifestyle and may contribute to the increased morbidity that frequently accompanies old age. Thus, modulation of the colon microflora is increasingly being studied as an effective, cost effective, and natural way to improve the health and well-being of elderly people as well as reduce the risks for various diseases. Recent human intervention studies have confirmed the potential of inulin-type fructans to beneficially modulate elderly gut microbiota and reduce the risk of disease (Tuohy 2007). A randomized, double-blind, controlled study conducted with 74 subjects aged 70 and over showed modulation of pro-inflammatory gene activation and serum levels of sCD14, leading the researchers to the conclusion that prebiotic addition can improve the low noise inflammatory process frequently observed in this sensitive population (Schiffrin et al. 2007).

20.7 Prebiotics as Therapeutic Agents

Environmental agents constantly challenge the human body to maintain overall health and well-being. It is now well documented that the bacterial microbiota resident in the human GI has a role not only in maintaining health but also in some disease (Fava et al. 2006; Guarner and Malagelada 2003; O'Hara and Shanahan 2006). This section provides an overview of current evidence establishing the beneficial effects of prebiotics on various diseases mainly based on human trials.

One of the most clear and significant disease preventing effects of prebiotics is the protection from pathogenic gastrointestinal infections. It is now commonly accepted that prebiotics promote the growth of probiotics, or "good" bacteria, which help displace pathogens from the mucosa, produce antimicrobial agents that harm pathogens, and compete with pathogens on binding sites and nutrients (O'Hara and Shanahan 2006). Ultimately, these functions are promoted by prebiotic intake which leads to an observed protection against food-borne pathogens. Human studies of such effects are ethically not approved and thus various in vitro data support this disease preventing effect of prebiotics (Bosscher et al. 2006; Dominguez-Vergara et al. 2009; Tuohy et al. 2005). One randomized placebo-controlled human study does depict the beneficial effect of a symbiotic preparation (containing a probiotic culture and oligofructose) in lowering intestinal permeability and pathogenic bacteria counts in patients admitted with high risk for developing sepsis (Jain et al. 2004). The beneficial effects of prebiotics in promoting health and preventing disease were also confirmed through a human study that enrolled 140 infants. This study found that 0.5 and 15 g of oligofructose and cereal, respectively, led to a statistically significant reduction in events of fever, frequency of vomiting, regurgitation, and abdominal discomfort (Saavedra and Tschernia 2002). Additionally, various studies have shown that prebiotics can beneficially affect patients with antibiotic-associated diarrhea, especially when the diarrhea arises from *C. difficile* (Macfarlane et al. 2006).

Prebiotics have also been demonstrated to be a possible therapeutic and preventive measure to modulate gut flora and consequently reduce the risk of colon cancer (Gibson 2004; Macfarlane et al. 2006; Rastall 2010; Sabater-Molina et al. 2009; Tuohy et al. 2005). Most of these studies are based on animal models and in vitro experiments and human studies are scarce. In general, it is accepted that prebiotics promote colon flora to metabolize carcinogenic compounds and secrete short chain fatty acids to the lumen; in this respect, butyrate has been a significant metabolite of interest (Duncan et al. 2004; Louis and Flint 2009). As reviewed by Rastall (2010), human studies have established that Inulin, FOS, and scFOS beneficially affect colorectal cell proliferation and genotoxicity in a way that justifies further research on the potential role of prebiotics in the prevention and treatment of colon cancer. The inflammatory nature of various gastrointestinal diseases, such as irritable bowel disorder (IBD), ulcerative colitis (UC), and Crohn's disease (CD), has also been studied as a potential target for the beneficial effects of prebiotics (Macfarlane et al. 2006; Rastall 2010; Tuohy et al. 2005). For example, sulfate reducing bacteria are believed to play a role in UC, accordingly some researchers believe that proliferation of bifidobacteria and production of butyrate in the colon can be induced through prebiotic consumption and could alleviate or treat UC. Moreover, the possible link between gut flora and obesity is now being enthusiastically studied, leading some researchers to the belief that the gut flora could be a novel therapeutic target with prebiotics being potential therapeutic agents or management tools (Champ and Hoebler 2009; Delzenne and Cani 2010; Duncan et al. 2008; Tuohy et al. 2009).

20.8 Emerging Prebiotics: From Understanding to Rational Design

The potential to affect human health and prevent disease through indigestible ingredients and gut flora management maintains a constant search for new and novel prebiotics. Among the various prebiotic candidates, a review of the current literature reveals that isomalto-oligosaccharides (IMO), Soy-oligosaccharides (SOS), Xylo-oligosaccharides (XOS), gluco-oligosaccharides, lactosucrose, and resistant starches can be classified as emerging prebiotics. Studies into these emerging prebiotics are currently limited to in vitro models or to small scale animal or human trials but show promising results in inducing the growth of Bifidobacteria and Lactobacilli and/or promoting formation of beneficial metabolites, e.g., butyrate, or reduction of fecal genotoxicity (Rastall 2010; Tuohy et al. 2005). For example, thermally produced resistant starch was found to have a bifidogenic and butyrogenic effect in an in vitro three stage continuous fermentation system inoculated with human feces (Lesmes et al. 2008). Furthermore, this study suggested that resistant starch crystalline polymorphism, arising from different thermal processing, could be a way to delineate different prebiotic effects on the human colon flora.

20.9 Conclusions

Prebiotics are cost-effective and efficient tools to beneficially promote the growth and/or activity of certain bacteria in the indigenous flora of the human GI, as established by independent human trials. Understanding the mechanism/s by which prebiotics exert their effects is an elemental step limiting further applications of prebiotics and establishing their beneficial effects. In the past, research of prebiotics and development of new prebiotics was based on a wide search of natural sources which were then screened through a trial and error approach both in vitro and in vivo. However, today's increased access to state-of-the-art instruments, in vitro gastrointestinal models, and advanced computerization tools are leading many researchers to adopt more complete and comprehensive approaches, e.g., metabolomics and metagenomics, as well as concepts of rational design. Thus, future research into prebiotics, emerging and novel prebiotics is becoming more mechanistic and science oriented, seeking to put innovations on sound scientific principles of rational design. This advanced "back to basics" approach combined with comprehensive analyses is likely to further expand and establish the use of prebiotics to affect humans in health and disease and support the modern lifestyle.

References

Belenguer A, Duncan SH, Calder AG, Holtrop G, Louis P, Lobley GE, Flint HJ (2006) Two routes of metabolic cross-feeding between Bifidobacterium adolescentis and butyrate-producing anaerobes from the human gut. Appl Environ Microbiol 72(5):3593–3599

Bosscher D, Van Loo J, Franck A (2006) Inulin and oligofructose as prebiotics in the prevention of intestinal infections and diseases. Nutr Res Rev 19(2):216–226

Bouhnik Y, Flourie B, DagayAbensour L, Pochart P, Gramet G, Durand M, Rambaud JC (1997) Administration of transgalacto-oligosaccharides increases fecal bifidobacteria and modifies colonic fermentation metabolism in healthy humans. J Nutr 127(3):444–448
Bu GH, Luo YK, Lu J, Zhang Y (2010) Reduced antigenicity of β-lactoglobulin by conjugation with glucose through controlled Maillard reaction conditions. Food Agric Immunol 21(2):143–156
Champ M, Hoebler C (2009) Functional food for pregnant, lactating women and in perinatal nutrition: a role for dietary fibres? Curr Opin Clin Nutr Metab Care 12(6):565–574
De Boever P, Deplancke B, Verstraete W (2000) Fermentation by gut microbiota cultured in a simulator of the human intestinal microbial ecosystem is improved by supplementing a soygerm powder. J Nutr 130(10):2599–2606
Delzenne NM, Cani PD (2010) Nutritional modulation of gut microbiota in the context of obesity and insulin resistance: potential interest of prebiotics. Int Dairy J 20(4):277–280
Den Hond E, Geypens B, Ghoos Y (2000) Effect of high performance chicory inulin on constipation. Nutr Res 20(5):731–736
Dominguez-Vergara AM, Vazquez-Moreno L, Montfort GRC (2009) Role of prebiotic oligosaccharides in prevention of gastrointestinal infections: a review. Arch Latinoam Nutr 59 (4):358–368
Duncan SH, Holtrop G, Lobley GE, Calder AG, Stewart CS, Flint HJ (2004) Contribution of acetate to butyrate formation by human faecal bacteria. Br J Nutr 91(6):915–923
Duncan SH, Lobley GE, Holtrop G, Ince J, Johnstone AM, Louis P, Flint HJ (2008) Human colonic microbiota associated with diet, obesity and weight loss. Int J Obes 32(11):1720–1724
Fava F, Lovegrove JA, Gitau R, Jackson KG, Tuohy KM (2006) The gut microbiota and lipid metabolism: implications for human health and coronary heart disease. Curr Med Chem 13(25):3005–3021
Flamm G, Glinsmann W, Kritchevsky D, Prosky L, Roberfroid M (2001) Inulin and oligofructose as dietary fiber: a review of the evidence. Crit Rev Food Sci Nutr 41(5):353–362
Fooks LJ, Fuller R, Gibson GR (1999) Prebiotics, probiotics and human gut microbiology. Int Dairy J 9(1):53–61
Gibson GR (2004) From probiotics to prebiotics and a healthy digestive system. J Food Sci 69(5): M141–M143
Gibson GR (2008) Prebiotics as gut microflora management tools. J Clin Gastroenterol 42(6): S75–S79
Gibson GR, Roberfroid MB (1995) Dietary modulation of the human colonic microbiota – introducing the concept of prebiotics. J Nutr 125(6):1401–1412
Gibson GR, Probert HM, Van Loo J, Rastall RA, Roberfroid MB (2004) Dietary modulation of the human colonic microbiota: updating the concept of prebiotics. Nutr Res Rev 17(2):259–275
Gmeiner M, Kneifel W, Kulbe KD, Wouters R, De Boever P, Nollet L, Verstraete W (2000) Influence of a synbiotic mixture consisting of Lactobacillus acidophilus 74–2 and a fructo-oligosaccharide preparation on the microbial ecology sustained in a simulation of the human intestinal microbial ecosystem (SHIME reactor). Appl Microbiol Biotechnol 53(2):219–223
Gopal PK, Sullivan PA, Smart JB (2001) Utilisation of galacto-oligosaccharides as selective substrates for growth by lactic acid bacteria including Bifidobacterium lactis DR10 and Lactobacillus rhamnosus DR20. Int Dairy J 11(1–2):19–25
Guarner F, Malagelada JR (2003) Gut flora in health and disease. Lancet 361(9356):512–519
Hur SJ, Lim BO, Decker EA, McClements DJ (2011) In vitro human digestion models for food applications. Food Chem 125(1):1–12
Jain PK, McNaught CE, Anderson ADG, MacFie J, Mitchell CJ (2004) Influence of synbiotic containing Lactobacillus acidophilus La5, Bifidobacterium lactis Bb 12, Streptococcus thermophilus, Lactobacillus bulgaricus and oligofructose on gut barrier function and sepsis in critically ill patients: a randomised controlled trial. Clin Nutr 23(4):467–475
Kaplan H, Hutkins RW (2000) Fermentation of fructooligosaccharides by lactic acid bacteria and bifidobacteria. Appl Environ Microbiol 66(6):2682–2684

Kaplan H, Hutkins RW (2003) Metabolism of fructooligosaccharides by Lactobacillus paracasei 1195. Appl Environ Microbiol 69(4):2217–2222

Knudsen KEB, Hessov I (1995) Recovery of inulin from Jerusalem-artichoke (helianthus-tuberosus l) in the small-intestine of man. Br J Nutr 74(1):101–113

Kong FB, Singh RP (2010) A human gastric simulator (HGS) to study food digestion in human stomach. J Food Sci 75(9):E627–E635

Krutmann J (2009) Pre- and probiotics for human skin. J Dermatol Sci 54(1):1–5

Lesmes U, Beards EJ, Gibson GR, Tuohy KM, Shimoni E (2008) Effects of resistant starch type III polymorphs on human colon microbiota and short chain fatty acids in human gut models. J Agric Food Chem 56(13):5415–5421

Louis P, Flint HJ (2009) Diversity, metabolism and microbial ecology of butyrate-producing bacteria from the human large intestine. FEMS Microbiol Lett 294(1):1–8

Macfarlane GT, Macfarlane S (2007) Models for intestinal fermentation: association between food components, delivery systems, bioavailability and functional interactions in the gut. Curr Opin Biotechnol 18(2):156–162

Macfarlane GT, Macfarlane S, Gibson GR (1998) Validation of a three-stage compound continuous culture system for investigating the effect of retention time on the ecology and metabolism of bacteria in the human colon. Microb Ecol 35(2):180–187

Macfarlane S, Macfarlane GT, Cummings JH (2006) Review article: prebiotics in the gastrointestinal tract. Aliment Pharmacol Ther 24(5):701–714

Minekus M, Marteau P, Havenaar R, Huisintveld JHJ (1995) A multicompartmental dynamic computer-controlled model simulating the stomach and small-intestine. ATLA, Altern Lab Anim 23(2):197–209

Molly K, Vandewoestyne M, Desmet I, Verstraete W (1994) Validation of the simulator of the human intestinal microbial ecosystem (Shime) reactor using microorganism-associated activities. Microb Ecol Health Dis 7(4):191–200

Moro G, Minoli I, Mosca M, Fanaro S, Jelinek J, Stahl B, Boehm G (2002) Dosage-related bifidogenic effects of galacto- and fructo-oligosaccharides in formula-fed term infants. J Pediatr Gastroenterol Nutr 34(3):291–295

Moro GE, Mosca F, Miniello V, Fanaro S, Jelinek J, Stahl B, Boehm G (2003) Effects of a new mixture of prebiotics on faecal flora and stools in term infants. Acta Paediatr 92:77–79

Nilsson U, Oste R, Jagerstad M, Birkhed D (1988) Cereal fructans – invitro and invivo studies on availability in rats and humans. J Nutr 118(11):1325–1330

O'Hara AM, Shanahan F (2006) The gut flora as a forgotten organ. EMBO Rep 7(7):688–693

Parracho H, McCartney AL, Gibson GR (2007) Probiotics and prebiotics in infant nutrition. Proc Nutr Soc 66(3):405–411

Perrin S, Warchol M, Grill JP, Schneider F (2001) Fermentations of fructo-oligosaccharides and their components by Bifidobacterium infantis ATCC 15697 on batch culture in semi-synthetic medium. J Appl Microbiol 90(6):859–865

Rastal RA (2010) Functional oligosaccharides: application and manufacture. Annu Rev Food Sci Technol 1:305–339 (Palo Alto: Annual Reviews)

Rivero-Urgell M, Santamaria-Orleans A (2001) Oligosaccharides: application in infant food. Early Hum Dev 65:S43–S52

Roberfroid M (2007) Prebiotics: the concept revisited. J Nutr 137(3):830S–837S

Saavedra JM, Tschernia A (2002) Human studies with probiotics and prebiotics: clinical implications. Br J Nutr 87:S241–S246

Sabater-Molina M, Larque E, Torrella F, Zamora S (2009) Dietary fructooligosaccharides and potential benefits on health. J Physiol Biochem 65(3):315–328

Sabharwal H, Sjoblad S, Lundblad A (1991) Affinity chromatographic identification and quantitation of blood-group a-active oligosaccharides in human-milk and feces of breast-fed infants. J Pediatr Gastroenterol Nutr 12(4):474–479

Sajilata MG, Singhal RS, Kulkarni PR (2006) Resistant starch – a review. Compr Rev Food Sci Food Saf 5(1):1–17

Salminen S, Salminen E (1997) Lactulose, lactic acid bacteria, intestinal microecology and mucosal protection. Scand J Gastroenterol 32:45–48

Schiffrin EJ, Thomas DR, Kumar VB, Brown C, Hager C, Hof MAV, Morley JE, Guigoz Y (2007) Systemic inflammatory markers in older persons: the effect of oral nutritional supplementation with prebiotics. J Nutr Health Aging 11(6):475–479

Schrezenmeir J, de Vrese M (2001) Probiotics, prebiotics, and synbiotics – approaching a definition. Am J Clin Nutr 73(2):361S–364S

Shadid R, Haarman M, Knol J, Theis W, Beermann C, Rjosk-Dendorfer D, Schendel DJ, Koletzko BV, Krauss-Etschmann S (2007) Effects of galacto-oligosaccharide and long-chain fructooligosaccharide supplementation during pregnancy on maternal and neonatal microbiota and immunity – a randomized, double-blind placebo-controlled study. Am J Clin Nutr 86 (5):1426–1437

Shimoni E (2005) Tailoring resistant starch polymorphs for optimized prebiotic effects. Ind Aliment Agric 122:15–18

Silvi S, Rumney CJ, Cresci A, Rowland IR (1999) Resistant starch modifies gut microflora and microbial metabolism in human flora-associated rats inoculated with faeces from Italian and UK donors. J Appl Microbiol 86(3):521–530

Tiihonen K, Ouwehand AC, Rautonen N (2010) Human intestinal microbiota and healthy ageing. Ageing Res Rev 9(2):107–116

Torres DPM, Goncalves MDF, Teixeira JA, Rodrigues LR (2010) Galacto-oligosaccharides: production, properties, applications, and significance as prebiotics. Compr Rev Food Sci Food Saf 9(5):438–454

Tuohy KM (2007) Inulin-type fructans in healthy aging. J Nutr 137(11):2590S–2593S

Tuohy KM, Probert HM, Smejkal CW, Gibson GR (2003) Using probiotics and prebiotics to improve gut health. Drug Discov Today 8(15):692–700

Tuohy KM, Rouzaud GCM, Bruck WM, Gibson GR (2005) Modulation of the human gut microflora towards improved health using prebiotics – assessment of efficacy. Curr Pharm Des 11(1):75–90

Tuohy KM, Costabile A, Fava F (2009) The gut microbiota in obesity and metabolic disease – a novel therapeutic target. Nutr Ther Metab 27(3):113–133

Van Loo J, Cummings J, Delzenne N, Englyst H, Franck A, Hopkins M, Kok N, Macfarlane G, Newton D, Quigley M, Roberfroid M, van Vliet T, van den Heuvel E (1999) Functional food properties of non-digestible oligosaccharides: a consensus report from the ENDO project (DGXII AIRII-CT94-1095). Br J Nutr 81(2):121–132

Yoo JY, Chen XD (2006) GIT physicochemical modeling – a critical review. Int J Food Eng 2 (4):1–7

Chapter 21
Host Genetics and Gut Microbiota

Keren Buhnik-Rosenblau, Yael Danin-Poleg, and Yechezkel Kashi

Abstract The gut microbiota consists of hundreds to thousands of bacterial species, and is strongly associated with the well-being of the host. Its composition differs among individual hosts, being affected by environmental factors such as food and maternal inoculation. Recently, the relative impact of the host's genetics has been uncovered in different mammal hosts. Here, we describe the effect of host genetic background on the composition of gut bacterial communities in a murine model, focusing on lactic acid bacteria (LAB) as an important group that contains a variety of strains associated with the improvement of various gut health disorders. Based on 16S-rDNA tRFLP analysis, variation was observed in fecal LAB populations of two genetic mouse lines BALB/C and C57BL/6J. *Lactobacillus johnsonii*, a potentially probiotic bacterium, appeared at significantly higher levels in the feces of C57BL/6J mice compared to BALB/C. The genetic inheritance of *L. johnsonii* levels was further tested in reciprocal crosses between the two mouse lines, where each mouse line served alternatively as the male or female parent. The two resultant groups of F1 offspring, having the same genetic content but exposed to different maternal microflora during and after birth, presented similar *L. johnsonii* levels, confirming that mouse genetics plays a major role in determining these levels, compared to the relatively lower maternal effect. Our findings suggest that mouse genetics has a major effect on the composition of the intestinal LAB population in general, and on the persistence of *L. johnsonii* in the gut in particular.

K. Buhnik-Rosenblau • Y. Danin-Poleg • Y. Kashi (✉)
Department of Biotechnology and Food Engineering, Technion-Israel Institute of Technology, Haifa 32000, Israel
e-mail: kashi@tx.technion.ac.il

E. Rosenberg and U. Gophna (eds.), *Beneficial Microorganisms in Multicellular Life Forms*, 281
DOI 10.1007/978-3-642-21680-0_21, © Springer-Verlag Berlin Heidelberg 2011

21.1 Introduction

The human digestive tract is a habitat for bacteria, which are tenfold more abundant than the total number of human cells. This intestinal microbiota is highly important for maintaining a healthy organism. Recent studies have shown a tight connection between the gut microbiota and different health disorders, such as inflammatory bowel diseases, atopic diseases, obesity, metabolic syndrome, intestinal cancers, and diabetes (Bajzer and Seeley 2006; Garrett et al. 2010; Mai and Draganov 2009; Qin et al. 2010; Turnbaugh et al. 2006; Vijay-Kumar et al. 2010; Wen et al. 2008). Colonization of the gastrointestinal tract begins at birth, when the newborn is first exposed to maternal microbiota, and continues thereafter via prolonged exposure to maternal and other environmental bacteria (Ducluzeau 1983; Mshvildadze et al. 2008). Eventually, the adult gut microbiota will consist of hundreds to thousands of bacterial species. The composition of the intestinal microbiota is known to differ among individual hosts (Costello et al. 2009; Dethlefsen et al. 2007; Eckburg et al. 2005; Ley et al. 2008b) as a result of variations in environmental factors, such as food consumption, exposure to maternal bacteria during infancy, antibiotic intake, or other stochastic factors (Alexander et al. 2006; Dethlefsen et al. 2006; Farano et al. 2003; Jakobsson et al. 2010; Ley et al. 2008a; Mshvildadze et al. 2008; Palmer et al. 2007; Turnbaugh et al. 2009).

21.2 Host Genetics and Gut Microbiota

The importance of the host's genetics in determining the intestinal microbiota has also been demonstrated. Three decades ago, Van de Merwe et al. (1983) found that the culturable fecal floras of monozygotic human twins were more similar than those of dizygotic twins, based on gram staining of representative colonies and characterization of cell morphologies. Over time, an increasing number of studies showed that the genotype of the human host affects the composition of gut microbiota, using modern molecular techniques mostly relaying on the fecal bacteria, extracted using culture independent methods. Zoetendal et al. (2001) analyzed the fecal–bacterial populations of human adults with varying degree of genetic relatedness using denaturing gradient gel electrophoresis (DGGE) of bacterial 16S rDNA amplicons and found that the similarity of DGGE profiles between monozygotic twins were significantly higher than those of unrelated individuals; in contrast, the DGGE profiles of marital partners, which are living in the same environment and share comparable feeding habits, showed low similarity. Three different species of wild monkeys were analyzed for their fecal microbiota using 16S rRNA sequencing, revealing that the microbial communities within the same species are more similar compared to those of different primate species (Yildirim et al. 2010). Microbial communities of those non-human primates were found to be distinct from human gut microbiota, reflecting host phylogeny.

A variety of murine models were studied as well, showing the effect of mouse genetics on its gut microbiota. The availability of different genetic mouse lines makes them an appropriate model. Indeed, altered bacterial populations were found to be consistently present in the gut of various genetic mouse lines. These observations are based on quantification of specific bacteria belonging to different representative groups (Alexander et al. 2006) by DNA fingerprinting of the bacterial population (Esworthy et al. 2010; Friswell et al. 2010) or by 16S rDNA pyrosequencing of the total microbiota (Benson et al. 2010). Kovacs et al. (2011) also found that mouse genetic background is a stronger determinant than mice gender in shaping the intestinal microbiota. Vaahtovuo et al. (2003) found that modulation of the microbiota by a course of antibiotic is followed by regeneration of the murine intestinal flora depending on the genotype of the host, supporting the effect of mouse genetic background in shaping the gut microbial communities. Recently, Benson et al. (2010) mapped 18 murine chromosomal segments, harboring quantitative trait loci (QTL) that showed significant linkage with relative abundance of specific microbial taxa, providing along with other studies clear evidence for the importance of host genetic control in shaping the diversity of individual microbiome in mammals.

21.3 Host Specific Genes That Influence the Gut Microbial Community

Bacterial communities are highly dynamic and therefore rapidly responding to any environmental change, which may be a result of genetic variation in a single gene of the host. Indeed, there is rapidly growing evidence for host specific genes affecting the composition of gut microbiota. Genes of the immune system have provided so far the largest amount of indications on their influence on the gut microbiota. Knockout of the gene encodes for the MyD88 protein, an adaptor for multiple innate immune receptors that recognize microbial stimuli, resulted in composition change of the distal gut microbiota (Wen et al. 2008); Immunodeficient (scid/scid) and immunocompetent (scid/+) mice monoassociated with segmented filamentous bacteria, presented different levels of these bacteria on the mucosa of the small intestine, after their weaning (Jiang et al. 2001). Mutations in the MEFV gene, which causes familial mediterranean fever (FMF) – autoinflammatory disorder, is associated with significant changes in gut bacterial community structure, characterized by loss of diversity and shifts within different bacterial phyla (Khachatryan et al. 2008). Different alleles of the major histocompatibility complex (MHC) genes were also found to affect the composition of gut microbiota (Toivanen et al. 2001).

The gut microbiota may be affected not only by genetic variation at genes belonging to the immune system. Ley et al. (2005) found that genetically obese mice, carrying mutation in the gene encodes for the leptin hormone (*ob*/*ob*), have fewer Bacteroidetes and more Firmicutes compared to their lean (+/+) siblings. Additionally, genetic variation at the intestinal surfaces, including the intestinal mucus, gut epithelial cells and extracellular matrix may also affect the gut microbiota composition. These intestinal surfaces are a target for bacterial adhesion, implying that variation in the surface-presented components may result in differential bacterial adhesion (Firon et al. 1985; Sharon 1987; Uchida et al. 2006). Along with the impact of host genetics on the composition of gut microbiota, genetic biodiversity among bacterial strains may influence bacterial adhesion in particular, and persistence in the gut in general (Barbas et al. 2009; Denou et al. 2008), contributing to the complexity of host-bacterium interactions.

21.4 Host Genetics and Bacterial Communities in Other Body Habitats

Host-bacterium interactions are not unique for the gut. Bacterial communities populate also other body habitats, such as the nose, oral cavity, urinary tract, vagina, and skin. Studies show connections between some of these microbiomes to the host genetics. Regarding the skin microbiome, Scharschmidt et al. (2009) showed that mice deficient in matriptase, a serine protease, present a shift in the skin microbiota. In the oral cavity, Nibali et al. (2011) found that a specific genotype of the IL6 gene is associated with high counts of the oral pathogen Aggregatibacter actinomycetemcomitans subgingivally. The vaginal microbiome was also found to be associated with polymorphism in different genes of the immune system, such as IL1, TLR4, and the MBL gene (Genc and Onderdonk 2011). Zhou et al. (2010) found differences in the frequencies of various microbial communities within the vaginal microbiota among women belonging to different ethnic groups, suggesting that host genetic factors may be more important in determining the species composition of vaginal bacterial communities than are cultural and behavioral differences. Yet, the gut microbiome is the most widely studied, being the most crowded and complicated of all other microbiomes.

21.5 Lactic Acid Bacteria

One important bacterial group colonizing the human body is the lactic acid bacteria (LAB), a heterogeneous group of gram-positive rods and cocci that belong to the phylum Firmicutes. There are indications of an association between oral administration of different LAB strains and improvement of gut health disorders, such as pouchitis, ulcerative colitis, infectious diarrhea, antibiotic-associated diarrhea, traveler's diarrhea, necrotizing enterocolitis, atopic eczema, and *Helicobacter pylori* infections (Guandalini 2010; Haller et al. 2010; Holubar et al. 2010;

Kalliomaki et al. 2010; Lionetti et al. 2010; Wolvers et al. 2010). *Lactobacillus* is the largest genus in the LAB, composed of over a 100 species and containing several probiotic strains (i.e., live microorganisms, which when administered in adequate amounts, confer a health benefit on the host) which are well-characterized. The genus *Lactobacillus* is highly divergent, but can nevertheless be divided into seven or eight groups (Claesson et al. 2007). One of them is the *acidophilus* group, which contains, among others, the bacterium *Lactobacillus johnsonii*. This species includes commensal as well as probiotic strains, such as *L. johnsonii* NCC533. The genome sequence of this strain reveals genes responsible for interactions with the host, as well as the absence of genes responsible for several important biosynthetic pathways, suggesting strong dependence on the host or on other microbes (Pridmore et al. 2004). Such dependence might partially explain the variation in the abundance of *L. johnsonii* in the guts of different animal hosts (Buhnik-Rosenblau et al. submitted for publication).

Here we describe our study (Buhnik-Rosenblau et al. 2011) that compare the intestinal LAB population of two genetically diverse mouse lines, C57BL/6J and BALB/C, held under the same environmental conditions (food, cages, etc.). The intestinal LAB population was analyzed by tRFLP followed by precise strain typing for *L. johnsonii* using variable number tandem repeat (VNTR) analysis. Our results suggest that mouse genetics plays a major role in determining *L. johnsonii* levels in particular, and the composition of the LAB population in general.

21.6 Profile of Fecal LAB Populations and *L. johnsonii* Levels in C57BL/6J and BALB/C Mice

We tested the variation in the gut microbiota between C57BL/6J and BALB/C mouse lines, concentrating on the fecal–bacterial sub-populations that grow on m-*Enterococcus* agar. Experiments were applied altogether on 50 C57BL/6J mice and 33 BALB/C mice in four biological replicates, over a 2-year period.

Fecal samples from the two mouse lines were collected and pooled from five individual 5- to 7-week-old mice to minimize the variation between individuals. tRFLP analysis of the bacterial population revealed different patterns for C57BL/6J and BALB/C mouse lines. These results were highly reproducible regardless of mouse origin: mice obtained from Harlan Laboratories Ltd. and self-bred mice presented similar patterns. Generally, the tRFLP patterns consisted of three major peaks: 74 bp, 189 bp, and 566 bp (Fig. 21.1). Minor peaks were excluded from the analysis in order to focus on the main bacterial species composing the bacterial population isolated from each line. To identify the bacterial species corresponding to each of the peaks, representative colonies were analyzed for their 16S terminal restriction DNA fragments (tRFs) followed by 16S rDNA sequencing. The 566 bp tRFLP peak was found to represent *Enterococcus faecalis*, the 189-bp peak

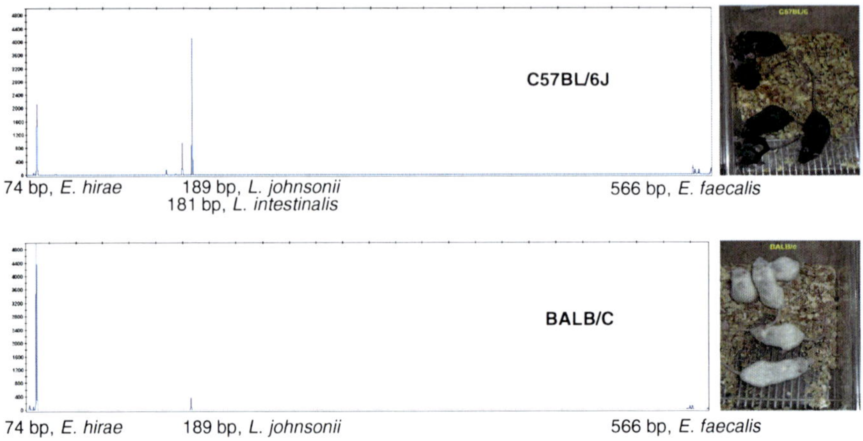

Fig. 21.1 tRFLP patterns of fecal LAB populations obtained from C57BL and BALB/C mice. Lactic acid bacteria (LAB) were grown on m-Enterococcus agar under anaerobic conditions. DNA fragments were observed using F-fluorescent-labeled primer for 16S PCR amplification followed by digestion with MspI restriction enzyme and fragment size analysis using ABI 3130 genetic analyzer. The size of specific fragments is indicated in bp

Lactobacillus johnsonii, and the 74-bp peak represented bacterial species belonging to the *Enterococcus faecium* cluster. Precise species identification based on sequencing of the housekeeping gene rpoA (Naser et al. 2005) revealed that the isolated species was *Enterococcus hirae*. The tRFLP patterns suggested that *L. johnsonii* is consistently present in bacterial populations originating from C57BL/6J mice, whereas it was nearly absent in BALB/C mice. In contrast, high levels of *E. hirae* were present in BALB/C mice and were lower or completely absent in C57BL/6J mice. *E. faecalis* was generally present at low levels in both mouse lines (Fig. 21.1). Males and females within each line showed similar patterns with slight non-significant differences. tRFLP patterns of bacterial populations grown under both aerobic and anaerobic conditions were similar, with the exception of a 181-bp peak, representing *Lactobacillus intestinalis*, which appeared mostly in C57BL/6J mice after anaerobic incubation. The most significant variation between mouse lines was in *L. johnsonii*, which was highly abundant in C57BL/6J relative to BALB/C mice, and we therefore chose to focus on this species in further experiments.

Quantitative estimation of the relative abundance of *L. johnsonii* in fecal samples of the two mouse lines, C57BL/6J and BALB/C, was performed by hybridization of *L. johnsonii*-specific probe (corresponding to gene LJ0244) to colonies grown on m-*Enterococcus* agar (Fig. 21.2a). Results were based on three biological replicates in which females and males were sampled separately. Hybridizations were performed in duplicate for each fecal sample. C57BL/6J mice were found to contain significantly higher levels of *L. johnsonii* in their feces than BALB/C mice (Fig. 21.2b). Both direct colony counts of *L. johnsonii* and their fraction of the total bacterial counts were tenfold higher in C57BL/6J mice than in

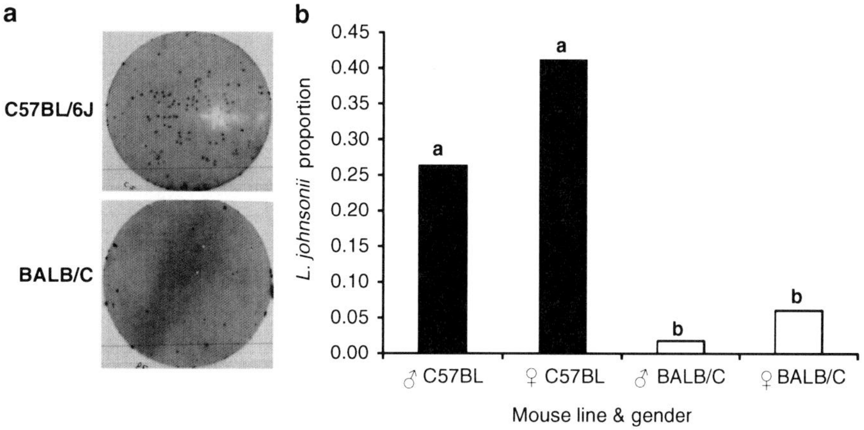

Fig. 21.2 (**a**) Colony hybridization of fecal LAB populations isolated from C57BL/6J and BALB/C mice targeting L. johnsonii. (**b**) Levels of Lactobacillus johnsonii in fecal samples of mouse lines C57BL (*black columns*) and BALB/C (*white columns*). Males and females were tested separately (five mice per group). Levels are expressed as proportion of L. johnsonii colonies out of total colonies grown on m-Enterococcus agar plates, and are given as average values of three and five independent biological replicates for males and females, respectively. L. johnsonii was enumerated using colony hybridization with L. johnsonii-specific probe. Columns headed by different letters are significantly different at $\alpha < 0.05$ by two-way ANOVA followed by Tukey HSD test

BALB/C mice (two-way ANOVA, $F = 6.079$, $p = 0.0025$ and $F = 17.867$, $p < 0.0001$, respectively). In contrast, the total bacterial counts were highly variable, with non-significant differences between mouse lines and genders ($F = 2.002$, $p = 0.1365$). Thus we chose to further refer to the parameter of *L. johnsonii* proportions, normalizing the variable total counts. The average proportion of *L. johnsonii* in C57BL/6J mice was 0.36, and 0.04 in BALB/C mice. Slight differences were observed between genders ($F = 4.25$, $p = 0.0487$), with the females always presenting a higher fraction of *L. johnsonii* compared to the males; nevertheless, line had the major effect ($F = 41.36$, $p < 0.0001$). Comparison of all means by Tukey HSD analysis further confirmed that the significant differences in *L. johnsonii* levels are indeed between the two mouse lines (Fig. 21.2b, $\alpha = 0.05$, $Q = 2.73$).

21.7 Strain Typing of *L. johnsonii* Isolates from C57BL/6J and BALB/C Mice

Since host-bacterium interactions are highly dependent on bacterial strain, we tested whether both mouse lines carry the same *L. johnsonii* clone by strain genotyping. Four representative *L. johnsonii* colonies were isolated from each mouse line and identified as *L. johnsonii* using species-specific PCR amplification of 23S rDNA. Further confirmation was achieved by sequencing of the 16S rDNA (1,534 bp) and of 302 bp of the LJ0244 gene. Precise typing was achieved using VNTR, which were found useful for fine discrimination among *L. johnsonii* from various hosts (Buhnik-Rosenblau et al. submitted for publication). Variation analysis of the eight isolates at 11 VNTR genomic loci revealed identical results, i.e., all isolates presented the same alleles in each of the tested loci. This genotyping identity was in contrast to the high variation found among 41 isolates of diverse origins, showing 2–10 alleles in the tested loci (Buhnik-Rosenblau et al. submitted for publication). Additional 21 *L. johnsonii* isolates from both mouse lines were tested for polymorphism at the two most polymorphic VNTR loci, giving identical alleles for all isolates at both loci (data not shown). These typing results supported the similarity of *L. johnsonii* clones isolated from both mouse lines.

21.8 *L. johnsonii* Levels in F1 Offspring of Reciprocal C57BL/6J × BALB/C Crosses

To confirm that *L. johnsonii* levels are mainly determined by host genetic factors rather than exposure to maternal microbiota, we tested C57BL/6J × BALB/C reciprocal crosses. Basically, the reciprocal hybrids are genetically similar (except for maternally inherited elements), whereas during and following birth, the F1 mice are exposed to either C57BL/6J or BALB/C maternal microbiota. The levels of *L. johnsonii* in fecal samples of the F1 hybrids were quantified in quadruplicate by colony hybridization (see above), and the average proportion of *L. johnsonii* was calculated. The obtained values were compared with *L. johnsonii* levels of the parental lines C57BL/6J and BALB/C. ANOVA was performed to test the difference in *L. johnsonii* levels among the genetic groups as non-significant differences were observed between genders (two-way ANOVA, $F = 3.63, p = 0.064$). Results showed significant differences among BALB/C, C57BL/6J and F1 hybrids of the reciprocal crosses ($F = 13.8513, p < 0.0001$). Comparison of all means by Tukey HSD analysis showed significantly lower *L. johnsonii* levels in the parental BALB/C line compared to the two F1 hybrids and the parental C57BL/6J line (Table 21.1,

Table 21.1 *Lactobacillus johnsonii* levels in fecal samples of F1 offspring and parental mouse lines C57BL and BALB/C, five mice per group

BALB/C	C57BL/6J	♀C57BL/6J × ♂BALB/C	♂C57BL/6J × ♀BALB/C
0.04 × 0.07b	0.36 × 0.18a	0.27 ± 0.16a	0.22 ± 0.14a

F1 offspring resulted from two separate reciprocal crosses: ♀BALB/C × ♂C57BL and ♂BALB/C × ♀C57BL. Levels are expressed as proportion of *L. johnsonii* colonies out of total colonies grown on m-*Enterococcus* agar plates, and are given as average values of four independent biological replicates. *L. johnsonii* was enumerated using colony hybridization with *L. johnsonii*-specific probe. Different letter represent significant difference at $\alpha < 0.05$

$\alpha = 0.05$, $Q = 2.67$). The levels of *L. johnsonii* were similar between hybrids of the cross ♀C57BL/6J × ♂BALB/C and the reciprocal cross ♀BALB/C × ♂C57BL/6J, presenting average values of 0.27 and 0.22, respectively (Table 21.1). Slightly higher levels of *L. johnsonii* were observed for the parental line C57BL/6J (0.36). These results indicated that mouse genetics is the major parameter determining the levels of *L. johnsonii*, whereas the maternal effect is relatively low. Comparison of *L. johnsonii* levels between the BALB/C parental line and the offspring of ♀BALB/C × ♂C57BL/6J, which differ only in their paternal genetics, showed highly significant differences, of almost an order of magnitude (Table 21.1, 0.04 and 0.22, respectively). All of the above suggest that mouse genetics plays a major role in determining the fecal levels of *L. johnsonii*.

21.9 General Discussion

An increasing number of studies are indicating a role for host genetics in determining gut microbiota composition. Here we show a major effect of the mouse's genetics on its fecal–bacterial population, focusing on a narrow spectrum: the lactic acid bacteria (LAB), at the species and strain level. C57BL/6J and BALB/C mice showed highly reproducible variation in their fecal LAB populations (Fig. 21.1), as analyzed by tRFLP. The same variation was observed between these inbred lines regardless of mouse origin (commercial source or self-breeding), indicating that the composition of the gut LAB population is genetically controlled. Similarly, a previous study observed variation in the cellular fatty acid profiles of the intestinal bacterial population between these two mouse lines, indicating that mouse genetics has an influence on the gastrointestinal microbiota (Vaahtovuo et al. 2005). Four bacterial species were found to dominate mouse feces in the bacterial spectrum tested here. *E. faecalis* was present in both mouse lines, whereas levels of *L. johnsonii*, *L. intestinalis*, and *E. hirae* were found to differ between the two mouse lines. Both *L. johnsonii* and *L. intestinalis* dominated in C57BL/6J vs. BALB/C mice. Quantitative analysis of *L. johnsonii* in mouse feces by colony hybridization using an *L. johnsonii*-specific probe demonstrated that the levels of *L. johnsonii* in BALB/C mice are an order of magnitude lower than in C57BL/6J mice (Fig. 21.2), further confirming the significant differences observed in *L. johnsonii* levels between inbred mouse lines with diverse genetic backgrounds. In contrast to C57BL/6J mice which presented high levels of both *L. johnsonii* and *L. intestinalis*, BALB/C mice presented high levels of *E. hirae*, indicating that these taxonomically related bacterial species might compete for the same niche or have similar metabolic functions.

The observed variation in the composition of intestinal bacterial species of genetically diverse hosts has been previously observed in humans (Dethlefsen et al. 2006, 2007), where it was claimed that the combination of bacterial species and strains is unique to each individual, calling for precise identification of the isolated bacterial strains. Strain identification is highly important, as different strains belonging to the same bacterial species might vary in their ability to persist in the gut of a specific host. As suggested by the "hologenome theory" (Zilber-Rosenberg and Rosenberg 2008), the host and its symbiont microflora (together defined as the "holobiont") are one unit of selection in the evolution. Thus, different bacterial strains might persist in the gut of each of the genetically diverged C57BL/6J and BALB/C mice. (Denou et al. 2008) previously showed that different *L. johnsonii* strains vary significantly in their gut residence time after oral feeding of mice having a single genetic background. Indications of strain-dependent persistence have also been found in other bacterial species (e.g., Barbas et al. 2009; Oh et al. 2010). In *Lactobacillus reuteri*, phylogeny analysis of strains revealed host-specific clusters. Moreover, in a feeding trial, administering a mixture of *L. reuteri* strains from various origins to germ-free mice resulted in better persistence of the mouse and rat isolates, suggesting host-specific ecological fitness up to the strain level (Oh et al. 2010). Strain typing was conducted on *L. johnsonii*, a potentially probiotic bacterium (Pridmore et al. 2004) that was found here to populate the gut of C57BL/6J mice at higher levels compared to BALB/C mice. Molecular typing of *L. johnsonii* using VNTR did not discriminate between strains isolated from the two mouse lines, in contrast to the high discrimination that we had achieved among other *L. johnsonii* strains isolated from various hosts in general, and mice in particular (Buhnik-Rosenblau et al. submitted for publication). VNTR, which are highly polymorphic genetic markers, have been found to be efficient for strain typing and for epidemiological studies in many bacterial species (e.g., Broza et al. 2009; Danin-Poleg et al. 2007; Mee-Marquet et al. 2009) as a result of their relatively high mutation rate (Van Ert et al. 2007). Thus, the fact that identical VNTR alleles were obtained here indicates that the isolates from both mouse lines are similar, and this suggests that the two mouse lines carry the same *L. johnsonii* strain. Nevertheless, only a comparison of the complete bacterial genomic sequences can give an unequivocal assertion of their similarity. Our results indicate that the same *L. johnsonii* strain might co-evolve with the two mouse lines, which in turn suggests that its differential persistence in the guts of C57BL/6J and BALB/C mice is a result of the genetic variation between these mouse lines rather than genetic variation in *L. johnsonii*. Maternal microbiota is a factor that may differentially influence *L. johnsonii* levels, as each BALB/C or C57BL/6J mouse is exposed to different maternal microbiota at the early stage of its life. The gut microbiota is known to be dynamic in the early stages of life as the young mammal is exposed to maternal microbiota starting at birth (Palmer et al. 2007), and to become relatively stable at later stages (Ley et al. 2006; Zoetendal et al. 1998). To test the effect of maternal microbiota on the bacterial populations of the two mouse lines, they were subjected to reciprocal crosses. The F1 offspring of the two crosses had the same genetics but differed in their maternally inherited genetic elements as well as in exposure to different maternal microbiota. The resultant reciprocal F1 groups presented similar levels of *L. johnsonii* (Table 21.1), suggesting that *L. johnsonii* levels are governed mainly by mouse genetics rather than by maternal microbiota or by other maternal effects. Both reciprocal F1 groups showed high levels of *L. johnsonii*, like those in the C57BL/6J mice and much higher than those in the BALB/C mice, suggesting that in the F1 mice, *L. johnsonii* is allowed to persist as in C57BL/6J mice. Furthermore, comparing mice with the same maternal parent (BALB/C) but different genomic content revealed significantly higher *L. johnsonii* levels in F1 offspring of the C57BL/6J male parent vs. those of the BALB/C male parent (i.e., BALB/C mice). Results clearly demonstrated that changing the paternal genetics results in a dramatic change in *L. johnsonii* levels, further confirming that the mouse's genetic background is the major parameter in determining fecal levels of *L. johnsonii*, whereas the maternal effect is relatively low. The use of F1 reciprocal crosses provided clear proof of the strong impact of mouse genetic background on *L. johnsonii* levels in BALB/C and C57BL/6J inbred lines. The findings presented here are further supported by numerous studies indicating the influence of host genetics on its gut microbiota, conducted with animal models and humans (Alexander et al. 2006; Jiang et al. 2001; Khachatryan et al. 2008; Vaahtovuo et al. 2003; Van de Merwe et al. 1983; Wen et al. 2008; Zoetendal et al. 2001). As mentioned above, 18 host quantitative trait loci were identified in an advanced intercross mouse line originating from a cross between C57BL/6J and an ICR-derived outbred line (HR), showing linkage with relative abundance of specific microbial taxa. Among these, the *L. johnsonii*/*L. gasseri* group was found to segregate with two genomic loci on murine chromosomes 7 and 14 (Benson et al. 2010). That study and others made use of novel DNA-sequencing techniques that provide a wide view of the gut microbiome (Frank and Pace 2008; Qin et al. 2010). However, concentrating on the narrow spectrum of LAB enables the isolation and characterization of potentially probiotic bacterial strains. The culture-dependant approach used here enabled isolation of a potentially probiotic strain belonging to *L. johnsonii*, a species that is already used in the industry. Furthermore, such an approach is expected to provide insight into the highly specific host-bacterium interactions down to the sub-species level of both host and bacteria. Indeed, our results present a highly specific interaction between mouse genetic background and a specific *L. johnsonii* strain.

In conclusion, our findings suggest that mouse genetics has a major effect in determining the composition of LAB populations in general, and the gut persistence of *L. johnsonii*, a potentially probiotic bacterial species, in particular. Maternal inoculation had relatively low effect on the levels of the tested bacteria. Finding LAB strains with specificity to host genetics in a murine model, as demonstrated here, could lead to the discovery of genes in both the bacteria and its host which may be involved in bacterial persistence in the gut. Such a discovery is expected to have a major impact, especially with respect to health-promoting bacteria (i.e., probiotics), and could lead to future development of probiotic products specifically oriented to the consumer's genetics, as part of the personalized medicine approach.

References

Alexander AD, Orcutt RP, Henry JC, Baker J, Bissahoyo AC, Threadgill DW (2006) Quantitative PCR assays for mouse enteric flora reveal strain-dependent differences in composition that are influenced by the microenvironment. Mamm Genome 17:1093–1104

Bajzer M, Seeley RJ (2006) Physiology – obesity and gut flora. Nature 444:1009–1010

Barbas AS, Lesher AP, Thomas AD, Wyse A, Devalapalli AP, Lee YH, Tan HE, Orndorff PE, Bollinger RR, Parker W (2009) Altering and assessing persistence of genetically modified E. coli MG1655 in the large bowel. Exp Biol Med 234:1174–1185

Benson AK, Kelly SA, Legge R, Ma F, Low SJ, Kim J, Zhang M, Oh PL, Nehrenberg D, Hua K, Kachman SD, Moriyama EN, Walter J, Peterson DA, Pomp D (2010) Individuality in gut microbiota composition is a complex polygenic trait shaped by multiple environmental and host genetic factors. Proc Natl Acad Sci USA 107:18933–18938

Buhnik-Rosenblau K, Danin-Poleg Y, Kashi Y (2011) Predominant effect of host genetics on levels of Lactobacillus johnsonii in mouse gut. Appl Environ Microbiol. In press

Buhnik-Rosenblau K, Matsko-Efimov V, Jung M, Shin H, Danin-Poleg Y, Kashi Y (2011) Submitted for publication

Broza YY, Danin-Poleg Y, Lerner L, Valinsky L, Broza M, Kashi Y (2009) Epidemiologic study of Vibrio vulnificus infections by using variable number tandem repeats. Emerg Infect Dis 15: 1282–1285

Claesson MJ, van Sinderen D, O'Toole PW (2007) The genus Lactobacillus – a genomic basis for understanding its diversity. FEMS Microbiol Lett 269:22–28

Costello EK, Lauber CL, Hamady M, Fierer N, Gordon JI, Knight R (2009) Bacterial community variation in human body habitats across space and time. Science 326:1694–1697

Danin-Poleg Y, Cohen LA, Gancz H, Broza YY, Goldshmidt H, Malul E, Valinsky L, Lerner L, Broza M, Kashi Y (2007) Vibrio cholerae strain typing and phylogeny study based on simple sequence repeats. J Clin Microbiol 45:736–746

Denou E, Pridmore RD, Berger B, Panoff JM, Arigoni F, Brussow H (2008) Identification of genes associated with the long-gut-persistence phenotype of the probiotic Lactobacillus johnsonii strain NCC533 using a combination of genomics and transcriptome analysis. J Bacteriol 190: 3161–3168

Dethlefsen L, Eckburg PB, Bik EM, Relman DA (2006) Assembly of the human intestinal microbiota. Trends Ecol Evol 21:517–523

Dethlefsen L, McFall-Ngai M, Relman DA (2007) An ecological and evolutionary perspective on human-microbe mutualism and disease. Nature 449:811–818

Ducluzeau R (1983) Implantation and development of the gut flora in the newborn animal. Ann Rech Vet 14:354–359

Eckburg PB, Bik EM, Bernstein CN, Purdom E, Dethlefsen L, Sargent M, Gill SR, Nelson KE, Relman DA (2005) Diversity of the human intestinal microbial flora. Science 308:1635–1638

Esworthy RS, Smith DD, Chu FF (2010) A strong impact of genetic background on gut microflora in mice. Int J Inflam 2010:986046

Farano S, Chierici R, Guerrini P, Vigi V (2003) Intestinal microflora in early infancy: composition and development. Acta Paediatr 91:48–55

Firon N, Duksin D, Sharon N (1985) Mannose-specific adherence of Escherichia-coli to Bhk cells that differ in their glycosylation patterns. FEMS Microbiol Lett 27:161–165

Frank DN, Pace NR (2008) Gastrointestinal microbiology enters the metagenomics era. Curr Opin Gastroenterol 24:4–10

Friswell MK, Gika H, Stratford IJ, Theodoridis G, Telfer B, Wilson ID, McBain AJ (2010) Site and strain-specific variation in gut microbiota profiles and metabolism in experimental mice. PLoS One 5:e8584

Garrett WS, Gordon JI, Glimcher LH (2010) Homeostasis and inflammation in the intestine. Cell 140:859–870

Genc MR, Onderdonk A (2011) Endogenous bacterial flora in pregnant women and the influence of maternal genetic variation. BJOG 118:154–163

Guandalini S (2010) Update on the role of probiotics in the therapy of pediatric inflammatory bowel disease. Expert Rev Clin Immunol 6:47–54

Haller D, Antoine JM, Bengmark S, Erick P, Rijkers GT, Lenoir-Wijnkoop I (2010) Guidance for substantiating the evidence for beneficial effects of probiotics: probiotics in chronic inflammatory bowel disease and the functional disorder irritable bowel syndrome. J Nutr 140: 690S–697S

Holubar SD, Cima RR, Sandborn WJ, Pardi DS (2010) Treatment and prevention of pouchitis after ileal pouch-anal anastomosis for chronic ulcerative colitis. Cochrane Database Syst Rev 6:CD001176

Jakobsson HE, Jernberg C, Andersson AF, Sjolund-Karlsson M, Jansson JK, Engstrand L (2010) Short-term antibiotic treatment has differing long-term impacts on the human throat and gut microbiome. PLoS One 5:e9836

Jiang HQ, Bos NA, Cebra JJ (2001) Timing, localization, and persistence of colonization by segmented filamentous bacteria in the neonatal mouse gut depend on immune status of mothers and pups. Infect Immun 69:3611–3617

Kalliomaki M, Antoine JM, Herz U, Rijkers GT, Wells JM, Mercenier A (2010) Guidance for substantiating the evidence for beneficial effects of probiotics: prevention and management of allergic diseases by probiotics. J Nutr 140:713S–721S

Khachatryan ZA, Ktsoyan ZA, Manukyan GP, Kelly D, Ghazaryan KA, Aminov RI (2008) Predominant role of host genetics in controlling the composition of gut microbiota. PLoS One 3:e3064

Kovacs A, Ben Jacob N, Tayem H, Halperin E, Iraqi FA, Gophna U (2011) Genotype is a stronger determinant than sex of the mouse gut microbiota. Microb Ecol 61:423–428

Ley RE, Backhed F, Turnbaugh P, Lozupone CA, Knight RD, Gordon JI (2005) Obesity alters gut microbial ecology. Proc Natl Acad Sci USA 102:11070–11075

Ley RE, Turnbaugh PJ, Klein S, Gordon JI (2006) Microbial ecology – human gut microbes associated with obesity. Nature 444:1022–1023

Ley RE, Hamady M, Lozupone C, Turnbaugh PJ, Ramey RR, Bircher JS, Schlegel ML, Tucker TA, Schrenzel MD, Knight R, Gordon JI (2008a) Evolution of mammals and their gut microbes. Science 320:1647–1651

Ley RE, Lozupone CA, Hamady M, Knight R, Gordon JI (2008b) Worlds within worlds: evolution of the vertebrate gut microbiota. Nat Rev Microbiol 6:776–788

Lionetti E, Indrio F, Pavone L, Borrelli G, Cavallo L, Francavilla R (2010) Role of probiotics in pediatric patients with Helicobacter pylori infection: a comprehensive review of the literature. Helicobacter 15:79–87

Mai V, Draganov PV (2009) Recent advances and remaining gaps in our knowledge of associations between gut microbiota and human health. World J Gastroenterol 15:81–85

Mee-Marquet N, Francois P, Domelier AS, Arnault L, Girard N, Schrenzel J, Quentin R (2009) Variable-number tandem repeat analysis and multilocus sequence typing data confirm the epidemiological changes observed with Staphylococcus aureus strains isolated from blood-stream infections. J Clin Microbiol 47:2863–2871

Mshvildadze M, Neu J, Mai V (2008) Intestinal microbiota development in the premature neonate: establishment of a lasting commensal relationship? Nutr Rev 66:658–663

Naser SM, Thompson FL, Hoste B, Gevers D, Dawyndt P, Vancanneyt M, Swings J (2005) Application of multilocus sequence analysis (MLSA) for rapid identification of Enterococcus species based on rpoA and pheS genes. Microbiology-Sgm 151:2141–2150

Nibali L, Madden I, Franch CF, Heitz-Mayfield L, Brett P, Donos N (2011) IL6–174 genotype associated with Aggregatibacter actinomycetemcomitans in Indians. Oral Dis 17:232–237

Oh PL, Benson AK, Peterson DA, Patil PB, Moriyama EN, Roos S, Walter J (2010) Diversification of the gut symbiont Lactobacillus reuteri as a result of host-driven evolution. ISME J 4: 377–387

Palmer C, Bik EM, DiGiulio DB, Relman DA, Brown PO (2007) Development of the human infant intestinal microbiota. PLoS Biol 5:1556–1573

Pridmore RD, Berger B, Desiere F, Vilanova D, Barretto C, Pittet AC, Zwahlen MC, Rouvet M, Altermann E, Barrangou R, Mollet B, Mercenier A, Klaenhammer T, Arigoni F, Schell MA (2004) The genome sequence of the probiotic intestinal bacterium Lactobacillus johnsonii NCC 533. Proc Natl Acad Sci USA 101:2512–2517

Qin J, Li R, Raes J, Arumugam M, Burgdorf KS, Manichanh C, Nielsen T, Pons N, Levenez F, Yamada T, Mende DR, Li J, Xu J, Li S, Li D, Cao J, Wang B, Liang H, Zheng H, Xie Y, Tap J, Lepage P, Bertalan M, Batto JM, Hansen T, Le Paslier D, Linneberg A, Nielsen HB, Pelletier E, Renault P, Sicheritz-Ponten T, Turner K, Zhu H, Yu C, Li S, Jian M, Zhou Y, Li Y, Zhang X, Li S, Qin N, Yang H, Wang J, Brunak S, Dore J, Guarner F, Kristiansen K, Pedersen O, Parkhill J, Weissenbach J, Bork P, Ehrlich SD, Wang J (2010) A human gut microbial gene catalogue established by metagenomic sequencing. Nature 464:59–65

Scharschmidt TC, List K, Grice EA, Szabo R, Renaud G, Lee CCR, Wolfsberg TG, Bugge TH, Segre JA (2009) Matriptase-deficient mice exhibit ichthyotic skin with a selective shift in skin microbiota. J Invest Dermatol 129:2435–2442

Sharon N (1987) Bacterial lectins, cell-cell recognition and infectious-disease. FEBS Lett 217: 145–157

Toivanen P, Vaahtovuo J, Eerola E (2001) Influence of major histocompatibility complex on bacterial composition of fecal flora. Infect Immun 69:2372–2377

Turnbaugh PJ, Ley RE, Mahowald MA, Magrini V, Mardis ER, Gordon JI (2006) An obesity-associated gut microbiome with increased capacity for energy harvest. Nature 444:1027–1031

Turnbaugh PJ, Ridaura VK, Faith JJ, Rey FE, Knight R, Gordon JI (2009) The effect of diet on the human gut microbiome: a metagenomic analysis in humanized gnotobiotic mice. Sci Transl Med 1:6ra14

Uchida H, Kinoshita H, Kawai Y, Kitazawa H, Miura K, Shiiba K, Horii A, Kimura K, Taketomo N, Oda M, Yajima T, Saito T (2006) Lactobacilli binding human A-antigen expressed in intestinal mucosa. Res Microbiol 157:659–665

Vaahtovuo J, Toivanen P, Eerola E (2003) Bacterial composition of murine fecal microflora is indigenous and genetically guided. FEMS Microbiol Ecol 44:131–136

Vaahtovuo J, Eerola E, Toivanen P (2005) Comparison of cellular fatty acid profiles of the microbiota in different gut regions of BALB/c and C57BL/6J mice. Antonie Van Leeuwenhoek 88:67–74

Van de Merwe JP, Stegeman JH, Hazenberg MP (1983) The resident faecal flora is determined by genetic characteristics of the host. Implications for Crohn's disease? Antonie Van Leeuwenhoek 49:119–124

Van Ert MN, Easterday WR, Huynh LY, Okinaka RT, Hugh-Jones ME, Ravel J, Zanecki SR, Pearson T, Simonson TS, U'Ren JM, Kachur SM, Leadem-Dougherty RR, Rhoton SD, Zinser G, Farlow J, Coker PR, Smith KL, Wang BX, Kenefic LJ, Fraser-Liggett CM, Wagner DM, Keim P (2007) Global genetic population structure of Bacillus anthracis. PLoS One 2:e461

Vijay-Kumar M, Aitken JD, Carvalho FA, Cullender TC, Mwangi S, Srinivasan S, Sitaraman SV, Knight R, Ley RE, Gewirtz AT (2010) Metabolic syndrome and altered gut microbiota in mice lacking toll-like receptor 5. Science 328:228–231

Wen L, Ley RE, Volchkov PY, Stranges PB, Avanesyan L, Stonebraker AC, Hu C, Wong FS, Szot GL, Bluestone JA, Gordon JI, Chervonsky AV (2008) Innate immunity and intestinal microbiota in the development of type 1 diabetes. Nature 455:1109–1113

Wolvers D, Antoine JM, Myllyluoma E, Schrezenmeir J, Szajewska H, Rijkers GT (2010) Guidance for substantiating the evidence for beneficial effects of probiotics: Prevention and management of infections by probiotics. J Nutr 140:698S–712S

Yildirim S, Yeoman CJ, Sipos M, Torralba M, Wilson BA, Goldberg TL, Stumpf RM, Leigh SR, White BA, Nelson KE (2010) Characterization of the fecal microbiome from non-human wild primates reveals species specific microbial communities. PLoS One 5:e13963

Zhou X, Hansmann MA, Davis CC, Suzuki H, Brown CJ, Schutte U, Pierson JD, Forney LJ (2010) The vaginal bacterial communities of Japanese women resemble those of women in other racial groups. FEMS Immunol Med Microbiol 58:169–181

Zilber-Rosenberg I, Rosenberg E (2008) Role of microorganisms in the evolution of animals and plants: the hologenome theory of evolution. FEMS Microbiol Rev 32:723–735

Zoetendal EG, Akkermans ADL, de Vos WM (1998) Temperature gradient gel electrophoresis analysis of 16S rRNA from human fecal samples reveals stable and host-specific communities of active bacteria. Appl Environ Microbiol 64:3854–3859

Zoetendal EG, Akkermans ADL, Akkermans-van Vliet WM, de Visser JAGM, de Vos WM (2001) The host genotype affects the bacterial community in the human gastrointestinal tract. Microb Ecol Health Dis 13:129–134

Part V
Evolution by Symbiosis

Chapter 22
Microbial Symbiont Transmission: Basic Principles and Dark Sides

Silvia Bulgheresi

Abstract For a microbial association to persist throughout generations, host progeny must either be capable of earning a microbial *fortune* from the environment (horizontal transmission) or inherit it from its parents (vertical transmission). The former modality relies on highly sophisticated molecular mechanisms of partners' recognition. The latter modality, instead, presupposes the microbial partners to be as deeply integrated into the host life cycle and to associate with its earliest developmental stages. Besides the common trends that just started to emerge, I discuss the under-explored aspects of bacterial transmission such as its cell biology, and how extracellular microbial symbionts – in turn – ensure their daughter cells the symbiotic lifestyle.

> If you love someone, set them free. If they come back they're yours. If they don't they never were.
>
> Richard Bach

22.1 Open and Closed Symbiotic Systems

Two major symbiont transmission modes exist, horizontal (or environmental) and vertical (usually, through the female germline) (reviewed in Bright and Bulgheresi 2010; Sachs et al. 2011). In the first case, host reproduction leads to aposymbiotic descendants, which at a later life stage are colonized by a facultative symbiont from the environment. The latter can be as large as the sea or as small as an insect carcass. Invariably, protein–sugar interactions take place and mediate partners'

S. Bulgheresi (✉)
Department of Genetics in Ecology, University of Vienna, Althanstrasse 14, 1090 Vienna, Austria
e-mail: silvia.bulgheresi@univie.ac

E. Rosenberg and U. Gophna (eds.), *Beneficial Microorganisms in Multicellular Life Forms*,
DOI 10.1007/978-3-642-21680-0_22,

recognition at the host site first contacted by the symbiont. This may persist at the infection site or be translocated from there to a symbiont-housing organ. Sometimes the environmental pool is replenished by symbiont release so that, in this case, the symbiont can not only enter but also exit the system. Host environmental exposure, albeit time-restricted, endows the symbiont to be subjected to purifying selection and the symbiotic life-style, in this case, does not lead to symbiont genome reduction (see for example Moran et al. 2008, 2009).

In the case of vertical transmission through the female germline, prior to host reproduction symbionts are typically translocated from the symbiont-housing organ to the female gonad, resulting in symbiotic descendants; once in the progeny, symbionts are further translocated to a specific tissue. Strictly, vertically transmitted symbioses may be regarded as closed systems in the sense that the symbiont is constantly *recycled*: it can neither enter the system from the environment, nor rejoin it. This condition of captivity does not allow the symbionts to purge their genetic repertoire. These irreversibly accumulate mutations resulting in gene inactivation and loss and ultimately in genome size reduction (Moran et al. 2008, 2009).

Between these two antithetic modes, a wealth of possibilities exists. For instance, if a non-obligate symbiont is vertically transmitted, occasional horizontal transmission of new symbionts can occur from another host population within the same species (*intra*specific host-switching), from a different host species (*inter*specific host-switching) or even from a free-living population. Vertically transmitted symbionts that are either facultative or that only recently became obligate display moderate genome size reduction, have many pseudogenes and mobile DNA (insertion sequences and phages). Increasing disconnection from a free-living pool will eliminate both of these kinds of genetic elements (Mira et al. 2001; Moran et al. 2004).

22.2 The Challenge of Horizontally Transmitted Symbioses: Recognizing Each Other

The choice of the appropriate partner can be based on two non-mutually exclusive molecular mechanisms: the presence or absence of one gene product or on minor variations of one gene product (reviewed in Chaston and Goodrich-Blair 2010; Mandel 2011). The most stunning example of the former molecular mechanism is the *Vibrio fischeri* Regulation of Symbiotic Colonization sensor (*rsc*S) gene (Mandel et al. 2009). It encodes a membrane-bound two-component sensor kinase that acts upstream of the response-regulator SypG. This transcriptional regulator – in turn – facilitates transcription of the exoploysaccharide locus *sypA-R* that is required for *V. fischeri* aggregation at the entrance of the developing light organ. Comparison of the genomes of the squid-specific *V. fischeri* strain with that of the pinecone fish-specific strain revealed that *rsc*S gene is exclusively found in the former strain. Nevertheless, ectopic expression of the *rsc*S gene in the pinecone-specific strain enables the heterologous symbiont to colonize the squid as efficiently as the squid-specific strain. Similarly, the *nilABC* operon confers on the bacterial symbiont *Xenorhabdus nematophila* the competence to colonize the insect-killing nematode *Steinernema carpocapsae* (Cowles and Goodrich-Blair 2008). Historically, this was the very first example of a single genetic locus, an operon in this case, affecting the host range of an animal-associated bacterium. It is predicted to encode an integral inner membrane protein, an outer membrane beta-barrel protein and an outer membrane lipoprotein, respectively (Cowles and Goodrich-Blair 2004). The phylogenetically related species *X. poinarii* and *X. bovienii*, together with six other examined so far, do not possess the operon encoding for these three structural proteins and thus they cannot colonize *S. carpocapsae*, unless they express it ectopically (Cowles and Goodrich-Blair 2008).

Comparison of the two above described specificity determinants, the *rcs*S gene and the *nil* operon, with their adjacent genetic (and phylogenetic) environment suggest that they both have been acquired by lateral gene transfer (Mandel et al. 2009; Cowles and Goodrich-Blair 2008). In line with the idea that environmental transmission resembles pathogenic infection, lateral gene transfer events appear also to have contributed to the evolution of pathogens such as *Yersinia pestis* (Hinnebusch et al. 2011).

Different from the above, legume–rhizobia recognition is primarily based on modifications of the Nod factor, a bacterial lipochito-oligosaccharide which interacts with legume receptors to induce species and/or strain-specific development of rhizobial-containing organs (nodules) (see for example Oldroyd and Downie 2008; Masson-Boivin et al. 2009). A single Nod factor modification is enough to mediate specificity: the absence of a sulfated group, for example, makes *Sinorhizobium meliloti* incapable of colonizing its native host alfalfa. On the other hand, it acquires the capability to nodulate the vetch (Faucher et al. 1988; Roche et al. 1991). Nevertheless, it should be noted that plant lectins could represent an *on–off* specificity system nested into "the logic of the Nod factor" as Mandel (2010) defined it. Indeed, if a pea lectin is ectopically expressed in the white clover, this can be colonized by pea-specific rhizobia, although the nodules are delayed and less efficient (Díaz et al. 1989; Van Eijsdenet al. 1995).

It has been argued that the presence/absence of a defined genetic element may be the specificity mechanism of choice in animal–microbe symbioses whereas "the logic of the Nod factor" prevails in plant–microbe symbioses (Mandel 2010). Nevertheless, it was recently found that the high sequence variability of a receptor, not bacterial but host-secreted, may mediate symbiont-specificity. Several isoforms of the Ca-dependent (C-type lectin, CTL) mannose-binding protein Mermaid are secreted by the marine nematodes *Laxus oneistus* and *Stilbonema majum*. These two species co-occur in the same microhabitat but carry two different 16 S rRNA phylotypes of sulfur-oxidizing bacteria on their surface. In particular, one type of carbohydrate recognition domain is exclusively expressed by *S. majum* and displays the highest agglutination activity toward *S. majum*-specific symbionts and the lowest toward *L. oneistus* symbionts. Also, the stony coral *Acropora millepora* and the lucinid mussel *Codakia orbicularis* were found to express several different isoforms of the Millectin and Codakin CTLs, but there are no reports yet of the functional significance of their diversity (Kvennefors et al. 2008; Gourdine and Smith-Ravin 2007). As we gain more and more evidence of selection for novel alleles (i.e. positive selection) at not only plant (De Mita et al. 2006, 2007) but also animal (Schwarz et al. 2008) loci encoding for surface receptors, we may discover that highly polymorphic host-symbiont recognition genes are not restricted to the former associations and that, even more importantly, are not necessarily a pathogenicity signature (Sachs et al. 2011).

22.3 The Challenge of Vertically Transmitted Symbioses: Deep Integration into Host Development

Perhaps the most meticulous way to ensure uninterrupted, trans-generational association of a bacterial symbiont has been described for the *Drosophila–Wolbachia* association (recently reviewed in Serbus et al. 2008; Fig. 22.1). Here, the symbiont is permanently present in the female germline stem cells (GSCs). Although – in this case – no symbiont translocation from or to the host reproductive organ is needed, persistent infection of the female GSC is all but a trivial task. GSC division produces a self-renewing daughter and a daughter that will differentiate into an oocyte (Fuller and Spradling 2007). *Wolbachia* are symmetrically distributed into both daughter cells. Although we largely ignore the underlying mechanisms, this equal partitioning enables *Wolbachia* to be stably transmitted into the oocytes throughout the life of the insect.

Better understood is the cell biology allowing *Wolbachia*, once inside the GSC daughter cell fated to become a mature oocyte, to be localized to the progeny germline. In early oogenesis, the prospective oocyte undergoes several rounds of mitosis with incomplete cytokinesis. All the resulting cells will become nurse cells, except the most posterior one that will give rise to the oocyte: *Wolbachia* are segregated therein by minus-end-directed motors (dyneins) traveling on microtubules. Later, as the oocyte matures, its cytoskeleton is rearranged so that plus-end-directed motors (kinesins) transport *Wolbachia* to the oocyte's posterior side. This guarantees symbiont integration into the germline of the progeny. Indeed, after the zygote nucleus multiplies into hundreds and after their migration to the periphery of the embryo, those found at the posterior pole are the first to be surrounded by plasma membranes and, most importantly, they are the ones fated to become germline cells nuclei. Therefore, after engaging with different microtubule-associated motor proteins, *Wolbachia* elegantly succeeds to be contained in the very first embryonic cells, the germline cells.

The integration of *Wolbachia* with, not merely the developmental program of the host, but its cell cytoskeleton and protein localization machineries goes even deeper. In contrast to the symmetric segregation observed in the GSCs, and also during the syncitial divisions, gastrulation and in symmetrically dividing epithelia,

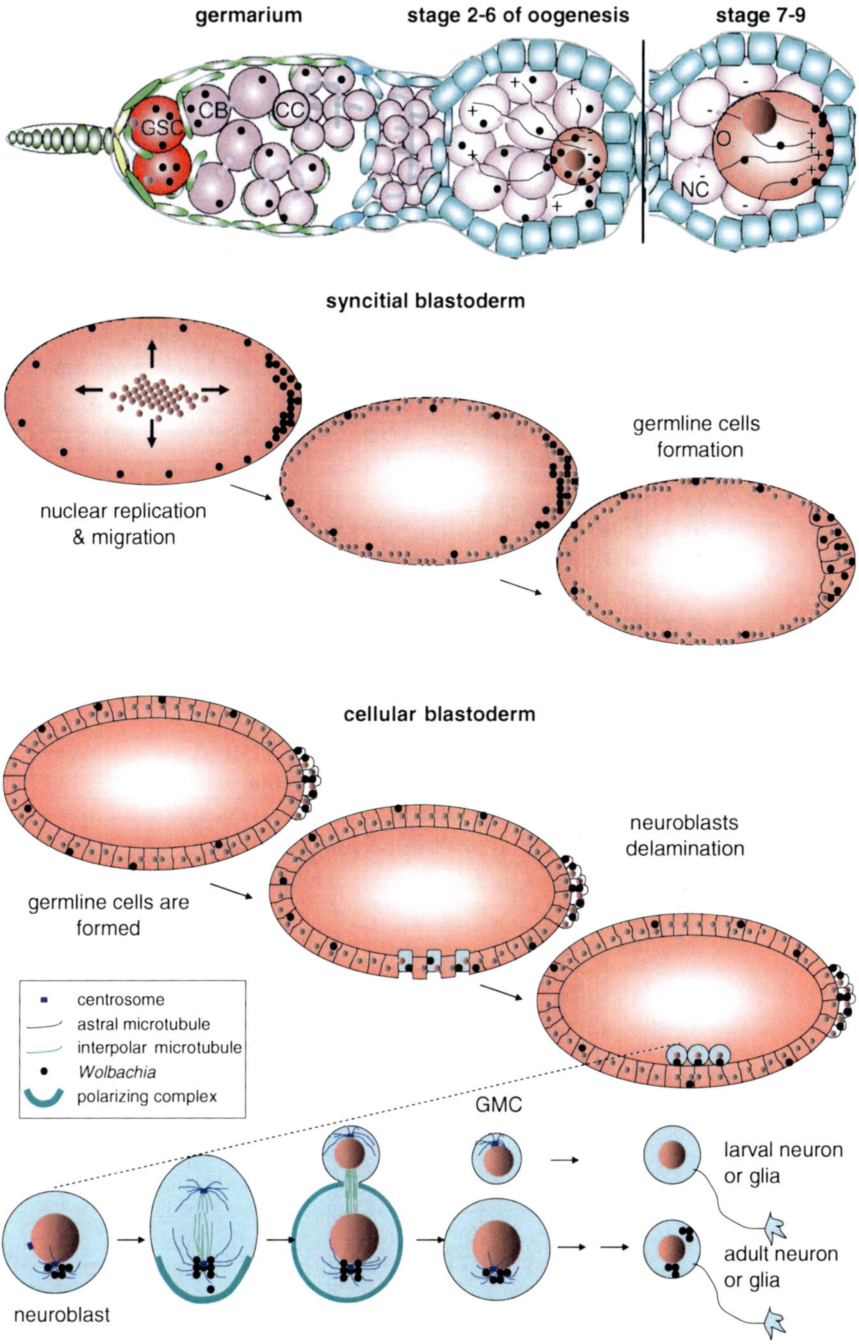

Fig. 22.1 Schematic view of *Wolbachia* localization throughout *Drosophila melanogaster* development. During germline stem cells (GSCs) mitosis, *Wolbachia* is partitioned between the

Wolbachia switch to a highly asymmetric segregation pattern in embryonic and larval neuroblasts (Albertson et al. 2009). Each neuroblast gives rise to a self-renewing daughter and a ganglion mother cell (GMC) that differentiates into larval glial and neuronal cells. Whereas the GMCs never inherit *Wolbachia*, the self-renewing neuroblasts retain the symbionts until insect adulthood. Although this asymmetric segregation probably results in overall less larval and brain cells being infected, it ensures that at least a few glia cells resulting from the final differentiation of the neuroblast are infected and reside in the adult central nervous system (reviewed in Dobson 2003). How can *Wolbachia* switch between two different segregation patterns that both rely on their tight association with astral microtubules and centrosomes? To undergo a switch from symmetric to asymmetric segregation, *Wolbachia* must either interact with different host factors at different ontogenetic stages (as exemplified by the alternative engagement to dynein and kinesin), or it is the function of the same host factors to change between stages (Albertson et al. 2009).

A recent, impressive finding is that *Wolbachia* segregation to the posterior pole of the oocyte occurs not only in arthropods, such as fruit flies, wasps, and mosquitoes (Serbus and Sullivan 2007; Serbus et al. 2008), but also in a host belonging to a different phylum, the filarial nematode *Brugia malayi* (Landmann et al. 2011). It is unclear whether *Wolbachia* accumulation at the posterior of fruit fly and nematode embryos to invade their germ cell precursors is due to convergent evolution or to the common developmental programs. If in the fruit fly, after occupying the future germline, *Wolbachia* must predispose their presence in selected adult neuronal lineages, in *B. malayi* the symbiont must manage accessing the adult chords which derive from the fusion of hypodermal cells. The first division of the *B. malayi* zygote is asymmetric and gives rise to an anterior AB blastomere and to a posterior P1 blastomere. Instead of invading the AB lineage, from which

---

Fig. 22.1 (continued) self-renewing stem cells and differentiating cytoblasts (CBs). As Wolbachia-bearing CBs progress through the germarium they become *Wolbachia*-infected egg chambers. The CB undergoes four rounds of mitosis with incomplete cytokinesis. This generates 16 interconnected cytocysts (CCs). Later on, all CCs will become nurse cells (NCs) except the most posterior one that will become the oocyte (O; red). After egg fertilization, the embryo undergoes multiple cycles of nuclear division so that hundreds of nuclei reside within the same cytoplasm (*syncitial blastoderm*). Subsequently, the nuclei migrate to the periphery and become associated with the cortex. Nuclei that migrate to the posterior region are surrounded by pole plasm and are fated to become germline cells (reviewed in Serbus et al. 2008). Embryonic neuroblasts are selected from a neuroectodermal layer and delaminate inwards. They divide asymmetrically along an apical–basal axis to regenerate a large self-renewing neuroblast and a small ganglion mother cell (GMC) (Albertson and Doe 2003; Kaltschmidt et al. 2000). The GMC undergoes an additional division round to give rise to neurons and glia (Goodman and Doe 1993). The evolutionarily conserved PAR protein complex is localized apically. Upon neuroblast delamination, the PAR complex mediates mitotic spindle orientation along the apical–basal axis (Kraut et al. 1996). Several rounds of asymmetric cell divisions follow. This figure is not to scale. Embryos are represented with their anterior pole to the left and the posterior to the right. In the schematic view of the neuroblast cell division apical is down and basal is up (Adapted from Niki et al. 2006 and Albertson et al. 2009) most hypodermal blastomeres – but no germ cells! – arise, *Wolbachia* are segregated to the posterior P1 blastomere. Only after embryo cellularization proceeds, the symbiont occupies a specific P1-derived blastomere (the C blastomere) which also produces some hypodermal cells. Landmann and colleagues (2010) speculate that during the evolution of the *Wolbachia*–nematode interaction, the bacteria followed conserved posterior determinants to ensure transmission to the germline, and only later on acquired an affinity for a specific blastomere to gain access to a defined adult somatic tissue. It is remarkable that in both the *B. malayi* zygote and the fruit fly neuroblasts *Wolbachia* colocalize with the PARtitioning defective (PAR) protein system (Goldstein and Macara 2007) and that this highly conserved and fundamental cell polarization system might be responsible for both segregation into fruit fly neuroblasts (Albertson et al. 2009) and to the posterior moiety of the nematode zygote (Landmann et al. 2010).

22.4 Turning Everything Upside Down: How Do Extracellular Symbionts Transmit the Host to Their Progeny?

Up to here, we only considered how the *host* copes with the problem of ensuring that its progeny will engage with beneficial microbes. But these, themselves, have to reproduce and, especially if extracellular, must guarantee their daughter cells the physical contact with their host. Several bacterial symbionts display unusual binary fission and/or cellular morphology, although it is not clear as to which extent this is due to their respective hosts. The most exotic alternatives to binary fission have been described for vertebrate gut symbionts (reviewed in Angert 2005) and marine invertebrate ectosymbionts (reviewed in Ott et al., 2004a, b). In both cases, bacteria are threatened to lose contact with their hosts either because of peristaltic movements and epithelial shedding (in the former case) or because of the highly turbulent nature of the marine environment (in the latter). *Metabacterium polyspora* is a spore-forming bacterium that resides in the gastrointestinal (GI) tract of guinea pigs, which feed on their feces. The coprophagous nature of these vertebrates is a key feature. As *M. polyspora* must – obligatorily – cycle through the GI tract, a life stage capable of surviving passage through the mouth and stomach of the host is required. This upper GI-resistant life stage is represented by the mature spores, which only germinate once they have reached the milder environment of the small intestine. Here, some undergo binary fission, whereas most start to sporulate (Angert and Losick 1998), will complete sporulation in the caecum and – finally – after passing the small intestine are defecated. Only if ingested by a guinea pig, the *M. polyspora* life cycle will start all over. What clearly sets this bacterium apart from other spore-forming ones is that a single cell can form multiple spores. This is possible because the asymmetric cell division producing the forespore takes place at both poles, and because each of the two forespores can produce multiple forespores. In this particular case, recruitment of *M. polyspora* from the environment is not sufficient *per se* to perpetuate the symbiosis. Instead, evolution had to co-opt sporulation – a developmental program usually activated by extreme, stressful conditions – to become an essential component of the bacterial life cycle. A component that ensures both symbiont propagation and the re-association of its descendants with the guinea pig gut i.e. symbiosis stability.

Another case of multiple spore-forming gut resident is the segmented filamentous bacterium (SFB or *Arthromitus*; Klaasen et al. 1993). Each SFB derives from a single, holdfast-bearing cell embedded among the microvilli protruding from the distal ileum. This grows, divides and develops into an up to 1 mm-long multicellular filament that is anchored to the surface of the gut epithelial cell (Davis and Savage 1974; Erlandsen and Chase 1974). Nevertheless, given that the gut epithelium is constantly sloughed off, it is not a stable substrate and the filament must manage to resettle somewhere else. It does so by producing intracellular offspring and releasing it, as it dies. This offspring can either be a holdfast-bearing daughter cell, capable of re-colonizing the epithelium of the same gut, or a spore, capable of enduring dispersal and colonizing the gut of a naïve host (Chase and Erlandsen 1976).

In contrast to *M. polyspora* and SFB, which both produce spores, the surgeonfish gut-associated, unusually large *Epulopiscium* spp. produce many intracellular active daughter cells (reviewed in Angert 2005). *Epulopiscium* spp. are among the closest relatives of *M. polyspora* and their viviparous reproduction highly resembles endospore production by the latter. Another peculiarity of *Epulopiscium* spp. is that offspring development follows a circadian cycle correlated with the host feeding habits, which results in developmentally synchronized populations. We do not know yet why this bacterium used the building blocks of a dormancy program to evolve viviparity as a reproduction strategy. This possibly maximizes the chances for a plant- and detritus-eating fish to recruit its anaerobic intestinal biota by coprophagy (Clements 1997).

All the alternatives to binary fission described so far come from vertebrate gut symbionts, but as already mentioned, also a number of ectosymbionts of marine invertebrates (reviewed in Ott et al. 2004a, b) appear to have extraordinary reproductive strategies. The non-septate filaments attached with one pole to the surface of the nematode *Eubostrichus dianae* can be up to 120 μm long and contain up to 50 nucleoids. As for the closely related *Eubostrichus parasitiferus*, its filamentous, crescent-shaped symbionts are attached to the nematode surface with both ends that makes it even more difficult to predict how they can coordinate multiplication with host attachment. Similarly, up to 20 μm long spirochetal ectosymbionts can be found on the posterior pole of devescovinid flagellates inhabiting low termite guts (Strassert et al. 2009). One end of the spirochetal cell is inserted into a deep pouch formed by the plasma membrane of the flagellate. No evidence of binary fission is available and it is a mystery how all these three bacterial symbionts reproduce.

Finally, a very atypical binary fission is carried out by the ectosymbionts of the two marine nematode species *L. oneistus* (Polz et al. 1994), and *Robbea* sp.3 (Bayer et al. 2009), and of the colonial ciliate *Kentrophoros* (Fenchel and Finlay 1989). All these symbionts are rods and form a monolayer covering the invertebrate surface.

Instead of setting their division planes perpendicular to their long axis, they all set it parallel i.e. they divide *longitudinally*. This allows both daughter cells, and not only one, to keep contact with the host surface. Nevertheless, it should be noted here that longitudinal division was also reported for *endo*symbionts of the gutless oligochaete *Olavius* (Giere and Krieger 2001; reviewed in Bright and Giere 2005) and of the deep-sea mussel *Bathymodiolus puteoserpentis* (Zielinski et al. 2009). It is unclear what drove these two symbionts to evolve such an atypical binary fission.

Unfortunately, all the above-mentioned symbionts displaying no or non-canonical binary fission are not culturable. Moreover, we ignore how their respective free-living counterparts, if existing, divide. It is possible that they lost their capacity to switch between canonical and non-canonical binary fissions depending on their free-living or symbiotic condition. More efforts are needed to culture these symbionts or - at least – to search and characterize the morphology and reproduction mode of their environmental counterparts, if existing. This research focus will be fundamental to assess the role of the host in shaping the symbiont life cycle.

22.5 Some Dark Sides of Microbial Transmission

In the case of horizontally transmitted symbiosis we have at least started to grasp the basic principles of partners' recognition. More obscure is the molecular base of tissue tropism of vertically transmitted symbionts in transit to their final destinations: how does a symbiont that just entered a developing host (or the germ cell giving rise to it!) find and recognize specific cell lineages that will house it in the adult? In the insect–*Wolbachia* association, apoptosis inhibition could be employed by the symbiont as a supplementary strategy (reviewed in Siozios et al. 2008), beside the already mentioned symmetric vs. asymmetric segregation switch (Sect. 23.2). The highly variable outer membrane protein (Wolbachia Surface Protein, WSP) can indeed activate Toll-mediated immune innate response, induce NO production and trigger inflammatory response by mammalian cells (Brattig et al. 2004; Porksakorn et al. 2007). Moreover, WSP can delay apoptosis in human cells typically involved in the innate immune response against pathogens (Bazzocchi et al. 2007). By structural and microevolutionary analysis, Baldo et al. (2010) predicted that the rapid evolutionary turnover of WSP loop motifs could help evading or inhibiting host immune response.

Another dark side of symbiont transmission is how ectosymbiotic ecdysozoans such as crustaceans, nematodes, and insects re-establish their symbiont coat upon molting (*intra*generational symbiont transmission). After they have shed their exoskeletons in order to grow and develop, do ecdysozoans need to re-recruit their symbionts environmentally, from scratch, by the same mechanisms used by the juvenile, aposymbiotic hosts? If so, is the genetic and cellular armoire needed for symbiosis establishment, at least partially, re-activated? Or, does a symbiont inoculum persist on the old exoskeleton and is it vertically transferred to the new exoskeleton? In this case partners' gene expression, metabolism and ultrastructure might be subjected to an *ontogenetic* rhythm similar to the circadian one observed in the squid–*Vibrio* symbiosis (Wier et al. 2010). Here, the rhythm is driven by the expulsion from the squid light-emitting organ of most of its luminous bacteria each day at dawn.

Metazoans such as cnidarians and flatworms, which not only may reproduce asexually but also regenerate symbiotic tissue, face an *intra*generational transmission challenge (similar to what is discussed above for ectosymbiotic ecdysozoans): how symbionts predispose their presence in freshly generated adult host tissue?

In the *Hydra viridis–Chlorella* association, the algal endosymbionts are transmitted by dividing gastrodermal cells in the developing bud. Since the algal symbionts are randomly distributed between daughter digestive cells, McAuley (1982) hypothesized that coordination of algal and digestive cells aids symbiosis perpetuation by ensuring the highest number of algae at host cell division. Although asexual reproduction by fragmentation or budding is very common in lower metazoans this is the only published study that deals with symbiont transmission in such cases of asexual reproduction: can we learn something from other systems? The trophosome of the mouth- and gutless deep sea tubeworm *Riftia pachyptila* consists of bacteriocytes (symbiont-housing cells). Although *Riftia* reproduces sexually, in the trophosome of adult tubeworms a steady proliferation of unipotent bacteriocyte stem cell takes place. However, no other host cells appear to differentiate into bacteriocytes (Pflugfelder et al. 2009). This is also true for all other known animal bacteriocytes such as those of aphids (Douglas and Dixon 1987; Miura et al. 2003; Braendle et al. 2003) or cockroaches (Lambiase et al. 1997): their determination and infection occurs only once in early host development and no aposymbiotic bacteriocytes are detectable after insect hatching. We do not know – at present – if proliferation of infected bacteriocytes is the only gateway to symbiotic tissue regeneration in the adult host. Alternatively, stem cells might differentiate into bacteriocytes continuously, throughout the host life cycle and be re-infected whenever symbiotic tissue is freshly produced. And – incidentally – another important and exciting research avenue will be the elucidation of which developmental factors make that symbiotic unit called “bacteriocyte” a bacteriocyte.

Acknowledgments I would like to thank Monika Bright, Ulrich Dirks, Salvador Espada, Harald R. Gruber-Vodicka, Niels R. Heindl, Wolfgang Miller, and Joerg A. Ott for inspiring discussions. The author was supported by the Austrian Science Fund (FWF) project P22470.

References

Albertson R, Doe CQ (2003) Dlg, Scrib and Lgl regulate neuroblast cell size and mitotic spindle asymmetry. Nat Cell Biol 5(2):166–170

Albertson R, Casper-Lindley C et al (2009) Symmetric and asymmetric mitotic segregation patterns influence Wolbachia distribution in host somatic tissue. J Cell Sci 122(Pt 24): 4570–4583

Angert ER (2005) Alternatives to binary fission in bacteria. Nat Rev Microbiol 3(3):214–224

Angert ER, Losick RM (1998) Propagation by sporulation in the guinea pig symbiont Metabacterium polyspora. Proc Natl Acad Sci USA 95(17):10218–10223
Bayer C, Heindl NR et al (2009) Molecular characterization of the symbionts associated with marine nematodes of the genus *Robbea*. Environ Microbiol Rep 1(2):136–144
Bazzocchi C, Comazzi S et al (2007) Wolbachia surface protein (WSP) inhibits apoptosis in human neutrophils. Parasite Immunol 29(2):73–79
Braendle C, Miura T et al (2003) Developmental origin and evolution of bacteriocytes in the aphid-Buchnera symbiosis. PLoS Biol 1(1):E21
Brattig NW, Bazzocchi C et al (2004) The major surface protein of Wolbachia endosymbionts in filarial nematodes elicits immune responses through TLR2 and TLR4. J Immunol 173(1): 437–445
Bright M, Bulgheresi S (2010) A complex journey: transmission of microbial symbionts. Nat Rev Microbiol 8(3):218–230
Bright M, Giere O (2005) Microbial symbiosis in Annelida. Symbiosis 38(1):1–45
Chase DG, Erlandsen SL (1976) Evidence for a complex life cycle and endospore formation in the attached, filamentous, segmented bacterium from murine ileum. J Bacteriol 127(1):572–583
Chaston J, Goodrich-Blair H (2010) Common trends in mutualism revealed by model associations between invertebrates and bacteria. FEMS Microbiol Rev 34(1):41–58
Clements KD (1997) Fermentation and gastrointestinal microorganisms of fishes. In: Mackie RI, White BA (eds) Gastrointestinal microbiology: gastrointestinal ecosystems and fermentations, vol 1. Chapman and Hall, New York, pp 156–198
Cowles CE, Goodrich-Blair H (2004) Characterization of a lipoprotein, NilC, required by Xenorhabdus nematophila for mutualism with its nematode host. Mol Microbiol 54(2):464–477
Cowles CE, Goodrich-Blair H (2008) The Xenorhabdus nematophila nilABC genes confer the ability of Xenorhabdus spp. to colonize Steinernema carpocapsae nematodes. J Bacteriol 190 (12):4121–4128
Davis CP, Savage DC (1974) Habitat, succession, attachment, and morphology of segmented, filamentous microbes indigenous to the murine gastrointestinal tract. Infect Immun 10(4):948–956
De Mita S, Santoni S et al (2006) Molecular evolution and positive selection of the symbiotic gene NORK in Medicago truncatula. J Mol Evol 62(2):234–244
De Mita S, Ronfort J et al (2007) Investigation of the demographic and selective forces shaping the nucleotide diversity of genes involved in nod factor signaling in Medicago truncatula. Genetics 177(4):2123–2133
Dobson SL (2003) Reversing Wolbachia-based population replacement. Trends Parasitol 19(3):128–133
Erlandsen SL, Chase DG (1974) Morphological alterations in the microvillous border of villous epithelial cells produced by intestinal microorganisms. Am J Clin Nutr 27(11):1277–1286
Faucher C, Maillet F et al (1988) Rhizobium meliloti host range nodH gene determines production of an alfalfa-specific extracellular signal. J Bacteriol 170(12):5489–5499
Fenchel T, Finlay BJ (1989) *Kentrophoros*: a mouthless ciliate with a symbiotic kitchen garden. Ophelia 30:75–93
Fuller MT, Spradling AC (2007) Male and female Drosophila germline stem cells: two versions of immortality. Science 316(5823):402–404
Giere O, Krieger J (2001) A triple bacterial endosymbiosis in a gutless oligochaete (Annelida): ultrastructural and immunocytochemical evidence. Invert Biol 120(1):41–49
Goldstein B, Macara IG (2007) The PAR proteins: fundamental players in animal cell polarization. Dev Cell 13(5):609–622
Goodman CS, Doe CQ (1993) Embryonic development of the drosophila central nervous system. Cold Spring Harbor Laboratory Press, Cold Spring Harbor

Gourdine JP, Smith-Ravin EJ (2007) Analysis of a cDNA-derived sequence of a novel mannose-binding lectin, codakine, from the tropical clam Codakia orbicularis. Fish Shellfish Immunol 22(5):498–509

Hinnebusch BJ, Rudolph AE et al (2002) Role of Yersinia murine toxin in survival of Yersinia pestis in the midgut of the flea vector. Science 296(5568):733–735

Kaltschmidt JA, Davidson CM et al (2000) Rotation and asymmetry of the mitotic spindle direct asymmetric cell division in the developing central nervous system. Nat Cell Biol 2(1):7–12

Klaasen HL, Koopman JP et al (1993) Intestinal, segmented, filamentous bacteria in a wide range of vertebrate species. Lab Anim 27(2):141–150

Kraut R, Chia W et al (1996) Role of inscuteable in orienting asymmetric cell divisions in Drosophila. Nature 383(6595):50–55

Kvennefors EC, Leggat W et al (2008) An ancient and variable mannose-binding lectin from the coral Acropora millepora binds both pathogens and symbionts. Dev Comp Immunol 32(12): 1582–1592

Lambiase S, Grigolo A, Laudani U, Sacchi L, Baccetti B (1997) Pattern of bacteriocyte formation in *Periplaneta americana* (L.) (Blattaria: Blattidae). Int J Insect Morphol Embryol 26:9–19

Landmann F, Foster JM et al (2010) Asymmetric Wolbachia segregation during early Brugia malayi embryogenesis determines its distribution in adult host tissues. PLoS Negl Trop Dis 4(7):e758

Mandel MJ (2010) Models and approaches to dissect host-symbiont specificity. Trends Microbiol 18(11):504–511

Mandel MJ, Wollenberg MS et al (2009) A single regulatory gene is sufficient to alter bacterial host range. Nature 458(7235):215–218

Masson-Boivin C, Giraud E et al (2009) Establishing nitrogen-fixing symbiosis with legumes: how many rhizobium recipes? Trends Microbiol 17(10):458–466

McAuley PJ (1982) Temporal relationships of host cell and algal mitosis in the green hydra symbiosis. J Cell Sci 58:423–431

Mira A, Ochman H et al (2001) Deletional bias and the evolution of bacterial genomes. Trends Genet 17(10):589–596

Miura T, Braendle C et al (2003) A comparison of parthenogenetic and sexual embryogenesis of the pea aphid Acyrthosiphon pisum (Hemiptera: Aphidoidea). J Exp Zool B Mol Dev Evol 295(1):59–81

Moran NA, Plague GR (2004) Genomic changes following host restriction in bacteria. Curr Opin Genet Dev 14(6):627–633

Moran NA, McCutcheon JP et al (2008) Genomics and evolution of heritable bacterial symbionts. Annu Rev Genet 42:165–190

Moran NA, McLaughlin HJ et al (2009) The dynamics and time scale of ongoing genomic erosion in symbiotic bacteria. Science 323(5912):379–382

Niki Y, Yamaguchi T et al (2006) Establishment of stable cell lines of Drosophila germ-line stem cells. Proc Natl Acad Sci USA 103(44):16325–16330

Oldroyd GE, Downie JA (2008) Coordinating nodule morphogenesis with rhizobial infection in legumes. Annu Rev Plant Biol 59:519–546

Ott JA, Bright M et al (2004a) Marine microbial thiotrophic ectosymbioses. Oceanogr Mar Biol Annu Rev 42:95–118

Ott JA, Bright M et al (2004b) Symbioses between marine nematodes and sulfur-oxidizing chemoautotrophic bacteria. Symbiosis 36(2):103–126

Pflugfelder B, Cary SC et al (2009) Dynamics of cell proliferation and apoptosis reflect different life strategies in hydrothermal vent and cold seep vestimentiferan tubeworms. Cell Tissue Res 337(1):149–165

Polz MF, Distel DL et al (1994) Phylogenetic analysis of a highly specific association between ectosymbiotic, sulfur-oxidizing bacteria and a marine nematode. Appl Environ Microbiol 60(12):4461–4467

Porksakorn C, Nuchprayoon S et al (2007) Proinflammatory cytokine gene expression by murine macrophages in response to Brugia malayi Wolbachia surface protein. Mediators Inflamm 2007:84318

Roche P, Debelle F et al (1991) Molecular basis of symbiotic host specificity in Rhizobium meliloti: nodH and nodPQ genes encode the sulfation of lipo-oligosaccharide signals. Cell 67(6):1131–1143

Sachs JL, Essenberg CJ et al (2011) New paradigms for the evolution of beneficial infections. Trends Ecol Evol 26(4):202–209

Schwarz JA, Brokstein PB et al (2008) Coral life history and symbiosis: functional genomic resources for two reef building Caribbean corals, Acropora palmata and Montastraea faveolata. BMC Genomics 9:97

Serbus LR, Sullivan W (2007) A cellular basis for Wolbachia recruitment to the host germline. PLoS Pathog 3(12):e190

Serbus LR, Casper-Lindley C et al (2008) The genetics and cell biology of Wolbachia-host interactions. Annu Rev Genet 42:683–707

Strassert JF, Desai MS et al (2009) The true diversity of devescovinid flagellates in the termite Incisitermes marginipennis. Protist 160(4):522–535

Wier AM, Nyholm SV et al (2010) Transcriptional patterns in both host and bacterium underlie a daily rhythm of anatomical and metabolic change in a beneficial symbiosis. Proc Natl Acad Sci USA 107(5):2259–2264

Zielinski FU, Pernthaler A et al (2009) Widespread occurrence of an intranuclear bacterial parasite in vent and seep bathymodiolin mussels. Environ Microbiol 11(5):1150–1167

Chapter 23
Hydra Go Bacterial

Thomas C.G. Bosch, Friederike Anton-Erxleben, René Augustin, Sören Franzenburg, and Sebastian Fraune

Abstract This chapter provides an overview of how the basal metazoan *Hydra* serves as model for untangling and dissecting the fundamental principles underlying complex host–microbe interactions.

23.1 Introduction

Recent work has shown that *Hydra*, long known to provide insight into the origin of developmental processes, has an elaborate innate immune system for sensing bacterial invaders and shaping the community of associated microbes. Because these immune responses function much the same way in vertebrates, *Hydra* can help to answer questions about the role of these pathways in other species, including humans. In contrast to vertebrates, *Hydra* is associated with only a few specific bacterial species. This together with the fact that *Hydra* has one of the simplest epithelia in the animal kingdom with only two cell layers and few cell types and the fact that not only a fully sequenced genome is available but also numerous genomic tools including transgenesis makes *Hydra* an ideal model to study all the inputs, outputs and the interconnections of an organism and its associated community of microorganisms (Fraune and Bosch 2007, 2010; Bosch et al. 2009; Fraune et al. 2009).

T.C.G. Bosch • F. Anton-Erxleben • R. Augustin • S. Franzenburg • S. Fraune
Zoologisches Institut, Christian-Albrechts-Universität Kiel Olshausenstrasse 40, 24098 Kiel, Germany
e-mail: tbosch@zoologie.uni-kiel.de

E. Rosenberg and U. Gophna (eds.), *Beneficial Microorganisms in Multicellular Life Forms*, 313
DOI 10.1007/978-3-642-21680-0_23, © Springer-Verlag Berlin Heidelberg 2011

23.2 *Hydra*, a Model in Developmental Biology and Comparative Immunology

Hydra belongs to the Cnidaria, one of the most basal eumetazoan phylum which is a sister taxon to all Bilateria. The animal represents a classical model organism in developmental biology which was introduced by Abraham Trembley as early as (1744). Because of its simple body plan, having only two epithelial layers (an endodermal and ectodermal epithelium separated by an extracellular matrix termed mesogloea), a single body axis with a head, gastric region and foot, and a limited number of different cell types, *Hydra* served for many years as model in developmental biology to approach basic mechanisms underlying *de novo* pattern formation, regeneration, and cell differentiation. The genome of *Hydra magnipapillata* is relatively large (1,300 Mb) (Chapman et al. 2010; reviewed in Steele et al. 2011). Since up to 40% of the whole genome are transposable elements, this was interpreted as "a very dynamic genome" in which recombination events might occur even without sexual recombination. Whether this in combination with horizontal gene transfer and trans-splicing allows the immortal, constantly regenerating and asexually proliferating polyps to quickly adapt to changing environmental conditions, remains a matter of debate. In addition to the *Hydra magnipapillata* genome, a large set of expressed sequence tags (ESTs) is available at www.compagen.org (Hemmrich and Bosch 2008). Adding to these relatively rich data sets, in the last years additional genome and transcriptome sequences could be retrieved from related basal metazoans such as corals and *Nematostella* shedding new and bright light on the ancestral gene repertoire (Putnam et al. 2007; Srivastava et al. 2008). The accumulated data show that Cnidaria possess most of the gene families found in bilaterians and, therefore, have retained many ancestral genes that have been lost in *D. melanogaster* and *C. elegans* (Kortschak et al. 2003; Miller et al. 2005; Technau et al. 2005). Since the genome organization and genome content of Cnidaria is remarkably similar to that of morphologically much more complex bilaterians, these animals offer unique insights into the content of the "genetic tool kit" present in the Cnidarian–bilaterian ancestor. For analytical purposes, an important technical breakthrough in studies using basal metazoans was the development of a transgenic procedure allowing efficient generation of transgenic *Hydra* lines by embryo microinjection (Wittlieb et al. 2006). This not only allows functional analysis of genes controlling development and immune reactions but also in vivo tracing of cell behavior.

23.3 Microbe–Epithelial Interactions in *Hydra*

In *Hydra*, a single layer of ectodermal epithelial cells covered by a glycocalix represents the physical barrier toward the environment and a single layer of endodermal epithelial cells separates the body from the content of the gastric cavity.

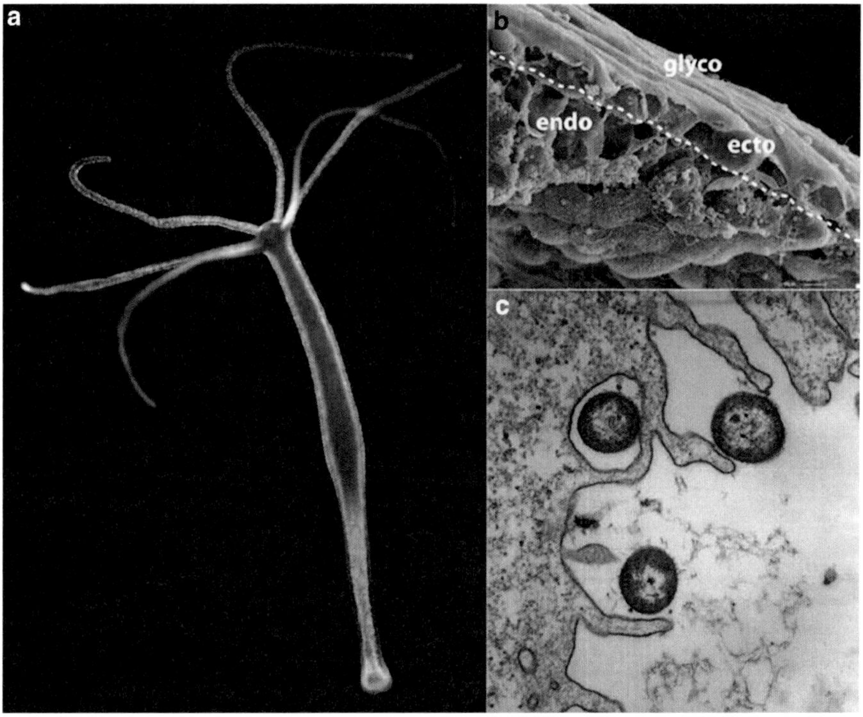

Fig. 23.1 Cnidarians are diploblastic animals (**a**) Live image of *Hydra oligactis* (Photo by S. Fraune) (**b**) Raster electron micrograph showing the ectodermal (*ecto*) and endodermal epithelium (*endo*); which are separated by an extracellular matrix (mesoglea – dashed line); a true mesoderm is missing. The apical part of *Hydra* ectodermal epithelial cells is covered by a glycocalix layer (*glyco*) (Photo by F. Anton-Erxleben). (**c**) Transmission electron micrographic of endodermal epithelial cells phagocyte bacteria from the gastric lumen (Panel C modified from Bosch et al. 2009)

The tube-like body structure resembles in several aspects the anatomy of the vertebrate intestine with the endodermal epithelium lining the gastric cavity and the ectodermal epithelium providing a permanent protection barrier to the environment (Fig. 23.1a–b). Although once considered simply a physical barrier, it is becoming increasingly evident that *Hydra's* epithelium is a crucial regulator of microbial homeostasis. Epithelial cells in both layers are multifunctional having both secretory and phagocytic activity, and most if not all innate immune responses are mediated by epithelial cells (Bosch et al. 2009; Jung et al. 2009). The interaction between bacteria and epithelium may differ in the ectoderm and endoderm because of morphological differences and the extent to which a glycocalix covers the ectodermal epithelial cells. Although there are no motile immune effector cells or phagocytes present in *sensu strictu*, endodermal epithelial cells not only contribute to digestion and uptake of food but also are capable of phagocytosing bacteria present in the gastric cavity (Fig. 23.1c). The detection of microbes by the host is achieved through families of pattern recognition receptors (PRRs) that recognize conserved molecular structures known as "Microbial Associated Molecular Patterns" (MAMPs). *Hydra´s* signaling receptors can be divided into two families: the Toll-like receptors (TLRs) and the nucleotide oligomerization domain (NOD)–like receptors (NLRs). The Toll-like receptor in *Hydra*, functions as a co-receptor with the MAMP recognizing Leucin-rich repeats and the signal transmitting TIR domain on two separated but interacting proteins (Bosch et al. 2009) (Fig. 23.2a). Homologous sequences to nearly all other components of the TLR pathway were identified (Bosch et al. 2009) including NFkB (Lange et al. 2011). RNAi knock-down experiments with *Hydra* TLR showed a drastic reduction of antimicrobial activity in the knock-down tissue compared to the wild type, which makes it apparent that antimicrobial activity relies directly on the activation of the TLR cascade (Bosch et al. 2009). In addition to recognizing MAMPs at the cell membrane, intracellular recognition of bacteria in *Hydra* is mediated by an unexpected large number (> 200) of cytosolic NOD-like receptors (Lange et al. 2011) (Fig. 23.2b). NLR proteins can signal through different multicomponent signal complexes to activate alternative pathways, including caspase activation and cell

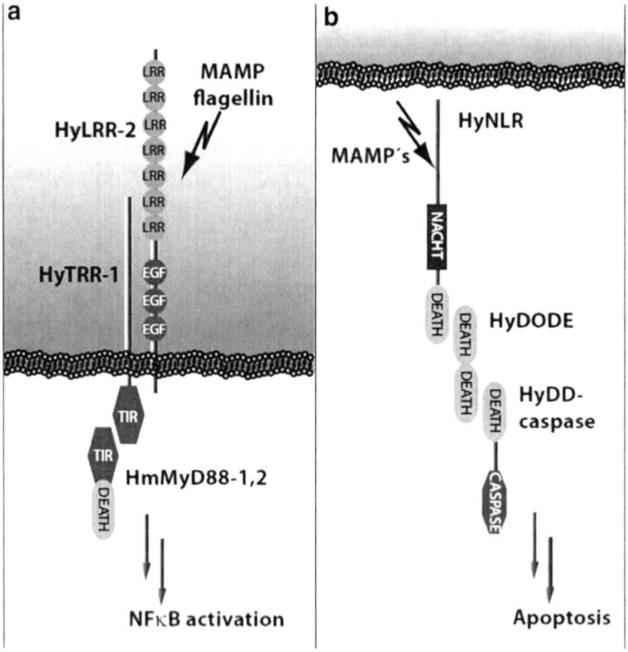

Fig. 23.2 Microbial associated molecular pattern recognition in *Hydra*. Scheme representing the molecules responsible for (**a**) extracellular and (**b**) cytosolic recognition and transmission of MAMPs in *Hydra* (**a**). The MAMP flagellin is detected by HyLRR-2 and the signal is transmitted into the cell via HyTRR-1. The signal is transduced further by conserved TLR-signalling pathway components. (**b**) Cytosolic MAMP´s are sensed by HyNLR´s and by homotypic interaction of death domain containing proteins programmed cell death is initiated (Modified form Lange et al. 2011)

death (Lange et al. 2011) (Fig. 23.2b). Upon detection of microbial infection and ligand binding, TLRs recruit adaptor proteins and cellular kinases that trigger downstream signaling cascades. These events require an adaptor protein MyD88 which acts through the adaptor TRIF, resulting in the transcriptional activation of downstream immune response genes. Accumulating evidence indicates that the inducible expression of antimicrobial factors depends on TLR or MyD88–dependent signaling. Up to now, we have isolated four families of antimicrobial peptides, including the Hydramacins, Arminins, Periculins, and serine protease inhibitors of the Kazal type (Bosch et al. 2009; Jung et al. 2009; Augustin et al. 2009a, b; Fraune et al. 2010). Interestingly, all antimicrobial peptides isolated so far are present in endodermal tissue only. That supports the view that the endodermal epithelium surrounding the gastric cavity is especially endangered by the regular uptake of food and that *Hydra´*s AMPs contribute to the chemical defense properties of this layer similar to AMPs in the human small intestine. In addition to endodermal epithelial cells, Periculin is also expressed in female germ cells and used for maternal protection of the embryo (Fraune et al. 2010).

23.4 The Microbial Community Is Largely Determined by the Host and not by the Environment

The observations summarized above demonstrate that *Hydra* has an effective innate immune system to interact with bacteria at the epithelial interface. What is the main function of this immune system? To keep out pathogens, or to allow the right community of microbes in? In principle, the community of microbes associated with *Hydra* could be determined by either the host or the environment. To test theories regarding the assembly of tissue-associated microbial communities and to sort out whether *Hydra´*s microbe composition depends primarily on the hosts or the environment, we have compared the microbiota of different *Hydra* species (Fraune and Bosch 2007). When analyzing different species we discovered that they differ greatly in their associated bacterial microbiota, although they were cultured under identical conditions. Comparing the cultures maintained in the laboratory for >30 years with polyps directly isolated from the wild revealed a surprising similarity in the associated bacterial composition. The significant differences in the microbial communities between the species and the maintenance of specific microbial communities over long periods of time strongly indicate distinct selective pressures within the epithelium (Fig. 23.3). When we constructed two phylogenetic trees, one of the *Hydra* hosts, and the other of their microbe communities, the microbe tree mirrored that for *Hydra*; the patterns of relationships among the microbe communities were identical to those among the *Hydra* species. This congruence establishes the *Hydra* host as the major force behind the makeup of microbe communities and shows that species specific microbe communities were shaped by divergences in host physiology over the course of evolution (Fraune and

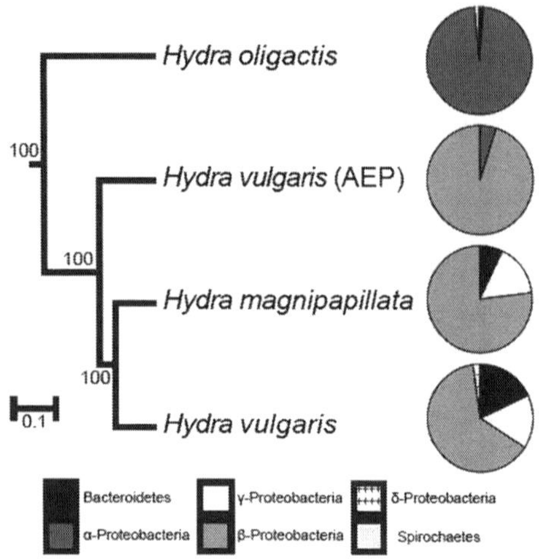

Fig. 23.3 *Hydra* polyps are colonized by species specific microbiota. Bacterial communities identified from different *Hydra* species displayed as overall composition of the analyzed bacterial communities (*pie diagrams*) and as a Jackknife environment cluster tree (weighted UniFrac metric). One hundred jackknife replicates were calculated, and each node was recovered with 99.9% (Scale bar: distance between the environments in UniFrac units) (Modified from Fraune et al. 2009)

Bosch 2007). New research in five closely related hominid species neatly supports our observations that the makeup of microbial communities is largely determined by the distinctive physiologies of closely related hosts (Ochman et al. 2010; PloS Biology).To decipher putative links between epithelial homeostasis and species-level bacterial phylotypes, we made use of the mutant strain sf-1 of *Hydra magnipapillata* which has temperature sensitive interstitial stem cells (Fraune et al. 2009). Treatment for a few hours at the restrictive temperature (28°C) induces quantitative loss of the entire interstitial cell lineage from the ectodermal epithelium while leaving both the ectodermal and the endodermal epithelial cells undisturbed. Intriguingly, 2 weeks after temperature treatment, when the tissue was lacking not only all interstitial cells as well as nematoblasts and most nematocytes, but in addition also had a reduced number of neurons and gland cells, the bacterial composition began to change drastically. Thus, changes in epithelial homeostasis cause significant changes in the microbial community, implying a direct interaction between epithelia and microbiota.

23.5 Apart from Their Antimicrobial Activities, AMPs Have Regulatory Functions in Host–Microbe Homeostasis

What is the driving force that leads to changes in microbiota composition? Because of their obvious ability to influence bacterial life, promising candidate molecules are AMPs. To investigate whether the ectotopic expression of an AMP may affect the number and composition of the colonizing microbiota at the ectodermal epithelial surface, transgenic *H. vulgaris* (AEP) expressing Periculin1a in ectodermal epithelial cells were generated. Comparing the bacterial load of these transgenic polyps with wild type control polyps revealed not only a significantly lower bacterial load in transgenic polyps overexpressing Periculin1a but also, unexpectedly, drastic changes in the bacterial community structure. Analyzing the identity of the colonizing bacteria showed that the dominant β-Proteobacteria decreased in number, whereas α-Proteobacteria were more prevalent. Thus, overexpression of Periculin causes not only a decrease in the number of associated bacteria but also a changed bacterial composition (Fraune et al. 2010). With the transgenic polyp overexpressing periculin we have created a new holobiont which is different from all investigated *Hydra* species. Thus, apart from their antimicrobial activities, AMPs are proposed to have additional regulatory functions. Future efforts will be directed toward analyzing the performance of this new Holobiont phenotype under different environmental conditions.

23.6 What Is It for? Toward Understanding the Functional Properties of the Microbiota

If all species are found to be associated with specific and beneficial microbes (Fraune and Bosch 2007, 2010), we need to find out why this is so. Do microbes give an advantage in certain environments or protection against certain infections? Are hosts deprived of their microbiota handicapped? To investigate the effect of absence of microbiota in *Hydra* we have produced gnotobiotic polyps which are devoid of any bacteria. While morphologically no differences could be observed to control polyps, *Hydra* lacking bacteria suffer from fungal infections unknown in normally cultured polyps (pers. observation). Thus, the beneficial microbes associated with *Hydra* appear to produce powerful antifungal compounds. Additional support for a fungal defense function of the bacteria associated with *Hydra* comes from analyzing the pathogenicity of oomycetes (*Saprolegnia* spec.). Many *Saprolegnia* species cause economic and environmental damage due to their ability to infect a wide range of plants and animals (Philips et al., 2008) causing saprolegniosis. Under standard culture conditions, *Saprolegnia* does not infect *Hydra*. Although zoospores are capable of attaching to the glycocalyx, germination appears to be inhibited. Under experimental conditions following, e.g., tissue dissociation and reaggregation in antibiotics-containing water, zoospores do not only attach but also germinate, infect and destroy the animals (pers. observation). Thus, severe defects in the epithelial barrier and the absence of an intact microbial community facilitate the germination of the spores pointing to a role of beneficial microbes in fungal defense. The biochemical characterization of the antifungal factors produced by certain bacteria associated with *Hydra* or other animal hosts might become an equally important effort to the discovery that toxins produced by fungi could kill bacteria that cause human disease. Another reason why hosts are always associated with specific microbiota may be found during development. As summarized elsewhere (Fraune and Bosch 2010), there is growing evidence that the microbial community affects animal development and in some cases is indispensable since it actually helps to orchestrate body plans early in life. For instance, although differentiation of the gut can be initiated in germ-free mice, signals from bacterial symbionts are required for the completion of differentiation. One of the hallmarks of *Hydra* is its rapid turnover of cells along the single body axis. The rate of proliferation is at least partly determined by Wnt signaling (Gee et al. 2010). Since a recent study from the Guillemin lab (Cheesman et al. 2010) has shown that in zebrafish one of the resident bacteria secretes a signal that acts on epithelial cells to promote the accumulation of b-catenin, a key component of the Wnt signaling pathway. Do microbes in *Hydra* promote epithelial proliferation, as they do in zebrafish?

23.7 Facts, Hypotheses, and Unresolved Issues

Any multicellular organism must be considered a metaorganism – comprised of the macroscopic host and synergistic interdependence with bacteria, archaea, fungi, viruses, and numerous other microbial and eukaryotic species including algal symbionts. The evolutionary dynamics within such a metaorganism and the involved molecular interactions are rather complex and often difficult to investigate experimentally. Here, we present *Hydra* as model to study host–microbe interaction in the context of a superorganismic organization. Molecular tools including transgenesis as well as rich genomic and transcriptomic resources make *Hydra* a valuable system, which can be manipulated experimentally. Because of its simplicity in body structure and the exclusive reliance on the innate immune system of epithelia, questions such as "How are these inter-kingdom alliances exactly coordinated, i.e., how do the interacting partners communicate, how do they coordinate their common interests and how do they solve potential conflicts of interests?" can now be addressed. The uncovered basic molecular machinery can be transliterated to more complex organisms including man since nearly all known molecules involved in innate immunity are present in *Hydra*. Equally pressing questions concern the stability and robustness of within-host microbial communities. The stability and function of these communities is enhanced by mutually beneficial interactions among the multiple species. However, how community stability is achieved is not yet understood. Here, a key point is to understand how far the overall function of the microbiota is influenced by individual as well as synergistic contributions of community members. The apparently crucial role of many microbes in development demonstrates that environmental and genetic information interact in determining the features of a phenotype. Now that genomic tools are available, it should be possible to estimate the extent to which overall host genome repertoires are shaped by the biotic microenvironment and to trace coevolutionary trajectories of host and symbiont genomes. In conclusion, comparative evolutionary approaches in key model organisms such as *Hydra* promise to allow the integration of information across many organisms, from multiple levels of organization and about entire systems to gain a new integrated understanding that incorporates more and more of the complexity that characterizes symbiotic systems.

Acknowledgment Research in our laboratory is supported in parts by grants from the Deutsche Forschungsgemeinschaft (DFG) and grants from the DFG Cluster of Excellence programs "The Future Ocean" and "Inflammation at Interfaces."

References

Augustin R, Siebert S, Bosch TCG (2009a) Identification of a kazal-type serine protease inhibitor with potent anti-staphylococcal activity as part of Hydra's innate immune system. Dev Comp Immunol 33:830–837

Augustin R, Anton-Erxleben F, Jungnickel S, Hemmrich G, Spudy B, Podschun R, Bosch TCG (2009b) Activity of the novel peptide arminin against multiresistant human pathogens shows the considerable potential of phylogenetically ancient organisms as drug sources. Antimicrob Agents Chemother 53:5245–5250

Bosch TCG, Augustin R, Anton-Erxleben F, Fraune S, Hemmrich G et al (2009) Uncovering the evolutionary history of innate immunity: the simple metazoan Hydra uses epithelial cells for host defence. Dev Comp Immunol 33:559–569

Chapman JA, Kirkness EF, Simakov O, Hampson SE, Mitros T et al (2010) The dynamic genome of Hydra. Nature 464:592–596

Cheesman SE, Neal JT, Mittge E, Seredick BM, Guillemin K (2010) Microbes and Health Sackler Colloquium: epithelial cell proliferation in the developing zebrafish intestine is regulated by the Wnt pathway and microbial signaling via Myd88. Proc Natl Acad Sci USA. doi:10.1073/pnas.1000072107

Fraune S, Bosch TCG (2007) Long-term maintenance of species-specific bacterial microbiota in the basal metazoan Hydra. Proc Natl Acad Sci USA 104:13146–13151

Fraune S, Bosch TCG (2010) Why bacteria matter in animal development and evolution. Bioessays 32:571–580

Fraune S, Abe Y, Bosch TCG (2009) Disturbing epithelial homeostasis in the metazoan Hydra leads to drastic changes in associated microbiota. Environ Microbiol 11:2361–2369

Fraune S, Augustin R, Anton-Erxleben F, Wittlieb J, Gelhaus C et al (2010) In an early branching metazoan, bacterial colonization of the embryo is controlled by maternal antimicrobial peptides. Proc Natl Acad Sci USA 107:18067–18072

Gee L, Hartig J, Law L, Wittlieb J, Khalturin K et al (2010) Beta-catenin plays a central role in setting up the head organizer in hydra. Dev Biol 340:116–124

Hemmrich G, Bosch TCG (2008) Compagen, a comparative genomics platform for early branching metazoan animals, reveals early origins of genes regulating stem-cell differentiation. Bioessays 30:1010–1018

Jung S, Dingley AJ, Augustin R, Anton-Erxleben F, Stanisak M et al (2009) Hydramacin-1, structure and antibacterial activity of a protein from the basal metazoan Hydra. J Biol Chem 284:1896–1905

Kortschak RD, Samuel G, Saint R, Miller DJ (2003) EST analysis of the cnidarian Acropora millepora reveals extensive gene loss and rapid sequence divergence in the model invertebrates. Curr Biol 13:2190–2195

Lange C, Hemmrich G, Klostermeier UC, López-Quintero JA, Miller DJ et al (2011) Defining the origins of the NOD-like receptor system at the base of animal evolution. Mol Biol Evol. doi:10.1093/molbev/msq349

Miller DJ, Ball EE, Technau U (2005) Cnidarians and ancestral genetic complexity in the animal kingdom. Trends Genet 21:536–539

Ochman H, Worobey M, Kuo C-H, Ndjango J-BN, Peeters M et al (2010) Evolutionary relationships of wild hominids recapitulated by gut microbial communities. Proc Natl Acad Sci USA. doi:10.1371/journal.pbio.1000546

Putnam NH, Srivastava M, Hellsten U, Dirks B, Chapman J et al (2007) Sea anemone genome reveals ancestral eumetazoan gene repertoire and genomic organization. Science 317:86–94

Srivastava M, Begovic E, Chapman J, Putnam NH, Hellsten U et al (2008) The Trichoplax genome and the nature of placozoans. Nature 454:955–960

Steele RE, David CN, Technau U (2011) A genomic view of 500 million years of cnidarian evolution. Trends Genet 27(1):7–13

Technau U, Rudd S, Maxwell P, Gordon PM, Saina M et al (2005) Maintenance of ancestral complexity and non-metazoan genes in two basal cnidarians. Trends Genet 21:633–639

Trembley A (1744) Mémoires, Pour Servir à l´Histoire d´un Genre de Polypes d´Eau Douce, *à Bras en Frome de Cornes*. Verbeek, Leiden

Wittlieb J, Khalturin K, Lohmann JU, Anton-Erxleben F, Bosch TCG (2006) Transgenic Hydra allow in vivo tracking of individual stem cells during morphogenesis. Proc Natl Acad Sci USA 103:6208–6211

Steele RE, David CN, Technau U (2011) A genomic view of 500 million years of cnidarian evolution. Trends Genet. Jan 27(1):7–13

Chapter 24
The Hologenome Concept

Eugene Rosenberg and Ilana Zilber-Rosenberg

Abstract The hologenome theory of evolution considers the holobiont (the animal or plant with all of its associated microorganisms) as a unit of selection in evolution. The hologenome is defined as the sum of the genetic information of the host and its microbiota. The theory is based on four generalizations, each of which is supported by a large body of empirical data: (1) All animals and plants establish symbiotic relationships with microorganisms; often the genetic information of the diverse microbiota exceeds that of the host. (2) Cooperation between the host and the microbiota contributes to the fitness of the holobiont. (3) Variation in the hologenome can be brought about by changes in either the host or the microbiota genomes; under environmental stress, the symbiotic microbial community can change rapidly by a variety of mechanisms including microbial amplification, horizontal gene transfer, and acquisition of new microorganisms from the environment. (4) Symbiotic microorganisms are transmitted between generations. These points taken together suggest that the genetic wealth of diverse microbial symbionts can play an important role both in adaptation and in evolution of higher organisms. During periods of rapid change in the environment, the diverse microbial symbiont community can aid the holobiont in surviving, multiplying, and buying the time necessary for the host genome to evolve. The distinguishing feature of the hologenome theory is that it considers all of the diverse microbiota associated with the animal or the plant as part of the evolving holobiont. The hologenome theory contains Lamarckian aspects within a Darwinian framework, accentuating both cooperation and competition within the holobiont and with other holobionts.

E. Rosenberg (✉) • I. Zilber-Rosenberg
Department of Molecular Microbiology and Biotechnology, Tel Aviv University, 69978 Israel
e-mail: eros@post.tau.ac.il; ilany60@netvision.net.il

E. Rosenberg and U. Gophna (eds.), *Beneficial Microorganisms in Multicellular Life Forms*,
DOI 10.1007/978-3-642-21680-0_24,

24.1 Introduction

Animals and plants arose from microorganisms and have remained in close association with them ever since. These associations, or symbioses, take many forms, mostly different levels of mutualism, where both the host and the symbiont benefit from the interaction, and to a much smaller degree – parasitism, where the symbiont benefits and the host suffers damage. These types of symbioses may change under different local conditions. Until recently, studies on symbiosis have concentrated on a single primary symbiont and its host. However, with the advent of molecular (culture-independent) techniques in microbiology during the last 15 years, it is now clear that all animals and plants live in close association with hundreds or thousands of different microbial species. In many cases, the number of symbiotic microorganisms and their combined genetic information far exceed that of their host. In the last few years, it has been demonstrated that these diverse microbiota with their large microbiomes play a remarkable role in the lives of animals and plants.

Evolutionary developmental biology is based on the principle that evolution arises from hereditable changes in development (Gilbert et al. 2010). In the past, the focus of these changes has been on the host genome (genetic and epigenetic) and occasionally on the genome of a specific primary symbiont (co-evolution). In this chapter, we shall first present the hologenome concept and then elaborate on the major points of the concept, using information presented in previous chapters as well as the recent literature.

24.2 The Hologenome Concept

The hologenome theory of evolution considers the holobiont with its hologenome, acting in consortium, as a unit of selection in evolution (Rosenberg et al. 2007; Zilber-Rosenberg and Rosenberg 2008; Sharon et al. 2010). The holobiont has been defined as the host organism and all of its symbiotic microbiota (Rohwer et al. 2002). The hologenome is the sum of the genetic information of the host and its microbiota. The hologenome theory posits that (1) all animals and plants harbor abundant and diverse microorganisms, acquiring from their host a sheltered and nutrient-rich environment, (2) these microbial symbionts affect the fitness of the holobiont and in turn are affected by it, (3) variation in the hologenome can be brought about by changes in the host genome, the microbial population genomes (microbiome), or both, (4) these variations, including those of the microbiome, can be transmitted from one generation to the next with fidelity and thus may also influence evolution of the holobiont. These four general principles taken together suggest that the genetic wealth of diverse microbial symbionts can play an important role both in adaptation and in evolution of higher organisms. During periods of rapid changes in the environment, the diverse microbial symbiont community can aid the holobiont in surviving, multiplying, and buying the time necessary for the host genome to evolve. A distinguishing feature of the hologenome theory is that it considers all of the diverse microbiota associated with the animal or the plant as part of the evolving holobiont. Thus, the hologenome theory fits within the framework of the "superorganism" proposed by Wilson and Sober (1989). The four principles underlying the hologenome theory will now be elaborated upon, using information presented in previous chapters as well as the recent literature.

24.3 All Animals and Plants Harbor Abundant and Diverse Microbial Symbionts

To begin, it is useful to provide some common definitions. The term "symbiosis" was first coined by Anton de Bary in the mid nineteenth century as "the living together of different species." This broad definition is generally accepted and easily comes to terms with the hologenome theory. The symbiotic system is usually constructed from a large partner termed the "host" and smaller partners called "symbionts." This arbitrary division by dimension between the host and the symbiont may not fit such systems as the holobiont because size can also be measured by cell number or by genome size, and in the case of many holobionts, the microbiota outnumber their host cells. In spite of these limitations, we will continue using the classical terms. Endo- and exosymbionts refer to those living inside or outside host cells, respectively.

The number of microbial symbionts of various hosts was initially estimated by viable counts. With the introduction of fluorescent microscopy coupled with staining techniques, we now know the viable counts are one to three magnitudes lower than the total counts. For example, human skin contains ca. 2×10^{10} and 1×10^{12} viable and total counts, respectively (Grice et al. 2009), and coral tissues contain 1×10^6 and 2×10^8 viable and total counts per cm^2, respectively (Koren and Rosenberg 2006). In most animals, including man, the largest numbers of symbionts are found in the digestive tract. Often, the number of symbiont cells greatly exceeds that of the host, e.g., humans contain ca. 10^{14} bacteria and 10^{13} host cells. In general, animals contain on an average 10^{9} bacteria per g wet weight, which interestingly is similar to rich soil.

Because the vast majority of microorganisms that have been observed on or in animal and plant tissues can not be cultured at present, current research on the diversity of microorganisms associated with a particular species relies primarily on culture-independent DNA-based technology (Eckburg et al. 2005). Although censuses of microorganisms associated with different animal and plant species are only in an early stage, certain interesting generalizations have emerged: (1) The diversity of microbial species associated with a particular animal or plant species is high (Table 24.1). (2) The host associated microbial community is very different from the community in the surrounding environment (Chelius and Triplett 2001;

Table 24.1 Examples of microbial species associated with specific animals and plants

Host	Minimum number of microbial species	Reference
Invertebrates		
Drosophila melanogaster	74	Mateos et al. (2007)
Tunicate	30	Mart´nez-Garc´a et al. (2007)
Marine sponge	1,694	Webster et al. (2001)
Coral *Oculina patagonica*	400	Koren and Rosenberg (2006)
Termite gut	367	Hongoh et al. (2005)
Earthworm gut	87	Wust et al. (2011)
Vertebrates		
Reindeer rumen	700	Sundset et al. (2007)
Great ape	8,914	Ochman et al. (2010)
Human gut	40,000	Frank and Pace (2008)
Human skin	1,200	Grice et al. (2010)
Human oral	1,179	Dewhirst et al. (2010)
Bovine rumen	341	Edwards et al. (2004)
Pig gut	375	Lesser et al. (2002)
Plants		
Leaf of the plant *Trichilia catigua*	617	Lambais et al. (2006)
Stems of soybean	200	Ikeda et al. (2010)
Seed of Norway spruce	46	Cankar et al. (2005)
Roots of *Zea mays*	74	Chelius and Triplett (2001)

Frias-Lopez et al. 2002; Rohwer et al. 2002; Sharp et al. 2007). (3) In some cases it has been shown that similar, but not identical, microbial populations are found on the same species that are geographically separated, while different populations are found on different species at the same location (Rohwer et al. 2002; Lambais et al. 2006; Fraune and Bosch 2007). (4) Different microbial communities often dominate different tissues of the same organism (Tannock 1995; Koren and Rosenberg 2006; Dethlefsen et al. 2007). (5) In several cases where a large diversity of associated bacterial species exists, certain bacterial groups dominate. For example, the human gut has been reported to contain thousands of bacterial species (Frank and Pace 2008), but only two divisions, Bacteroidetes and Firmicutes, make up 99% of the total bacterial population (Ley et al. 2006a).

The association of microorganisms with hosts can take many different forms. Some may be transitory and have little effect on adaptation or evolution of the holobiont. At the other extreme, there are several examples of well-studied long-lasting interactions (e.g., the rumen system) between the host and the microorganism, which can lead to total dependence of one on the other. Between these two extremes lies a gradient of interactions of varying strengths, including pathogenesis (see Chap. 20). It should be noted that the study of the interactions between hosts and their associated microorganisms is complicated by the fact that most associated microorganisms have not been cultured and that most of the interactions involve more than one microorganism with the host, e.g., the human gut microbiota (Ley et al. 2006a) and the co-aggregating bacteria in the human mouth

(Foster and Kolenbrander 2004). The idea that microbial diversity can play a critical role under conditions of fluctuating environments has been referred to as the insurance policy hypothesis (Yachi and Loreau 1999).

Let us consider some factors that determine the diversity of microorganisms associated with the holobiont. We shall first consider those characteristics that would result in high diversity. Many microorganisms are specialists. Given that hosts provide a variety of different niches that can change with the developmental stage of the host, the diet and other environmental factors, a diverse microbial community is established, with different microbial strains filling the different niches. Another factor that contributes to bacterial diversity is bacteriophages. It has been established that high concentrations of bacteriophages are present in animal and plant tissues (Breitbart et al. 2003; see Chap. 15). If any microorganism becomes too abundant, it may be lysed by bacteriophages. This concept, referred to as the "kill the winners" hypothesis (Thingstad and Lignell 1997), is supported by mathematical models of the bacteria: bacteriophage dynamics (Weitz et al. 2005). On the other hand, there exist opposing forces that limit the number of strains that can survive and become established in the holobiont, notably the innate and adaptive immune systems. The innate or nonspecific immune system is the first line of defense and includes physical barriers, antimicrobial molecules, enzymes, specific binding proteins for microbial attachment (e.g., the peptidoglycan-binding protein and lectin complement system), production of reactive oxygen species and phagocytes (Iwanaga and Lee 2005). Interestingly, resident symbiotic bacteria are also part of the innate immune system – by occupying potential adhesion sites and by producing antibiotics (Ritchie 2006). The adaptive or the specific immune system in vertebrates includes specific recognition of "foreign" microorganisms, generation of responses to eliminate these microorganisms, and development of immunological memory to hasten the response to subsequent infections with the same microbe. In essence, the immune system of the host is responsible for both limiting the types of microorganisms that can survive within the host and recognizing and accommodating the normal microbiota, thereby regulating the kinds of microorganisms that can reside in the holobiont. Also, it is important to note that plants have evolved myriad phytochemicals, whose purpose is to prevent infection by harmful microorganisms (Wallace 2004) and enable coexistence with beneficial ones (Smith et al. 1999; Stougaard 2000; Wilkinson 2001).

24.4 Cooperation Between the Host and the Microbiota Contributes to the Fitness of the Holobiont

Natural selection is the central concept of Darwinian theory– the fittest survive and spread their advantageous traits through populations, but being the "fittest" is not an absolute property because it varies with environmental influences. Considering the holobiont as a unit of selection in evolution, we argue that the cooperation between the normal microbiota and the host generally leads to improved fitness. In addition, the genetic diversity of the microbiota can extend the range of environments in which the holobiont can compete successfully. Several chapters in this book elaborate on some of the ways in which microorganisms contribute to the fitness of their host. Table 24.2 summarizes some of these symbioses as well as others taken from the literature.

Table 24.2 Contribution of symbionts to the fitness of the holobiont

Holobiont–microbiota	Microbial contribution	References
General		
All eukaryotes–mitochondria	Respiration	Margulis (1993)
Plants–chloroplasts	Photosynthesis	Chap. 23
Invertebrates		
Aphid–*Buchnera*	Provides essential amino acids and other nutrients	Chap. 2
Aphid–microbiota	Growth at high temperature; resistance to parasites	Russell et al. (2003)
Termite–microbiota in hindgut	Nitrogen fixation; humic acid and lignocellulose digestion	Chap. 1
Stinkbug midgut–*Burkholderia*	More efficient food utilization	Kikuchi et al. (2007)
Drosophila–Lactobacilli	Mating preference	Chap. 4
Arthropods–*Wolbachia*	Effects fertility and sex determination	Veneti et al. (2005)
Squid light organ–*Vibrio fischeri*	Camouflage against predators	McFall-Ngai (1999)
Squid nidamental gland–microbiota	Protection of eggs and embryos against pathogens	Barbieri et al. (2001)
Corals–microbiota	Photosynthesis; protection against pathogens; nitrogen fixation; chitin digestion	Chaps. 10 and 11
Sponges–microbiota	Breakdown of complex polymers; nitrogen cycling; protection against pathogens	Taylor et al. (2007)
Vertebrates		
Cow rumen–microbiota	Provides all nutritional needs from cellulose;	Dehority (2003)
Human gut–microbiota	Stimulation of immune system; protection against pathogens; angiogenesis and muscle thickness; vitamin synthesis;fiber breakdown; fat storage and obesity; metabolism; modulates brain development and behavior	Chaps. 14–22, Xu et al. (2007), Hooper et al. (2002), Ley et al. (2006), O'Hara and Shanahan (2006), and Heijtza et al. (2011)
Land plants–microbiota		
	Protection against pathogens	Chap. 8
	Promote growth	Chap. 6
	Nitrogen fixation	Chap. 5
	Supply minerals from soil	Wang and Qui (2006)

In several well-studied cases, neither the host nor the primary symbiont can survive without the other (absolute mutualism). For example, in the aphid–*Buchnera* symbiosis, the primary endosymbiotic bacterium has lost many genes required for independent growth, during evolution whereas the aphid partner depends on essential amino acids lacking in its diet that are synthesized and furnished by the symbiont (Baumann et al. 2006). While the aphid–*Buchnera* primary endosymbiosis is an example of absolute dependency, most of the symbiosis systems, as indicated in Table 24.2, are not based on life or death interactions, but rather the microbial partners that contribute in different degrees to the holobiont's well-being.

One of the classical techniques to establish how a microbe affects its host is to produce germ-free animals and plants. An early example was the demonstration by Beijerinck (1901) that the bacterium *Rhizobium* is responsible for nodule formation and nitrogen-fixation in leguminous plants. Sterile (germ-free) seeds neither produce the characteristic nodules nor fix nitrogen in the absence of the bacterium. Over 15,000 plant legume species form specific associations with nitrogen-fixing *Rhizobia* that involve the formation of root nodules. Studies on the legume–*Rhizobia* symbiosis have demonstrated that biochemical cross-talk is a prerequisite for nodule formation and that the hybrid molecule leghemoglobin (part plant, part bacterial) maintains the low oxygen tension required for nitrogen fixation (Jones et al. 2007).

Germ-free mice exhibit significant differences in gut development and function as compared with mice grown conventionally, i.e., possessing normal gut microbiota. The germ-free mice demonstrate enlarged caeca (Wostmann 1981), a slow digested food transit time (Abrams and Bishop 1967), altered kinetics of epithelia turn-over in the small intestine (Savage et al. 1981), an increased caloric intake Q5 (Wostmann et al. 1983), and a greater susceptibility to infection (Silva et al. 2004). The influence of microbiota on energy metabolism in germ-free conventionalized mice was observed within two weeks of the introduction of microbiota (B¨ackhed et al. 2004). It included microbial fermentation of polysaccharides not digested by the host, absorption of the microbially produced short-chain fatty acids, more efficient absorption of the monosaccharides from the intestine, conversion of breakdown products in the liver to more complex lipids and microbial regulation of host genes that promote fat deposition in adipocytes. These events were accompanied by lower food intake and higher metabolic rate. The cross-talk between microbiota and its mouse host regarding energy metabolism involves at least two demonstrated mouse factors (B¨ackhed et al. 2007). In addition, it has been shown in mice (Ley et al. 2005; Turnbaugh et al. 2006) and humans (Ley et al. 2006b) that obesity is correlated with a higher proportion of microorganisms from the Bacteroidetes division as compared to those from the Firmicutes division. Moreover, in obese humans a gradual transition was observed from the obese microbiota to the lean microbiota during the course of a restrictive energy intake (Ley et al. 2006b).

Although there are a wide variety of different mechanisms by which microorganisms benefit their host, certain generalizations can be derived from the literature:

1. Symbionts protect animals against pathogens (e.g., Silva et al. 2004; Reshef et al. 2006; Ritchie 2006; Jaenike et al. 2010; Turnes et al. 1989; Scarborough et al. 2005).
2. Symbionts aid in the digestion of complex polysaccharides (e.g., B¨ackhed et al. 2004; Dehority 2003; Taylor et al. 2007; Coyne et al. 2005; Martens et al. 2009; Brune and Stingl 2006).
3. Symbiotic prokaryotes provide utilizable nitrogen via nitrogen fixation to animals, plants, fungi, and protists that live on carbohydrate-rich diets (e.g., Fiore et al. 2010; Nardi et al. 2002; Mohamed et al. 2008; Kneip et al. 2007; Lilburn et al. 2001; Iniguez et al. 2004).
4. Symbionts often provide essential amino acids and vitamins to animals that live on nutritionally poor or unbalanced diets (e.g., Gündüz and Douglas 2009; Dale and Moran 2006; Skaljac et al. 2010; Ben-Yosef et al. 2010).

Pathogenic microorganisms are traditionally characterized as those that harm their animal or plant host by cell or tissue damage and sometimes death. Although pathogens represent only a small minority of the microorganisms that are associated with higher organisms, they have been studied extensively and much of what we know about host–microbial interactions has emerged from research on infectious diseases. One of the most important recent discoveries in this area is that many of the genes necessary for pathogen–host and symbiont–host interactions are located on similar mobile genetic elements (Hentschel et al. 2000), referred to as pathogenic and symbiotic islands, respectively. As discussed in the next section, genetic variability brought about by these elements plays an important role in the interactions of pathogens and symbionts with their hosts.

24.5 Genetic Variation in Holobionts

Variation is the raw material for evolution. According to the hologenome theory of evolution, genetic variation can arise from changes in either the host or the symbiotic microbiota genomes. Variation in host genome occurs during sexual reproduction and development, by recombination, chromosome rearrangements, mutation, and epigenetic variations. Variation in the microbiota occurs not only by these mechanisms but also by three other processes that are unique to the hologenome theory of evolution: microbial amplification, acquisition of novel strains, and horizontal gene transfer. These latter three processes can occur rapidly under environmental demand and are important elements in the adaptation, development, and evolution of animals and plants.

The first, microbial amplification is the most rapid and easy to understand mode of variation in holobionts. It involves changes in the relative numbers of the diverse types of associated microorganisms that can occur as a result of changes in environmental conditions. The holobiont is a dynamic entity with certain microorganisms multiplying and others decreasing in number as a function of local conditions. An increase in the number of a particular microbe is equivalent to gene amplification. Considering the large amount of genetic information encoded in the diverse microbial population of holobionts, microbial amplification is a powerful mechanism for adapting to changing conditions. In fact, changes of symbiont populations as a function of external factors are well documented in many biological systems. For example, when *Drosophila melanogaster* was transferred from a molasses-based medium to a starch-based medium, *Lactobacillus plantarum* was amplified ten-fold (Sharon et al. 2010). In cold blooded organisms, such as corals, bacterial species rise and fall with changes in temperature (Koren and Rosenberg 2006). Children on a high fiber diet had a high abundance of bacteria from the genus *Prevotella* and *Xylanibacter*, known to contain a set of bacterial genes for cellulose and xylan hydrolysis, whereas children on a high carbohydrate diet had abundant *Shigella* and *Escherichia* (De Filippo et al. 2010). Further support for amplification of certain bacteria following a change in diet comes from a study of infant gut microbiota (Koenig et al. 2010) in which ingestion of solid table foods caused a change in infant gut microbiota with sustained increase in the abundance of Bacteroidetes. In a study performed on humanized gnotobiotic mice, it was observed that a 1 day change in diet from high fiber to high fat brought about an immediate change in microbiota (Turnbaugh et al. 2009).

The second mechanism for introducing variation into holobionts is acquiring new symbionts from the environment. Animals and plants come in contact with billions of microorganisms during their lifetimes. It is reasonable to assume that occasionally, as a random event, one of these microorganisms will find a niche and become established in the host. Under the appropriate conditions, the novel symbiont may become more abundant and affect the phenotype of the holobiont. Unlike microbial amplification, acquiring new symbionts can introduce entirely new genes into the holobiont. Applied examples of such amplification, or in some circumstances of acquisition of novel strains – are prebiotics (see Chap. 21) and probiotics (see Chap. 22). By introducing specific strains of bacteria known to contribute to the health of the holobiont, one can achieve recovery from *Clostridium difficile*-associated diarrhea in humans (MacConnachie et al. 2009) and changes in metabolic characteristics (Laitinen et al. 2009). The third mechanism is the microbe–microbe interaction of horizontal gene transfer by which new traits can be transferred from microorganisms not generally associated with the holobiont to resident microbes. An example of the latter is the transfer of genes coding for porphyranases, agarases and associated proteins from a marine bacteria member of the Bacteroidetes to the human gut bacterium *Bacteroides plebeius* in Japanese population (Hehemann et al. 2010).

We suggest that once a beneficial genetic variation in a holobiont has occurred as a result of changes in the microbiota (in a single specific symbiont or in multiple symbionts), two general pathways may be possible for ensuring that any useful genetic information is conserved in future holobiont generations: (1) The microbial genes can be inserted into the host genome, as in the transfer of carotenoid biosynthetic genes from a fungus to aphids (Moran and Jarvik 2010), and/or (2) the host and microbe can undergo secondary changes that stabilize and benefit the interactional symbiosis. This latter kind of adaptation can occur in primary symbiosis as in corals and their algae (Fallowski et al. 1984) or in secondary symbiosis as in the bovine rumen which fosters the growth of anaerobic cellulose-degrading microorganisms, which benefit the host not only by conversion of the cellulose to utilizable fatty acids, but also by satisfying its protein and vitamin requirements (Dehority 2003).

24.6 Transmission of Symbionts Between Holobiont Generations

The hologenome theory of evolution relies on ensuring the continuity of partnerships between holobiont generations. Accordingly, both host and symbiont genomes must be transmitted with accuracy from one generation to the next. The precise modes of vertical transmission of host genomes are well understood and need not be discussed here. However, in recent years, it has become clear that microbial symbionts can also be transmitted from parent to offspring hosts by a variety of methods (see Chap. 24). McFall-Ngai (2002), in an insightful review on the influence of bacteria on animal development, divided the modes for maintaining symbionts faithfully between generations into two categories, transovarian and environmental transmission, while correctly acknowledging that there are numerous intermediate cases. We suggest that the numerous intermediate cases, in fact, best describe the large variety in modes of transmission which are known at present to reconstitute plant and animal holobionts. It is this continuum of modes of transmission from direct to indirect that makes it impractical to place them in specific categories. The molecular mechanisms that mediate symbiont attraction and accumulation, interpartner recognition and selection, as well as symbiont confrontation with the host immune system have recently been reviewed (Bright and Bulgheresi 2010).

Mitochondria and chloroplasts, which can be considered (extreme) symbionts, are transmitted by the most direct mode, namely, cytoplasmic inheritance. Direct transmission from parent to offspring also occurs with other symbionts where the microorganisms are in or on the reproductive cells. For example, in the aphid–Buchnera symbiosis, bacteria are intracellularly situated in bacteriocytes and are transferred to and transmitted via the eggs (Baumann et al. 2006). Direct contact is another slightly less direct mode of transmission demonstrated in mammals in which many of the symbionts are derived during passage through the birth canal or subsequently by close physical contact with parent or family and community members. Support for vertical transmission of microbiota in humans comes from studies which showed a greater similarity of microbiota within family members when compared with between families (Zoetendal et al. 2001) and to similarity between microbiota of vaginally delivered infants and their mother's virginal microbiota (Dominguez-Bello et al. 2010). Moreover, it has also been observed that a correlation exists between mother's BMI, weight and weight gain during pregnancy and infant's microbiota (Collado et al. 2010) implying a possible effect on fetal and child metabolic development. The conserved transmission of microbiota from parent to offspring for many generations has been used as a window into human migration (Devi et al. 2006).

Another slightly less direct mode of transmission used by many animals is feeding feces from the adult to the baby. For example, in the termite hindgut–microbiota symbiosis, feces from adult termites (containing abundant microorganisms) are fed to newly hatched juveniles by workers in the colony (Abe et al. 2000). In the case of the baby koala (Joey), the mother produces "pap," which is a form of feces, when the Joey is about 6 months old and begins weaning from milk to gumleaves. The Joey leaves the pouch to eat the pap which contains the bacteria necessary to digest gum leaves. In the bovine rumen–microbiota symbiosis, the offspring acquires the microbiota by feeding on grass that is contaminated with feces and sputum from their parents, as well as by passage through the birth canal (Dehority 2003).

A less direct, but precise, mode of transmission is exemplified in the squid light organ–*Vibrio fischeri* symbiosis where the high specificity of the light organ for *V. fischeri* has evolved together with the need to acquire the motile bacteria from the surrounding seawater. The adult squid releases large amounts of *V. fischeri* into the water at dawn every day, ensuring that sufficient symbionts are available to colonize the hatchlings (McFall-Ngai 1999). Furthermore, the squid provides a habitat in which only *V. fischeri* that emits light is able to maintain a stable association (McFall-Ngai 1999; Visick et al. 2000). Thus, even in transfer via the environment (often referred to as horizontal transfer), the holobiont is reconstituted faithfully.

Some animals and most plants can develop from cells other than gametes, namely, from somatic cells (Buss 1987). The most striking example is vegetative reproduction in plants. When a fragment of a plant falls to the earth, it may root and grow into a fully developed plant. In such cases, it will clearly contain some of the symbionts of the original plant (direct transfer). In addition, it will most likely incorporate rhizosphere fungi (mycorrhiza) and other microorganisms from the soil adjacent to the parent. Symbionts of the metazoan Hydra demonstrate both specificity and accuracy of transmission (Fraune and Bosch 2007). The Hydra reproduces similarly to plants, namely, vegetatively (by budding) and sexually. The report showed, first, that two different species of Hydra were colonized by different communities of microorganisms and, second, in both cases the two species of Hydra were populated with similar microorganisms both in the laboratory and in nature, even after more than 30 years of maintaining the animals in the laboratory.

Summarizing this section, we suggest that regardless of the mechanism used, there is now growing evidence that the microbial component of the holobiont is transferred from generation to generation. The large varieties in modes of transmission have an interesting implication: individuals can acquire and transfer symbionts throughout their lives, and not just during their reproductive phase. This means that the parents, grandparents, nannies, siblings, spouses, or any organism that is in close contact with an offspring can transfer symbionts and thereby influence the holobiont of the next generation.

24.7 The Hologenome Theory Contains Aspects of Both Lamarckism and Neo-Darwinism

The only methodical evolution theory preceding Darwin was the one presented by Jean-Baptiste Lamarck, a renowned French botanist, zoologist and philosopher of science, published in 1809 in his book "Philosophie Zoologique" (discussed in Burkhardt 1972). Lamarck believed in spontaneous creation of living creatures and in their slow evolution – a linear continuous evolvement from simple to complex creatures. He also proposed that environmental forces lead to change in organisms. Lamarck believed that these environmentally induced changes were then passed on to future generations. Thus, the main features of his theory, referred to today as "Lamarckism," are:

1. Use and disuse – individuals lose characteristics they do not use and develop characteristics that are useful.
2. Inheritance of acquired characteristics – individuals transmit acquired characteristics to their offspring.

Interestingly, Darwin believed, as did Lamarck and many others at that time, that an organism can transmit traits it acquired during its lifetime to its offspring. But with the advent of neo-Darwinism at the beginning of the twentieth century, Lamarckism was discredited and largely ignored throughout most of the last century. There were two major scientific arguments for rejecting Lamarckism. First, the evolutionary theorist August Weismann argued that inheritance takes place only by means of germ cells and that germ cells cannot be affected by any somatic cells of the body acquired during their lifetime (Weismann 1893). Second, Mendelian genetics considers that variation, the raw material for Darwinian evolution, occurs by random mutations in the population. The modern synthesis of evolution theory that emerged during the twentieth century and continued to disregard Lamarck was in essence an amalgamation of neo-Darwinism, molecular genetics, and population dynamics. Since the 1980s, however, Lamarckism is being considered with growing interest by mainstream evolution thought (Gould 1999). Jablonka and Lamb (2005) in their book *Evolution in four Dimensions* comprehensively discussed inheritance of acquired characteristics via epigenetic systems, including DNA methylation, self-sustaining feedback loops, prions, chromatin-marking, and RNA interference. These mechanisms include the inheritance of changes that are not DNA sequence based and therefore argue for withdrawal from the strict genotype–phenotype separation dogma of neo-Darwinism.

Microbial amplification and acquisition of novel microbes into holobionts closely fit the Lamarckian first principle of "use and disuse." The holobiont loses characteristics (microbes) it does not use and gains characteristics (microbes) that are useful. As described above, these acquired microbes can be transmitted to offspring, thus satisfying the second principle of Lamarckism. One of the initial reasons for rejecting Lamarckism was the so-called Weismann barrier (Pollard 1984); genetic information cannot pass from somatic cells to germ cells and on to the next generation. Weismann and his contemporaries were unaware, of course, of the abundant microbial community associated with multicellular organisms and the enormous genetic information they possess.

24.8 Unresolved Questions and Future Research

The concept of the holobiont, including all of the associated microorganisms, is relatively new. By its very nature, even a simple holobiont, such as a Hydra, is an enormously complex dynamic system. Following are three of the many unresolved questions:

1. *What are the physiological functions of specific symbionts within the complex microbiota?* Most of what we currently know about the microbial community of holobionts has been derived from analyses of the 16S rRNA gene sequences of the total DNA of animals and plants. Although these data can be used to determine the abundance of different operational taxonomic units (OTUs), it fails to provide definitive information on the physiological functions of the symbionts. For example, certain strains of *Escherichia coli* are pathogenic, while others are beneficial. Accordingly, it will be useful in the future to culture many of the abundant symbionts and study their properties.
2. *How does the holobiont ensure that genetic information in beneficial symbionts is maintained?* Two general pathways may be possible for ensuring that any useful genetic information is conserved in future holobiont generations: (1) The microbial genes can be inserted into the host genome, as in the transfer of carotenoid biosynthetic genes from a fungus to aphids (Moran and Jarvik 2010), and/or (2) the host and microbe can undergo secondary changes that stabilize and benefit the interactional symbiosis. This latter kind of adaptation can occur in primary symbiosis as in corals and their algae (Fallowski et al. 1984) or in secondary symbiosis as in the bovine rumen which fosters the growth of anaerobic cellulose-degrading microorganisms, which benefit the host not only by conversion of the cellulose to utilizable fatty acids, but also by satisfying its protein and vitamin requirements (Dehority 2003). What determines, during evolution, which functions in the holobiont will be taken on by the host and

which by the symbiont? Which microbial genes will be inserted into the host genome and which will be kept external within the microbiota?

3. *What are the roles of viruses in holobionts?* Viruses are abundant in humans (see Chap. 15) and probably most other holobionts. As was initially the case with bacteria, viruses are considered agents of disease; however, they may, at least in principle, benefit the holobiont by carrying useful genetic information and ensuring the diversity of the microbial population.

As we gain a better understanding of the holobiont it should be possible to alter the microbiota by pre- and pro-biotics in order to improve health.

References

Abe T, Bignell DE, Higashi M (eds) (2000) Termites: evolution, sociality, symbioses, ecology. Kluwer Academic Publishers, Dordrecht

Abrams GD, Bishop JE (1967) Effect of normal microbial flora on gastrointestinal motility. Proc Soc Exp Biol Med 126:301–304

Bäckhed F, Ding H, Wang T, Hooper LV, Koh GY, Nagy A, Semenkovich CF, Gordon JI (2004) The gut microbiota as an environmental factor that regulates fat storage. Proc Natl Acad Sci USA 101:15718–15723

Bäckhed F, Manchester JK, Semenkovich CF, Gordon JI (2007) Mechanisms underlying the resistance to diet-induced obesity in germ-free mice. Proc Natl Acad Sci USA 104:979–984

Barbieri E, Paster BJ, Hughes D, Zurek L, Moser DP, Teske A, Sogin ML (2001) Phylogenetic characterization of epibiotic bacteria in the accessory nidamental gland and egg capsules of the squid Loligo pealei (Cephalopoda: Loliginidae). Environ Microbiol 3:151–167

Baumann P, Moran NA, Baumann L (2006) Bacteriocyte-associated endosymbionts of insects. In: Dworkin M, Rosenberg E, Schleifer KH, Stackebrandt E (eds) The prokaryotes, vol 1. Springer, New York, pp 403–438

Beijerinck MW (1901) Über oligonitrophile mikroben, centralblatt für bakteriologie, parasitenkunde, infektionskrankheiten und hygiene. Abteilung II 7:561–582

Ben-Yosef M, Aharon Y, Jurkevitch E, Yuval B (2010) Give us the tools and we will do the job: symbiotic bacteria affect olive fly fitness in a diet-dependent fashion. Proc Biol Sci 277:1545–1552

Breitbart M, Hewson I, Felts B, Mahaffy JM, Nulton J, Salamon P, Rohwer F (2003) Metagenomic analyses of an uncultured viral community from human feces. J Bacteriol 185:6220–6223

Bright M, Bulgheresi S (2010) A complex journey: transmission of microbial symbionts. Nat Rev Microbiol 8:218–230

Brune A, Stingl U (2006) Prokaryotic symbionts of termite gut flagellates: phylogenetic and metabolic implications of a tripartite symbiosis. Prog Mol Subcell Biol 41:39–60

Burkhardt RW (1972) The inspiration of Lamarck's belief in evolution. J Hist Biol 5:413–438

Buss LW (1987) The evolution of individuality. Princeton University Press, Princeton

Cankar K, Kraigher H, Ravnikar M, Rupnik M (2005) Bacterial endophytes from seeds of Norway spruce (Picea abies L. Karst). FEMS Microbiol Lett 244:341–345

Chelius MK, Triplett EW (2001) The diversity of Archaea and bacteria in association with the roots of *Zea mays* L. Microb Ecol 41:252–263

Collado MC, Isolauri E, Lairinen K, Sahminen S (2010) Effect of mother's weight on infant's microbiota acquisition, composition, and activity during early infancy: a prospective follow-up study initiated in early pregnancy. Am J Clin Nutr 95(5):1023–1030

Coyne MJ, Reinap B, Lee MM, Comstock LE (2005) Human symbionts use a host-like pathway for surface fucosylation. Science 307:1778–1781
Dale C, Moran NA (2006) Molecular interactions between bacterial symbionts and their hosts. Cell 126:453–465
De Filippoa C, Cavalieria D, Di Paolab M et al (2010) Impact of diet in shaping gut microbiota revealed by a comparative study in children from Europe and rural Africa. Proc Natl Acad Sci USA 107:14691–14696
Dehority BA (2003) Rumen microbiology. Nottingham University Press, Nottingham
Dethlefsen L, McFall-Ngai M, Relman DA (2007) An ecological and evolutionary perspective on human-microbe mutualism and disease. Nature 449:811–818
Devi SM, Ahmed I, Khan AA et al (2006) Genomes of *Helicobacter pylori* from native Peruvians suggest a mixture of ancestral and modern lineages and reveal a western type cag-pathogenicity island. BMC Genomics 7:191
Dewhirst FE, Chen T, Izard J, Paster BJ, Tanner ACR, Yu WH, Lakshmanan A, Wade WG (2010) The human oral microbiome. J Bacteriol 192:50012–50017
Dominguez-Bello MG, Costellob EK, Knight R (2010) Delivery mode shapes the acquisition and structure of the initial microbiota across multiple body habitats in newborns. Proc Natl Acad Sci USA 107:11971–11975
Eckburg PB, Bik EM, Bernstein CN, Purdom E, Dethlefsen L, Sargent M et al (2005) Diversity of the human intestinal microbial flora. Science 308:1635–1638
Edwards JE, McEwan NR, Travis AJ, Wallace RJ (2004) 16S rDNA library-based analysis of ruminal bacterial diversity. Antonie van Leeuwenhoek 86:263–281
Fallowski PG, Dubinsky Z, Muscatine L, Porter JW (1984) Light and the bioenergetics of a symbiotic coral. Bioscience 34:705–709
Fiore CL, Jarett JK, Olson ND, Lesser MP (2010) Nitrogen fixation and nitrogen transformations in marine symbioses. Trends Microbiol 10:455–463
Foster JS, Kolenbrander PE (2004) Development of a multispecies oral bacterial community in a saliva-conditioned flow cell. Appl Environ Microbiol 70:4340–4348
Frank DN, Pace NR (2008) Gastrointestinal microbiology enters the metagenomics era. Curr Opin Gastroenterol 24:4–10
Fraune S, Bosch TCG (2007) Long-term maintenance of species-specific bacterial microbiota in the basal metazoan Hydra. Proc Natl Acad Sci USA 104:13146–13151
Frias-lopez J, Zerkle AL, Bonheyo GT, Fouke BW (2002) Partitioning of bacterial communities between seawater and healthy, black band diseased, and dead coral surfaces. Appl Environ Microbiol 68:2214–2228
Gilbert SF, McDonald E, Boyle N et al (2010) Symbiosis as a source of selectable epigenetic variation: taking the heat for the big guy. Philos Trans R Soc Lond B Biol Sci 365:671–678
Gould SJ (1999) A division of worms. Nat Hist 108:18–26
Grice EA, Kong HH, Conlan S (2009) Topographical and temporal diversity of the human skin microbiome. Science 324:1190–1192
Gündüz E, Douglas AE (2009) Symbiotic bacteria enable insect to use a nutritionally inadequate diet. Proc R Soc 276:987–991
Hehemann JH, Correc G, Barbeyron T et al (2010) Transfer of carbohydrate-active enzymes from marine bacteria to Japanese gut microbiota. Nature 464:908–914
Heijtza RD, Wange S, Anuard F et al (2011) Normal gut microbiota modulates brain development and behavior. Proc Natl Acad Sci USA. doi:10.1073/pnas.1010529108
Hentschel U, Steinert M, Hacker J (2000) Common molecular mechanisms of symbiosis and pathogenesis. Trends Microbiol 8:226–231
Hongoh Y, Deevong P, Inoue T, Moriya S, Trakulnaleamsai S, Ohkuma M, VongKaluny C, Noparatnaraporn N , Kudo T (2005) Intra- and interspecific comparisons of bacterial diversity and community structure support coevolution of gut microbiota and termite host. Appl Environ Microbiol 71:6590–6599

Hooper LV, Midtvedt T, Gordon JI (2002) How host-microbial interactions shape the nutrient environment of the mammalian intestine. Ann Rev Nutr 22:283–307
Ikeda S, Okubo T, Anda M et al (2010) Community- and genome-based views of plant-associated bacteria: plant–bacterial interactions in soybean and rice. Plant Cell Physiol 51:1398–1410
Iniguez AL, Dong YM, Triplett EW (2004) Nitrogen fixation in wheat provided by *Klebsiella pneumoniae* 342. Mol Plant Microbe Interact 17:1078–1085
Iwanaga S, Lee BL (2005) Recent advances in the innate immunity of invertebrate animals. J Biochem Mol Biol 38:128–150
Jablonka E, Lamb MJ (2005) Evolution in four dimensions: genetic, epigenetic, behavioral, and symbolic variation in the history of life. MIT Press, Cambridge
Jaenike J, Unckless R, Cockburn SN, Boelio LM, Perlman SJ (2010) Adaptation via symbiosis: recent spread of a *Drosophila* defensive symbiont. Science 329:212–215
Jones KM, Kobayashi H, Davies BW, Taga ME, Walker GC (2007) How rhizobial symbionts invade plants: the *Sinorhizobium medicago* model. Nat Rev Microbiol 5:619–633
Kikuchi Y, Hosokawa T, Fukatsu T (2007) Insect–microbe mutualism without vertical transmission: a stinkbug acquires a beneficial gut symbiont from the environment every generation. Appl Environ Microbiol 73:4308–4316
Kneip C, Lockhart P, Voss C, Maier UG (2007) Nitrogen fixation in eukaryotes–new models for symbiosis. BMC Evol Biol 7:55. doi: 10.1186/1471-2148-7-55
Koenig JE, Spor A, Scalfone N et al (2010) Succession of microbial consortia in the developing infant gut microbiome. Proc Natl Acad Sci USA 108(Suppl 1):4578–4585
Koren O, Rosenberg E (2006) Bacteria associated with mucus and tissues of the coral *Oculina patagonica* in summer and winter. Appl Environ Microbiol 72:5254–5259
Laitinen K, Poussa T, Isolauri E et al (2009) Probiotic and dietary counseling contribute to glucose regulation during and after pregnancy: a randomised controlled trial. Br J Nutr 101:1679–1687
Lambais MR, Crowley DE, Cury JC, B¨ull RC, Rodrigues RR (2006) Bacterial diversity in tree canopies of the Atlantic forest. Science 312:1917
Leser TD, Amenuvor JZ, Jensen TK, Lindecrona RH, Boye M, Møller K (2002) Culture-independent analysis of gut bacteria: the pig gastrointestinal tract microbiota revisited. Appl Environ Microbiol 68: 673–690
Ley RE, B¨ackhed F, Turnbaugh P, Lozupone CA, Knigh RD, Gordon JI (2005) Obesity alters gut microbial ecology. Proc Natl Acad Sci USA 102:11070–11075
Ley RE, Peterson DA, Gordon JI (2006a) Ecological and evolutionary forces shaping microbial diversity in the human intestine. Cell 124:837–848
Ley RE, Turnbaugh PJ, Klein S, Gordon JI (2006b) Human gut microbes associated with obesity. Nature 444:1022–1023
Lilburn TG, Kim KS, Ostrom NE, Byzek KR, Leadbetter JR, Breznak JA (2001) Nitrogen fixation by symbiotic and free-living spirochetes. Science 292:2495–2498
MacConnachie AA, Fox R, Kennedy DR, Seaton RA (2009) Faecal transplant for recurrent *Clostridium difficile*-associated diarrhoea: a UK case series. QJM 102:781–784
Margulis L (1993) Symbiosis in Cell Evolution: Microbial Communities in the Archean and Proterozoic Eons. 2nd edn. W.H. Freeman and Co., New York
Martens EC, Roth R, Heuser JE, Gordon JI (2009) Coordinate regulation of glycan degradation and polysaccharide capsule biosynthesis by a prominent human gut symbiont. J Biol Chem 284:18445–18457
Martínez-García M, Díaz-Valéz M, Wanner G, Ramos-Esplá A, Antón J (2007) Microbial community associated with the colonial ascidian Cyctodytes dellechiajei. Environ Microbiol 9:521–534
Mateos M, Castrezana SJ, Nankivell BJ, Estes AM, Markow TA, Moran NA (2006) Heritable endosymbionts of Drosophila. Genetics 174:363–376
McFall-Ngai MJ (1999) Consequences of evolving with bacterial symbionts: insights from the squid–Vibrio association. Annu Rev Ecol Syst 30:235–256

McFall-Ngai MJ (2002) Unseen forces: the influence of bacteria on animal development. Dev Biol 242:1–14

Mohamed NM, Colman AS, Tal Y, Hill RT (2008) Diversity and expression of nitrogen fixation genes in bacterial symbionts of marine sponges. Environ Microbiol 10:2910–2921

Moran NA, Jarvik T (2010) Lateral transfer of genes from fungi underlies carotenoid production in aphids. Science 328:624–627

Nardi JB, Mackieb RI, Dawson JO (2002) Could microbial symbionts of arthropod guts contribute significantly to nitrogen fixation in terrestrial ecosystems? J Insect Physiol 48:751–763

Ochman H, Worobey M, Kuo, CH et al. (2010) Evolutionary relationships of wild hominids recapitulated by gut microbial communities. PLoS Biol. 8, e1000546

O'Hara AM, Shanahan F (2006) The gut flora as a forgotten organ. EMBO Rep 7:688–693

Pollard JW (1984) Is Weismann's barrier absolute? In: Ho MW, Saunders PT (eds) Beyond neo-Darwinism: introduction to the new evolutionary paradigm. Academic, London, pp 291–315

Reshef L, Koren O, Loya Y, Zilber-Rosenberg I, Rosenberg E (2006) The coral probiotic hypothesis. Environ Microbiol 8:2067–2073

Ritchie KB (2006) Regulation of microbial populations by coral surface mucus and mucus-associated bacteria. Mar Ecol Prog Ser 322:1–14

Rohwer F, Seguritan V, Azam F, Knowlton N (2002) Diversity and distribution of coral-associated bacteria. Mar Ecol Prog Ser 243:1–10

Rosenberg E, Koren O, Reshef L, Efrony R, Zilber-Rosenberg I (2007) The role of microorganisms in coral health, disease and evolution. Nat Rev Microbiol 5:355–362

Russell JA, Latorre A, Sabater-Mũnoz B, Moya A, Moran NA (2003) Side-stepping secondary symbionts: widespread horizontal transfer across and beyond the Aphidoidea. Mol Ecol 12:1061–1075

Savage DC, Siegel JD, Snellen JE, Whitt DD (1981) Transit time of epithelial cells in the small intestines of germfree mice and ex-germfree mice associated with indigenous microorganisms. Appl Environ Microbiol 42:996–1001

Scarborough CL, Ferrari J, Godfray HCJ (2005) Aphids protected from pathogen by endosymbiont. Science 310:1781–1783

Sharon G, Segal D, Ringo JM et al (2010) Commensal bacteria play a role in mating preference of *Drosophila melanogaster*. Proc Natl Acad Sci USA 107:20051–20056

Sharp KH, Eam B, Faulkner DJ, Haygood MG (2007) Vertical transmission of diverse microbes in the tropical sponge *Corticium* sp. Appl Environ Microbiol 73:622–629

Silva AM, Barbosa FHF, Duarte R, Vieira LQ, Arantes RME, Nicoli JR (2004) Effect of *Bifidobacterium longum* ingestion on experimental salmonellosis in mice. J Appl Microbiol 97:29–37

Skaljac M, Zanic K, Ban SG, Kontsedalov S, Ghanim M (2010) Co-infection and localization of secondary symbionts in two whitefly species. BMC Microbiol 10:142

Smith KP, Handelsman J, Goodman RM (1999) Genetic basis in plants for interactions with disease-suppressive bacteria. Proc Natl Acad Sci USA 96:4786–4790

Stougaard J (2000) Regulators and regulation of legume root nodule development. Plant Physiol 124:531–540

Sundset MA, Praesteng KE, Cann IK, Mathiesen SD, Mackie RI (2007) Novel rumen bacterial diversity in two geographically separated sub-specie of reindeer. Microb Ecol 54:424–438

Tannock G (1995) Normal microflora. Chapman & Hall, London

Taylor MW, Radax R, Steger D, Wagner M (2007) Sponge associated microorganisms: evolution, ecology and biotechnological potentials. Microbiol Mol Biol Rev 71:295–347

Thingstad TF, Lignell R (1997) Theoretical models for the control of bacterial growth rate, abundance, diversity and carbon demand. Aquat Microb Ecol 13:19–27

Turnbaugh PJ, Ley RE, Mahowald MA, Magrini V, Mardis ER, Gordon JI (2006) An obesity-associated gut microbiome with increased capacity for energy harvest. Nature 444:1027–1031

Turnbaugh PJ, Ridaura VK, Faith JJ et al (2009) The effect of diet on the human gut microbiome: a metagenomic analysis in humanized gnotobiotic mice. Sci Transl Med 1(6):6ra14

Turnes MS, Hay ME, Fenical W (1989) Symbiotic marine bacteria chemically defend crustacean embryos from a pathogenic fungus. Science 246:116

Veneti ZL, Reuter M, Montenegro H, Hornett EA, Charlat S, Hurst GD (2005) Interactions between inherited bacteria and their hosts: the Wolbachia paradigm. The influence of Cooperative Bacteria on Animal Host Biology (McFall-Ngai MJ, Henderson B & Ruby E-G, eds), pp. 119–141. Cambridge University Press, New York

Visick KL, Foster J, Doino J, McFall-Ngai MJ, Ruby EG (2000) *Vibrio fischeri* lux genes play an important role in colonization and development of the host light organ. J Bacteriol 182:4578–4586

Wallace RJ (2004) Antimicrobial properties of plant secondary metabolites. Proc Nutr Soc 63:621–629

Wang B, Qui YL (2006) Phylogenetic distribution and evolution of mycorrhizas in land plants. Mycorrhiza 16:299–363

Webster NS, Wilson KJ, Blackall LL , Hill RT (2001) Phylogenetic diversity of bacteria associated with the marine sponge Rhopaloeides odorabile. Appl Environ Microbiol 67:434–444

Weismann A (1893) The germ-plasm: a theory of heredity. Charles Scribner's Sons/Electronic Scholarly Publishing, New York

Weitz JS, Hartman H, Levin SA (2005) Coevolutionary arms races between bacteria and bacteriophage. Proc Natl Acad Sci USA 102:9535–9540

Wilson DS, Sober E (1989) Reviving the superorganism. J Theor Biol 136:337–356

Wilkinson DM (2001) Mycorrhizal evolution. Trends Ecol Evol 16:64–65

Wostmann BS (1981) The germ-free animal in nutritional studies. Annu Rev Nutr 1:257–297

Wostmann BS, Larkin C, Moriarty A, Bruckner-Kardoss E (1983) Dietary intake, energy metabolism and excretory losses of adult male germfree Wistar rats. Lab Anim Sci 33:46–50

Wust PK, Horn MA, Drake HL (2011) Clostridiaceae and Enterobacteriaceae as active fermenters in earthworm gut content. ISME 5:92–106

Xu J, Mahowald MA, Ley RE (2007) Evolution of symbiotic bacteria in the distal human intestine. PLOS Biol 5:1574–1586

Yachi S, Loreau M (1999) Biodiversity and ecosystem productivity in a fluctuating environment: the insurance hypothesis. Proc Natl Acad Sci USA 96:1463–1468

Zilber-Rosenberg I, Rosenberg E (2008) Role of microorganisms in the evolution of animals and plants: the hologenome theory of evolution. FEMS Microbiol Rev 32:723–735

Zoetendal EG, Akkermans ADL, van Vliet WM et al (2001) A host genotype affects the bacterial community in the human gastrointestinal tract. Microb Ecol Health Dis 13:129–134

Index

A
Aciduric, 191
Acetate CoA transferase gene, 250
Acetogenesis, reductive, 14
  coexistence, 14
Acidic fermentation, 247
Acidogenic, 191
Actinobacteria, 214
Actinorhizal bacteria, 74
Adaptation, 65
Adaptive immune systems, 327
ADP-ribosylation, 45
Aerobic, 214
Aggregatibacter actinomycetemcomitans, 193
Agrobacterium, 125
Agrobacterium radiobacter K84, 126
Alfano, J.R., 125, 126
Algal endosymbionts, 308
Allorhizobium, 73, 75
Alpha-aminoadipate reductase (alpha-AAR), 155
Alpha-proteobacteria, 75, 76
Alveolates, 152
Aminocyclopropane carboxylate (ACC) deaminase, 97
Anabaena, 74
Anaerostipes spp., 248
Ancient DNA, 198
Animal models, 271
Antibiotic production, 126
Antibiotics, 213
Antigens, 225
Anti-inflammatory, 216
Antimicrobial peptides, 317
Aphids, 29, 30
Apicomplexa, 157
Aplanochytrium, 154
Archaea, 163, 168, 169
Archaeological, 197
Armantifilum, 9
Arminins, 317
Assortative mating, 58
  genetic basis, 60
  medium-induced, 64
Atherosclerosis, 193
Azobacteroides, 8, 19
Azorhizobium, 73, 75, 76
  caulinodans, 76
Azospirillum brasilense, 92

B
Baby koala (Joey), 333
Bacilli
  enterococcus, 214
  lactobacillus acidophilus, 216
  lactobacillus casei, 216
  lactobacillus farciminis, 216
  lactobacillus paracasei, 216
  lactobacillus plantarum, 216
  lactobacillus reuteri, 216
Bacillus, 127, 128
Bacteria, 165, 166, 170, 270
Bacterial commensalism, 257
Bacterial communities, 105, 111, 177, 178, 184, 284
Bacterial community compositions, 113
Bacterial interactions, 177, 178
Bacterial microbiota, 274
Bacterial persistence, 291
Bacterial populations, 283
Bacterial survival, 100

DOI 10.1007/978-3-642-21680-0,

Bacterial virulence, 253, 257
Bacteriocyte, 308
Bacteriophages, 327
Bacteroides plebeius, 331
Bacteroidetes, 7, 19, 132, 214, 226, 326, 329
  bacteroides, 213
  prevotella, 214
Bacteroids, 77, 82
Bactrocera oleae, 27, 33
Beneficial effects, 273, 275
Beta-proteobacteria, 76
Bifidobacteria, 213, 216, 245, 246
Bile acid malabsorption, 215
Biobreeding diabetes prone (BB-DP) rats, 234
Biochar, 131
Biocontrol, 123, 124
Biocontrol fungi (BCF), 125
Biofilm, 190
Biological control, 90
Blattabacterium, 17
Bothrosome, 155
Bowel habits, 273
Bradyrhizobium, 73–76, 79, 82
  B. japonicum, 74, 76, 77, 79, 82
Bristol stool form, 212
Brugia malayi, 304
Buchnera, 30
Burkholderia, 73, 75, 76, 98
  B. phymatum, 76
  B. silvatlantica, 98
  B. tropica, 98
  B. unamae, 98
Butyrate production, 247–249

C
Caduceia, 9
Callosobruchus maculates, 66
Caries, 191
Cellulolytic bacteria, 246
Cellulolytic ruminococci, 246
Cellulosomal *scaC* gene, 250
Ceratitis capitata, 27, 33
Checkerboard DNA–DNA hybridization, 196
Chironomids, 43–54
Chironomus
  developmental stage, 44
  egg masses, 45
Chitinases, 126
Chloroplasts, 332
Cholera enterotoxin (CT), 44
Cholera epidemics, 48
Cholera toxin *(ctxA)*, 47, 50, 53
Ciliate, 157, 158
Citrobacter rodentium, 237
Clone libraries, 226
Closed symbiotic system, 299–300
Clostridia, 226
  clostridium coccoides, 214
  clostridium diffcile, 214
  clostridium histolyticum, 214
  clostridium lituseburense, 214
  clostridium thermosuccinogenes, 214
  clostridum perfringens, 213
  Collinsella, 214
  coprobacillus, 214
  coprococcus, 214
  eubacterium rectale, 214
  faecalibacterium prausnitzii, 214
  lachnospiraceae, 214
  ruminococcus bromii, 214
  ruminococcus torques, 214
  Veillonella, 214
Cnidaria, 314
Cnidarians, 315
Coadhesion, 190
Coaggregation, 191
Coccoid bodies, 154, 157
Co-evolution, 324
Co-existence, 184
Coliforms, 213
Colitis, 247
Colon microbiome, 269
Colorectal cancer, 243, 244, 247
Commercial inoculants, 94
Community, 113
Competition, 105, 110, 112, 178, 180, 184
Compost, 129
Constipation, 212
Co-occurrence, 180
Cooperation, 178, 184
Coral, 163–171
Corallochytrea, 155
Cryptocercus, 5
CT. *See* Cholera enterotoxin
Cultivation, 213
Culturability, 244–245
Culture-dependant approach, 291
Cupriavidus, 73, 75, 76
  C. taiwanensis, 76
Cuticular hydrocarbons, 58
Cyanobacteria, 74
Cytokinins, 90
Cytryn, E., 130, 131

D
DAPG. *See* 2,4-Diacetylphloroglucinol
Defaunation, 6
Dental hygiene, 194
Devescovina, 9
Devosia, 73, 75
    D. neptuniae, 75
DGGE, 213
Diabetes, 193
2,4-Diacetylphloroglucinol (DAPG), 129, 130
Diarrhea, 212
Diet, 244
Dietary starch, 248
Digestion models, 271
Dinoflagellates, 144
Dipterans legs, 46
Disease, 265, 274
Divergent selection, 61
Diversity, 105, 106, 109–111, 244–245
DNA-binding transcription factor, 50
Drosophila, 58, 61
    commensal bacteria, 63
    courtship, 59
    melanogaster, 58
    pseudoobscura, 61
    willistoni, 61
Dysbiosis, 225

E
Early colonizer, 190
Ecological competence, 128
Ecological model, 180
Ecological plaque hypothesis, 191
Ectoplasmic extensions, 155
    ectoplasmic net, 155, 156
Ectosymbiotic ecdysozoans, 307
Elad,Y., 131
Elusimicrobia, 7, 19
Emerging prebiotics, 276
Enamel, 190
Endomicrobia, 7
Endomicrobium, 9, 19
Endophytic bacteria, 74
Endophytic colonization, 100
Endosymbionts, 18
    genome, 18
Ensifer, 73–78
    medicae, 74, 76
    meliloti, 76, 78–80
Enterobacteriaceae, 33, 35
Epigenetic, 324
Epigenetic variations, 330
Epithelium, 314, 315, 317
EPS. *See* Exopolysaccharides
Epulopiscium spp., 306
Escherichia, 331
Escherichia coli, 227
Eubacterium hallii, 248
Exopolysaccharides (EPS), 91
Exosymbiont, 325
ExPEC. *See* Extraintestinal pathogenic
        E. coli strains
Extracellular symbionts, 305–307
Extraintestinal pathogenic *E. coli* strains
        (ExPEC), 254–259

F
Facultative aerobes, 213
Faecalibacterium prausnitzii, 227, 248
Feed effciency, 204, 206, 207
FFAR. *See* Free fatty acid receptor
Field inoculation, 94
Fingerprinting, 214
Firmicutes, 7, 214, 225, 326
FISH, 214
Flagellates, 12
    cospeciaton, 10
    parabasalid, 12
Flavobacterium, 132
Fluoride, 191
Frankia, 74
Free fatty acid receptor (FFAR), 247
Functional clubs, 250
Fungal community, 109
Fungal infections, 319
Fungi, 163, 166–168
Fusobacterium nucleatum, 193

G
Ganglion mother cell (GMC), 304
Gastrointestinal diseases, 275
Gene transfer agents (GTAs), 146
Genome, 73, 75–77
Germ-free mice, 329
Germline stem cells (GSCs), 302–303
Gibberellins, 90
Gingival crevicular fluid (GCF), 194
Gingivitis, 193
Gluconacetobacter, 97
Gluconacetobacter diazotrophicus, 97
Glycocalix, 314, 315
GMC. *See* Ganglion mother cell
Graber, E.R., 131

Green fluorescent protein (GFP), 46
GSCs. *See* Germline stem cells
GTAs. *See* Gene transfer agents
Gut community, 245–246, 249–250
Gut flora, 266
Gut health disorders, 284
Gut microbial community, 247
Gut microbiology, 272
Gut microbiome, 233
Gut microbiota, 244, 246, 274
  bacterial, 6
  co-cladogenesis, 10
  flagellates, 5
  microaerophiles, 16
Gut microflora, 273
Gycoprotein matrix, 48

H

Haemagglutinin protease (HAP), 47, 48
Halofolliculina, 158
HapR, 47, 50–52
HAP secretion regulation, 49–50
Health, 265
Health and well-being, 265
Helicostoma, 157
Hemicelluloses, 12
Herbaspirillum, 95
Holobionts, 32, 33, 153, 157, 158
Hologenome theory, 290
Horizontal gene transfer, 76, 77
Horizontally transmitted symbioses, 300–302
Host genetics, 282, 284, 289, 291
H. seropedica, 96
Human colon, 247–249
Human colonic bacteria, 246
Human digestive tract, 282
Human microbiome, 231–232
Human milk derived oligosaccharides, 266
Human trials, 276
Hybridization, 196
Hydra, 333
Hydramacins, 317
Hydrogenosomes, 12
Hydrolytic enzymes, 155
Hygiene hypothesis, 233

I

IAA. *See* Indole-3-acetic acid
IBD. *See* Inflammatory bowel disease
Iirritable bowel syndrome, 243
Immune response, 195
Immune system, 283, 284, 327
Independent studies, 273
Indole-3-acetic acid (IAA), 90
Induced systemic resistance, 124
Infant, 274
Inflammation, 192
Inflammatory bowel disease (IBD), 233, 243
Innate, 327
Inoculated roots, 92
Inoculation, 91
Insurance policy hypothesis, 327
Inter-species quorum sensing, 49
Interspecific host-switching, 300
Interstitial stem cells, 318
Intestinal microbial communities, 243–251
Intestine, 265
Intraspecific host-switching, 300
Inulin-type fructans, 274
Invertebrates, 326
In vitro gastrointestinal models, 276
In vitro GI models, 272
In vitro systems, 271

J

Joe, A., 125, 126

K

Kautsky, L., 130
Kill the winners hypothesis, 327
Klebsiella, 33, 35, 36, 64
Kolton, M., 131

L

Labyrinthula, 157
Labyrinthulidae, 154
Labyrinthuloides, 153, 154
Labyrinthulomycetes, 153
Lactic acid bacteria, 284
Lactobacilli, 213
Lactobacillus, 191
  L. plantarum, 64
  strains, 234
Leaky gut hypothesis, 236–237
Legume-rhizobia, 82
Legume-*Rhizobium*, 79
Legumes, 73–75, 77–83, 93
  Aeschynomene, 77, 79
  alfalfa, 78
  bean, 78
  clover, 78
  Discolobium, 77

Medicago truncatula, 74
Neptunia, 77
pea, 78
peanut, 77
Sesbania, 77
soybean, 74, 78
Stylosanthes, 77
Leptin hormone, 283
Lignin, 11
Lignocellulose, 10
Lipopolysaccharides (LPS), 91
Loper, J.E., 125, 126
LPS. *See* Lipopolysaccharides

M
MAMPs. *See* Microbe-associated molecular patterns
Mandelbaum, R.T., 130
Maternal effect, 289, 291
Maternal microbiota, 282, 288, 290
Mavrodi, D.V., 125, 126
Mavrodi, O.V., 125, 126
Mechanistic studies, 271
Meller-Harel, Y., 131
Mermaid, 301
Mesomycetozoa, 155
Mesorhizobium, 73, 75, 76
M. loti, 76, 79
Metabacterium polyspora, 305
Metabolic interactions, 183
Metabolic modeling, 178, 183, 184, 186
Metabolic models, 179
Metabolic network, 178–180
Metagenomics, 250
Metazoans, 308
Methanogenesis, 14
coexistence, 14
Methanogenic, 213
methanobrevibacter smithii, 213
Methanogens, 7
Methods, 270
Methylobacterium, 73, 75, 76
M. nodulans, 75, 76
Mice, 224, 285–287, 289–291
Microarray, 196
Microarray analysis, 232
Microbe-associated molecular patterns (MAMPs), 127, 128
Microbial amplification, 330
Microbial colonisation, 244
Microbial communication, 191
Microbial communities, 111
Microbial transmission, 307–308
Microbiota, 212
Minz, D., 130
Mitochondria, 332
Mixotricha, 9
Motility, 216
Mouse, 224
Mutualism, 140, 324
Mycorrhiza, 109, 333

N
Neomonada, 155
Next generation sequencing, 232
N_2 fixation, 73, 81
N_2 fixation genes
fix, 81
nif, 81
NifA, 81
nifB, 81
nifD, 81
nifE, 81
nifH, 81
nifK, 81
NifL, 81
nifN, 81
nifV, 81
NFκB, 316
Niche exclusion, 126
Nitricoxide, 90
Nitrogen fixation, 31, 90
Nitrogen-fixing activity, 17
NLRs. *See* NOD-like receptors
Nod factors, 78–80
NOD-like receptors (NLRs), 316
Nodulation, 75, 77–83
Nodulation competitiveness, 82
Nodulation genes, 78–81
nod, 78–80
nodA, 78
nodB, 78
nodC, 78
nodD, 79
nodD$_2$, 79
nodH, 80
nodL, 80
nodP, 80
nodQ, 80
noe, 78
nol, 78
nolA, 79
nolR, 79

Nodule, 73–75, 77–82
determinant nodules, 74, 78
indeterminant nodules, 74, 78
Non-biting midges, 44
Non-O1 heat-stable enterotoxin, 47
Non-targeted metagenomics, 250
Nostoc, 74
Novel prebiotics, 275
Nutritional habits, 228

O

Ocean acidification, 163–171
Ochrobactrum, 73, 75
cytisi, 75
lupini, 75
Ofek, M., 130
Oligosaccharide, 268
Open symbiotic system, 299–300
Oral, 189
Oral cavity, 190
Oral microbiome, 198
Oxygen gradients, 15

P

Paenibacillus, 99
Paenibacillus polymyxa, 99
PANGEA, 235
Parasitism, 324
Paraurонema, 158
PARtitioning defective (PAR) protein system, 305
Pasternak, Z., 132
Pathogen, 225
Paulsen, I.T., 125, 126
PCR, 196
Periculins, 317, 319
Periodontitis, 192
Persistence, 284, 290
Personalized medicine approach, 291
PGPR. *See* Plant-growth promoting rhizobacteria
pH, 164–171
PHAs. *See* Poly-beta-hydroxyalkanoates
Phenazine, 129
Phylotypes, 189, 244, 245
Phytochemicals, 327
Phytohormones, 90
Plant-growth promoting rhizobacteria (PGPR), 74
Plant membrane localized pattern recognition receptors (PRRs), 127
Plant-microbe interactions, 106
Plant roots, 105, 107, 110
Plants, 326
Plaque, 189
Poly-beta-hydroxyalkanoates (PHAs), 91
Polysaccharides, 267
Polyunsaturated fatty acids (PUFA), 155, 157, 158
Porphyromonas gingivalis, 193
Post-infectious IBS (PI-IBS), 215
Prebiotic action, 269
Prebiotic effects, 270, 273
Prebiotics, 265, 266, 270, 274, 276
Preterm births, 193
Prevotella, 331
Probiotic bacterial, 291
Probiotics, 212, 285
Proteases, 126
Proteobacteria, 214
α-Proteobacteria, 146
β-Proteobacteria, 319
campylobacter jejuni, 215
enterobacteriaceae, 213
Proteolytic activities, 48
Protist, 152–155, 157–158
PRRs. *See* Plant membrane localized pattern recognition receptors
Pseudomonads, 125
Pseudomonas, 98, 127, 128
P. aurantiaca, 98
P. fluorescens st, 126
P. putida, 98
P. stutzeri, 98
Pseudotrichonympha, 19
pSym, 78
P symbionts, 28–30
PUFA. *See* Polyunsaturated fatty acid
Pulvilum, 46
Pyro-sequencing, 227

Q

Quantitative real-time PCR (qPCR), 214
Quorum sensing (QS), 143
Quorum-sensing signals, 49–53

R

Ralstonia, 75, 76
Rav David, D., 131
Reciprocal crosses, 288, 290, 291
Recycling, 17
Regulatory T cells, 227
Reproductive isolation, 58
Resistant starch (RS), 246

Reticulorumen, 203–208
Retrievable enamel chip model, 195
Rhizobia, 73–75, 77–83, 93
Rhizobium, 73, 75–80, 329
  etli, 76
  leguminosarum, 76
Rhizosphere, 77, 105–113
Rhizosphere competence, 125
Root bacterial communities, 110, 112
Root bacterial communities composition, 112
Root-bacterial interactions, 110
Root depositions, 106–108
Root microbial communities, 112
Roseburia/Eubacterium rectale, 247
Roseobacter, 142
RS. *See* Resistant starch
Rumen, 203–205, 207
Rumen bacteria, 246
Ruminants, 203, 206, 208
Ruminococci, 246, 250
Ruminococcus bromii, 246

S
Salivary pellicle, 190
Saprophytic, 152, 154
SAR. *See* Systemic acquired resistance
Seeds, 110, 113
Segmented filamentous bacterium (SFB), 306
Septicemia, 254, 255, 257, 259
Sequencing, 196
SFB. *See* Segmented filamentous bacterium
Shigella, 331
Shinella, 73, 75
  S. kummerowiae, 75
Siderophores, 126
Signal molecules, 49
Silber, A., 131
Sinorhizobium, 73, 75, 76
Small intestinal bacterial overgrowth, 213
Small intestine, 244
Speciation, 65
Specific plaque hypothesis, 191
Spirochaetes, 6
Spiroplasma, 63
Squid light organ–*Vibrio fischeri* symbiosis, 333
16S rRNA, 214
  analyses, 236
  analysis, 246, 247
S symbionts, 28, 29
Sterile insect technique, 27, 36
Stinkbugs, 30
Strain typing, 287, 290
Stramenopile, 152–155, 157
Streptococci, 190
  S. mutans, 191
Sucrose, 197
Superorganism, 325
Symbiosis, 74, 76, 79, 80, 82, 325
  digestive, 10
  nutritional, 16
Symbiotic, 76
Symbiotic islands (SIs), 76
Symbiotic plasmids, 77
  pSyms, 76, 77
Systemic acquired resistance (SAR), 127, 128
Systemic resistance, 127

T
Tannerella forsythia, 193
Targeted metagenomics, 250
TaxCollector, 235
Temporal stability, 228
Tephritidae, 27, 32
Tephritids, 27
Termites, 30–32
  cospeciaton, 10
Therapeutic and preventive measure, 275
Thomashow, L.S., 125, 126
Thraustochytridae, 155
Thraustochytrids, 152, 153, 155, 157, 158
Thraustochytrium, 153, 154
TNO's intestinal model (TIM), 272
Toll-like receptors (TLRs), 316, 317
Tooth, 190
Transcript analysis with the aid of afffnity capture (TRAC), 214
Transgenesis, 313, 320
Transmission of symbionts, 332
Treatments, 228
tRFLP, 285, 286, 289
Trichoderma, 125, 127, 131
Trichomitopsis, 12
Trichonympha, 9, 12, 19
Tsechansky, L., 131
Tsetse, 28, 32
Type 1 diabetes, 231–237
Typing, 288

U
Upgrading, 17
Uric acid, 17
Urinary tract infection (UTI), 254, 257–259

V
VBNC. *See* Viable but non-culturable
Veillonella parvula, 191
Vertebrates, 326
Vertically transmitted symbioses, 302–305
Viable but non-culturable (VBNC), 46
Vibrio–chironomids association, 45
Vibrio cholerae, 43–53
  vs. Chironomids, 51, 52
  serogroups, 45
Vibrio fischeri, 300
Virulence-associated genes, 47
Virulence factor expression, 50, 51
Visceral sensation, 216

W
Waste stabilization ponds (WSP), 46
Weller, D.M., 125, 126
Wolbachia, 30, 32, 63

X
Xenorhabdus nematophila, 301
Xylanibacter, 331

Y
Yersinia pestis, 301

Z
Zonulin, 236

Printed by Publishers' Graphics LLC USA
MO20120403-235
2012